W9-BRE-740

Acknowledgments

I want to give a sincere and warm "thank you" to all the individuals who took time out of their busy schedules to help me turn this book into a reality. The hard work and dedication of these manufacturers, public relations firms, associations, and government bodies helped me to create a worthwhile tool certain to benefit construction professionals, consumers, and the environment. I certainly appreciate the time they spent to supply information and photos, answer questions, review sections (so they would stay on track), and proof text. There isn't enough room here for all of you personally to be listed — but check out the appendix! Again, many thanks for all your help!

To the following, I would like to give a special thanks:

To Renee Rewiski, Editor of Building and Remodeling News, who over the years has given me many opportunities to write articles on alternative materials, many of which found their way into this book; to Huck DeVenzio, Advertising Manager for Hickson Corporation, for all his suggestions in the Permanent Wood Foundation section and the Fire-Retardant section, both in Chapter 2; to Donald R. Moody, P.E., President of Residential Steel Partnership (RSP) for his contribution to "The History and Future of Steel" in Chapter 4; to Geoffrey Stone, AISI Project Manager for Residential Construction, for the information given on behalf of the American Iron & Steel Institute (AISI); to Marilyn H. LeMoine, Manager, Publications Department, and Lora K. Metzner, Senior Writer for APA-The Engineered Wood Association, for sharing their comments and bringing to my attention some factual information about plywood and engineered wood products used in Chapter 4; to Jack Davidson, Manager of B.C. Shake & Shingle Association, for his input and the information he shared in the roofing section of Chapter 5; to EIMA for sharing the information on EIFS used in Chapter 6; to Barbara French, Marketing Manager of Keystone Retaining Wall Systems, Inc., for input and the wonderful illustrations used in Chapter 12; to Dianne Walsh Astry, Director of American Lung Association, Minneapolis Affiliate, for the information provided on asbestos, lead, and formaldehyde used in Chapter 13; to Jill Manlove Mayfield, Marketing Coordinator, Planning, Environmental and Conservation Services Department for the City of Austin, Texas, for the definition of "Green" Building used in Chapter 13; to Heidi Harting-Rex, Administrative Director of the Center for Resourceful Building Technology (CRBT), for administering the written interview with Steve Loken, founder of the organization, in Chapter 13; to Tom Craig, Certified Plan Examiner, City of Spokane, Washington, for his time in reviewing certain issues concerning the use of the alternative materials described in this book; and to Karen Craig for her hard and dedicated work in organizing, editing, and proofreading our fifth book together — they're getting easier as we get better organized!

Thanks also to Laurence Jacobs, Editor of Craftsman Book Company, for publishing my first book with them.

I would also like to thank my wife, Kimberly, for her support throughout all my projects, including this one which, for no particular reason, seemed never to end!

This indeed was a community effort and I thank you all for your help and support on a very worthwhile project.

Trademarks

Access Able™
Sachwin Products, Inc.

ACQ® Preserve
CST — Chemical Specialties, Inc.

Adjust-A-Box®
RACO

Adjust-A-Brace®
RACO

Affinity®
ABTco, Inc. (ABT Building Products Corporation)

AFM®
AFM Corporation

Alpan®
Alpan, Inc.

Alterna™
JELD-WEN, Inc.

American Lung Association®
American Lung Association

AMOFOAM®
Amoco Oil Company

AMOWRAP®
Amoco Oil Company

Anaglypta®
Crown Berger Ltd.

AquaTile®
ABTco, Inc. (ABT Building Products Corporation)

BCI®
Boise Cascade

Blow-In-Blanket®
Ark Seal International, Inc.

Blow-In-Blanket System®
Ark Seal International, Inc.

Blue™
The Dow Chemical Company

BOCA®
Building Officials and Code Administrators International, Inc.

Bomacron®
Bomanite Corporation

Bomanite®
Bomanite Corporation

BoWrench®
Cepco Tool Company

Brock Deck™
Royal Crown Ltd.

Cable•Rail™
Feeney Wire Rope & Rigging, Inc.

Canterbury Architectural Moulding®
ABTco, Inc. (ABT Building Products Corporation)

Capri®
Stillwater Products, Inc.

Carefree Decking System™
U.S. Plastic Lumber, Ltd.

Carpenter's Steel Floor Joist®
HL Stud Corporation

Carpenter's Steel "Hud" Stud®
HL Stud Corporation

Carpenter's Steel Stud®
HL Stud Corporation

Carpenter's Steel System®
HL Stud Corporation

Cedar Breather™
Benjamin Obdyke Incorporated

Cedarline™
Giles & Kendall, Inc.

Cedarlite™
Monier, Inc.

Challenge®
JELD-WEN, Inc.

Classicast®
Focal Point Architectural Products, Inc.

Classic-Craft®
Thema-Tru Corporation

Clip Strip®
Royal Crown Ltd.

C-LOC®
Crane Plastics Company

Clorox®
The Clorox Company

Clutterbuster™
Hinge-It Corporation

Cocoon™
GreenStone Industries, Inc., a Louisiana-Pacific Company

Colorlok®
Masonite Corporation

Colortones™
Gloucester Co., Inc.

COM-PLY®
APA — The Engineered Wood Assn.

ComfortBase®
Homasote Company

Comfort Line®
Comfort Line Inc.

ComfortTherm™
Johns Manville Corp.

Composibor™
Louisiana-Pacific Corporation

Conform®
American ConForm Industries, Inc.

Contour-All®
Focal Point Architectural Products, Inc.

Corian®
DuPont Company

Counter Snap™
O'Berry Enterprises, Inc.

Cross Vinylattice®
Cross Industries, Inc.

Cultured Stone®
Stone Products Corporation

Custom®
Custom Building Products

Debri-Shield™
Genova Products, Inc.

Dec-Klip®
Ben Manufacturing, Inc.

Deckmaster®
DeckMaster

Deck Strap™
Supreme Decking Inc.

Deco-Roof™
Stillwater Products, Inc.

Dekmate™
Dek-Block Ontario Ltd. (Canadian Dekbrands)

DesignWood®
Thompson Industries, Inc.

Diamond Snap-Form™
AFM Corporation

Diamond Snap-Ties™
AFM Corporation

DRICON®
Hickson Corporation

Duratron®
Image Carpet, Inc.

DuraFlex®
ResinArt East, Inc.

EC Deck Clip™
U.S. Plastic Lumber, Ltd.

Easy Corner™
Triad Group

Easy-ply
Homasote Company

EB-TY™
Blue Heron Enterprises LLC

Eco-shake®
Re-New Wood, Incorporated

Elite®
JELD-WEN, Inc.

Enviro-Tech®
Image Carpets, Inc.

Eurorack™
Hinge-It Corporation

E-Z Deck®
ZCL Mfg. Canada

Fancy Cuts®
Shakertown 1992, Inc.

Faster Plaster™
Flexi-Wall Systems

Fiber-Classic®
Therma-Tru Corporation

Fiber Frame™
Comfort Line Inc.

Fiberock™
United States Gypsum Company

FIBREX®
Fibrex Insulations, Inc.

FiRP™
Fiber Technologies Incorporated

Flexi-Wall®
Flexi-Wall Systems

Flushmate®
Sloan Value Company

FocalFlex Material®
Focal Point Architectural Products, Inc.

Focal Point®
Focal Point Architectural Products, Inc.

Fold-Form®
Lite-Form International

Folding Lite-Form®
Lite-Form International

4-Way®
Homasote Company

Frostguard®
Traco A Three Rivers Aluminum Co.

Fusion Finish
*ABTco, Inc. (ABT Building Products
Corporation)*

Gang-Lam®
Louisiana-Pacific Corporation

Gibraltar®
Wilsonart International Inc.

Glass Fusion™
TerraGreen Ceramics, Inc.

GRABBER®
John Wagner Associates, Inc.

Grasscrete®
Grass Concrete International, Ltd.

GreenStone™
*GreenStone Industries, Inc., a
Louisiana-Pacific Company*

Hardibacker®
James Hardie & Coy Pty Limited

Hardisoffit™
James Hardie & Coy Pty Limited

**HEALTH HOUSE® A Project of the
American Lung Association®**
*The Minneapolis Affiliate of the
American Lung Association*

Heat Mirror™
Southwall Technologies

Helping Hands™
*Creative Remodeling
Services, Inc. (C.R.S., Inc.)*

Hinge-It®
Hinge-It Corporation

HY-LITE®
HY-LITE Block Windows

IBP Glass Block Grid System™
Innovative Building Products, Inc.

IMSI®
*Insulated Masonry Systems,
International, Ltd.*

IMSI Structure/Coat®
*Insulated Masonry Systems
International, Ltd.*

INFLOOR®
Maxxon Corporation

Insta-Seal®
Insta-Foam Products, Inc.

Iron Woods™
Timber Holdings Ltd.

Jam-It™
Hinge-It Corporation

JELD-WEN®
JELD-WEN, Inc.

KapSeal™
Keystone Retaining Wall System, Inc.

Keystone®
Keystone Retaining Wall System, Inc.

Kitchen Classics™
Sterling Plumbing Group, Inc.

Kuehn™
Kuehn Bevel

Lincrusta®
Crown Berger Ltd.

Lite-Form®
Lite-Form International

Magna-Latch®
D&D Technologies

MaxiPlank™
Maxitile, Inc.

Maxxon™
Maxxon Corporation

Metaltite™
Metaltite Corporation, Inc.

Met-Tile®
Met-Tile, Inc.

MiraDRI®
MiraDRI Moisture Protection Products

MIRAFLEX™
Owens Corning Fiberglas Corp.

Mister Miser®
Mister Miser Urinal

Natural Cork Planks Plus™
Natural Cork, Ltd. Co.

Natural Impressions™
Holz Dammers Moers (HDM) USA

Nature Guard®
Louisiana-Pacific Corporation

NO COAT™
NO COAT Products

Noxon®
Noxon Inc.

Ondura®
Nuline Industries

Pau Lope
Greenheart-Durawoods Inc.

Perfect-Fit™
Vance Industries, Inc.

Perform Guard®
AFM Corporation

Phenoseal®
Gloucester Co., Inc.

Phoenix Recycled Plastics™
Phoenix Recycled Plastics

PinePlus™
*ABTco, Inc. (ABT Building
Products Corporation)*

PINK*PLUS*®
Owens Corning Fiberglas Corp.

Plaster In A Roll™
Flexi-Wall Systems

PlayGuard®
Carlisle Tire & Wheel Company

Plyboo™
Smith & Fong Company

Pole-Wrap™
Pole-Wrap

Power Beam®
Anthony Forest Products Co.

PowerLog®
Anthony Forest Products Co.

PrimeMoulding™
ABTco, Inc. (ABT Building Products Corporation)

proVent™
ado Products

QuickConnect™
Best Dressed Homes Company

QuikJack™
Cepco Tool Company

QUICK'N'EASY CORNER®
NO COAT Products

R-Control®
AFM Corporation

Re-Bath®
Re-Bath Corporation

Resistron®
Image Carpets, Inc.

Re-Source Lumber®
Re-Source Building Products Ltd.

Reward Wall Systems, Inc.™
Reward Wall Systems, Inc.

Rigid Roll™
TrimLine Roof Ventilation Systems Inc.

Romex®
General Cable Company

Shakertown®
Shakertown 1992, Inc.

SHEETROCK™
United States Gypsum Company

Shower Tower™
The Swan Corporation

Sink Undermounter™
Vance Industries, Inc.

SmartBlock™
American ConForm Industries, Inc.

SmartLap™
Louisiana-Pacific Corporation

SmartPanel™
Louisiana-Pacific Corporation

SmartSystem™
Louisiana-Pacific Corporation

SpaceJoist™
Truswal Systems Corporation

SpaceMaker™
Trus Joist MacMillan

SpecLam™
AFM Corporation

SpringLine®
Noxon Inc.

SoftPave®
Carlisle Tire & Wheel Company

Solar SensorLight™
Alpan, Inc.

Solartex®
EnviroWorks Inc.

Sound Barrier Floor Systems™
Homasote Company

Standard Building Code©
Southern Building Code Congress International, Inc.

Starter Trac™
Plastic Components, Inc.

Step Warmfloor™
Electro Plastics, Inc.

Sterling®
Sterling Plumbing Group, Inc.

Strong-Tie®
Simpson Strong-Tie Company, Inc.

Sturd-I-Floor®
APA — The Engineered Wood Assn.

Style-Mark™
Style-Mark Inc.

STYROFOAM®
The Dow Chemical Company

SuperTop®
Carlisle Tire & Wheel Company

Supreme Decking®
Supreme Decking Inc.

Swanstone®
The Swan Corporation

SwanTile™
The Swan Corporation

Tex-Plus™
ZCL Mfg. Canada

The Nailer®
The Millennium Group, Inc.

Therma-Floor®
Maxxon Corporation

ThermaSteel™
Thermastructure XT, Corp.

The Soft Bathtub®
International Cushioned Products Inc.

The Sun Tunnel™
The Sun Tunnel

3-D®
ICS 3-D Panel Works, Inc.

Tide®
Procter & Gamble

TimberStrand®
Trus Joist MacMillan

TimberTech™
Crane Plastics Company

Titebond II®
Franklin International

TRAWOOD®
Traco A Three Rivers Aluminum Co.

TREX®
TREX Company, LLC

Trim-Fit™
Vance Industries, Inc.

TrimLine®
TrimLine Roof Ventilation Systems Inc.

Triple Crown®
Royal Crown Ltd.

Tropic Top™
Tropic Top

Trowel & Seal™
Custom Building Products

Tru-Close®
D&D Technologies

Two Bit Snapper™
Noxon Inc.

Tubwal®
The Swan Corporation

Ultraflex®
NO COAT Products

ULTRA-LATH™
Plastic Components, Inc.

UltraOak™
ABTco, Inc. (ABT Building Products Corporation)

Uniform Building Code™
International Conference of Building Officials

Unique® **Refinishers**
Unique Refinishers, Inc.

VersaDek™
Versadek Industries

Vikrell™
Sterling Plumbing Group, Inc.

WALLMATE®
The Dow Chemical Company

WeatherLock™
Owens Corning Fiberglas Corp.

WECU-Crackless™
Wicander Enterprises, Inc.

WECU-Soundless™
Wicander Enterprises, Inc.

WECU Soundless+™
Wicander Enterprises, Inc.

WENCO®
JELD-WEN, Inc.

Werzalit®
JELD-WEN, Inc.

Wheatboard™
PrimeWood, Inc.

Williamsburg®
Colonial Williamsburg Foundation

Wilsonart®
Wilsonart International Inc.

Wind-lock™
Wind-Lock Corporation

Windows®
Microsoft Corporation

Wolmanized®
Hickson Corporation

WonderBoard®
Custom Building Products

Contents

1

Alternative Building Materials

▮▮

As you drive down the road, what do you see as you look around your community? Do you see new buildings and homes going up, old structures being restored, and others coming down? Have you ever wondered what could be done to save some of these buildings from destruction? And what about new construction — have you noticed anything new or unusual about building procedures or the types of materials used?

Personally, I'm always interested in seeing what kind of construction is going on. After all, that's my business. But at the same time I really hate to see structures torn down. It particularly bothers me when a builder's high-quality workmanship is wiped out in the blink of an eye. I guess you'd call that progress — here today, gone tomorrow. But the other side of the coin is that an enormous volume of valuable resources — the *materials* manufactured and used for these structures (which still have plenty of life) — are destroyed, discarded, dumped into landfills or, as is the case in my area, sent to a waste-to-energy plant. Why are we so eager to wipe out and destroy our past? Is it really "progress" or is it that we simply haven't taken the time to understand the dramatic impact such actions have on the quality of life in and around our communities?

There are alternatives to everything we do in life. The decisions we make today will affect someone tomorrow or even a few years down the road. That's why it's important to try to make choices that won't have a negative impact in the future. Unfortunately,

it isn't always easy to make those choices. As a builder and/or remodeler, you need to ask yourself if what you're choosing to do will improve rather than harm your community. You *can* make a difference!

Don't get me wrong, I'm not preaching against progress. But I do firmly believe that the small stuff we take for granted often has the biggest impact on our lives. So if we can move a building to a new location and save it for someone else to use, that's also a form of progress. If we can refurbish or rebuild it using conventional materials or some of the *alternative materials* that are mentioned in this book, that's even better. That's a kind of progress that's not destructive. It doesn't promote waste. If everyone throughout the world worked toward the goal of nondestructive progress, just think of the impact it would have on our quality of life! Think of the resources we could save!

Making Choices

Let's talk about the alternative materials that I just mentioned. What does the word "alternative" mean to you? To me, in its simplest form, it means "choices." In the construction trade, that means choices between kinds of materials and work techniques. As a contractor or builder, you have a responsibility to make good, well-informed choices for your customer. If you've been doing things the same way your dad did, and his dad before him, then making new choices may come hard. But to be fair to yourself and others, you need to take the time to examine new products coming to the market. Be open to change, even though these products may require different or new methods of installation.

I bet you think I'm going to tell you to get up to speed to stay ahead of the competition. Normally, I would — but the fact is that your customers may beat me to it! Today's customers do their homework and they're smart shoppers. If you're going to stay in the game, you need to be one step ahead of *them*! Do the research, find the products — it could make a difference on which contractor the customer selects for their project.

Be careful though. Alternative products are simply flooding the market, and manufacturers do such a great job of promoting their products that it's easy to be misled. It takes a few years to weed out products that don't have good market appeal or that don't really have the stamina that they're supposed to have. You need to do your homework.

That's where I got the idea to write this book. My purpose is to provide you with information about the alternative products on the market. Unfortunately, I can't cover all of them — there are just too many. But I've tried to cover the ones you're most likely to use. Some of them I've used myself, others I haven't. Some now have established performance records and others are still fairly new to the industry. I hope to give you a basic understanding of how to use and install these products so that you can make an informed decision about whether to consider one of them for your next project. It's that simple!

If you're interested in one of the newer products, you can contact the manufacturers or supply houses to get samples, documentation and references. Talk to other professionals who've installed the product, and, if possible, visit a job site or two and personally view the results. To be successful, you need to educate yourself about the latest building technology.

Why Use Alternative Materials?

Do you ever wonder what happens to the plastic milk containers and the white plastic bags (post-consumer waste) that you give to the recycler? Well, some of today's construction products contain components made from recycled plastics, while others include sawdust from mills (post-industrial waste) and similar by-products of other industries throughout the country. Local mandatory recycling programs, recycling drives, recycling vendors and waste-to-energy plants all generate waste that can be reused in new materials and products. When you use alternative materials made from recycled waste products, you help the environment. Tell this to your customers.

Some products are manufactured using alternative methods rather than alternative materials. Alternative manufacturing methods often make use of materials that at one time would have been considered waste. Mill ends (2 × 4 studs made up of short 2 × 4s with glued finger joints) are an example of a product produced by alternative methods. Other products are made from blends or combinations of virgin materials and manufactured materials that increase the product's life expectancy. These also help the environment. Products that last longer are less likely to need replacing, and so cost less in the long run. The result is a reduced need for new materials as well as fewer materials discarded into landfills.

Using alternative materials, or products that take advantage of alternative manufacturing methods, not only helps protect and restore our natural resources, but also creates employment, which is healthy for our economy. It can also make your business more profitable. As you can see, everyone benefits!

Can You Save Money or Time?

Do alternative materials and procedures save time and money? I'm not sure there's an easy answer here. Most of the products I've worked with cost about the same as conventional materials — some a bit more. This may be because some of these products just haven't been on the market that long.

For products made with recycled materials, it's possible that a shortage of available recyclable materials may develop. This doesn't mean there isn't enough to be recycled, but not enough people are coming forward and doing their part to recycle. When demand for the product is high and natural resources diminish, then the price goes up. I suspect in this case it won't be the demand that will drive product costs up, but a shortage of recyclable materials. My question is, how much of a demand can we put on recyclable resources before they dry up?

Generally, after a product has been on the market a few years, demand for it increases. Also, new facilities are built to produce the products from recycled materials, and manufacturers make production more efficient. Together, those factors begin to drive the cost down.

As for saving time, it's just like any new product that comes to the market. There's a learning curve; it takes time to understand the characteristics and the feel of the product. This means (in some cases) that it'll take longer to install these products, at least for the first couple of jobs. You can probably reduce that labor time once you've had the opportunity to install the product a few times. I know when I first started using TREX decking (a product of TREX Company, LLC), it took me a while to get used to it, especially handling a board 16 feet long. Once I'd worked with it a few times and understood its characteristics, I was able to cut my labor time, in some cases by almost 50 percent. But don't count on any alternative product necessarily saving you 50 percent on your labor. You may be able to save on the job costs for the overall project, but some savings may come from labor, others from less maintenance, while others come from fewer callbacks.

It's probably too early to tell whether or not installing alternative products is cost effective, especially for products that haven't been on the market that long. Only you can make that decision on your jobs. The records you maintain could provide valuable information for the manufacturer and also serve as a selling tool for potential customers. Not only that, it will help you to understand the products better and to evaluate whether you're really helping the environment and saving money with the new materials. Set your own track record when it comes to these products!

Do They Meet Building Codes?

Do alternative materials meet building codes? Most of them do — but not necessarily! What does that mean? Not all new products have the track record required for acceptance by building code departments, and some products are accepted by building departments in some areas and not in others. That's why it's important when researching alternative products to request information from the manufacturer on the product's physical and mechanical properties. They should have a report on product testing by an independent, nationally recognized, certified testing corporation. In some cases, this documentation will be enough to get the product accepted.

Better yet, check to see if the product has been evaluated and listed with the National Evaluation Service, Inc. (NES). If so, they'll have an evaluation report that you could provide as evidence of code compliance to your local building department. Don't assume that the building department in your area knows of the product you plan to use. When you approach them, it may actually be the first they've heard of this product. If this is the case, the flags will go up immediately. Be prepared to show some type of documentation or an evaluation report on the product. The National Evaluation Service Secretariat is located at:

National Evaluation Service, Inc.
900 Montclair Road, Suite A
Birmingham, AL 35213-1206
205-599-9888
jheaton@sbcci.org (e-mail)
www.nateval.org (Web site)

Local building departments follow their own adopted codes or one of the three widely-accepted model building codes:

National Building Code
Building Officials & Code Administrators
International (BOCA)
4051 West Flossmoor Road
Country Club Hills, IL 60478-5795
708-799-2300

Uniform Building Code
International Conference of Building
Officials (ICBO)
5360 Workman Mill Road
Whittier, CA 90601
562-699-0541

Standard Building Code
Southern Building Code Congress
International (SBCCI)
900 Montclair Road
Birmingham, AL 35213
205-591-1853

Each of these organizations supports an evaluation service. If the manufacturer plans to market the product on a national level, they'll submit the product to NES. But if they plan to market only in their own geographic area, they may go to their model code evaluation service to have the product evaluated at a certified testing laboratory and quality agency.

The evaluation reports are available to the member building department jurisdictions and other users. Some organizations may charge for this report, but others will send it to you automatically through a subscription service. If you want to use a product in an area that's not under the model code where the evaluation took place, get an evaluation report for the building official in your area. Of course, it doesn't guarantee the product will be accepted, but it may meet local codes.

Here are the numbers for the evaluation services of the three model code agencies:

BOCA Evaluation Services, Inc.
708-799-2305
boca@aecnet.com (e-mail)
www.boca-es.com/~boca-es (Web site)

ICBO Evaluation Service, Inc.
562-699-0543
es@icbo.org (e-mail)
http://www.icbo.org (Web site)

SBCCI Public Safety Testing & Evaluation
  Services, Inc.
205-599-9800
rfazel@sbcci.org (e-mail)

You'll get along better with inspectors if they know beforehand what you plan to do. In other words, don't pull any surprises on the inspector in the field. Get your materials or method of construction approved *before you start work*. The building department staff knows that new products hit the market every day. They do have some latitude and can help you both comply with the codes and satisfy your customer. It's in your best interests to keep an open mind, work with the departments involved, be willing to compromise, and encourage your customer to do the same.

Here's the bottom line. Work *with* the building department. And, of course, know the products you want to use. Do your homework and get the proper

documentation supplied by the manufacturer or the evaluation report supplied by the model code agency's evaluation service.

Will Your Customer Benefit from These Products?

Can alternative materials be beneficial to your customers? Of course! Anything you do that makes your customers feel more comfortable within their environment provides a benefit. But it's important to find out what your customer needs, then use the right products to fit those needs. You could inadvertently select a product that causes health problems for the customer. Certain paints, carpets, caulks, and adhesives contain chemicals that could affect asthmatics or people with certain allergies. It's your responsibility to ask your customer and check with the manufacturer on any new products you're considering.

In other words, work with your customer and know the products you want to install. Remember, you walk away from the completed project. Your customer, on the other hand, has to live within the environment you create. Walking away isn't an option for them.

Selling Alternative Materials

When selling alternative materials and construction techniques, you need to know your customers as well as the products. It could make the difference in whether or not you close a sale. So, what types of customers do you work for? Looking back on my own customers (and I assume they're pretty typical), I've outlined the most common types of customers you'll be dealing with on a regular basis:

- The customer is looking for a basic, affordable structure; will sacrifice looks and conveniences to get it.

- The customer is looking for the biggest bang for the buck; will sacrifice quality.

- The customer is interested in premium products, but for reasons other than a premium item; wants a product that will hold up under a disaster.

- The customer is willing to pay more for a premium product to get a product that's a little unusual.

▌The customer is willing to pay more for a product in order to achieve maximum energy efficiency for their home, to save on their utility bill!

Perhaps you could add a few additional types to this list. If so, then you're on the right track to really knowing your customers. It may seem trivial, but this knowledge can really help as you prepare your bids. You can select just the right product at just the right price. It really could make a difference whether or not you get the job!

Reusing Materials

What about salvaged materials? Earlier I mentioned all the old buildings that are being torn down. While many are salvaged for their materials, others are simply dismantled and hauled off to the dump. Why not salvage all that can be salvaged and reused, and then — *and only then* — discard what *has* to be discarded? Reusing building materials is actually very cost-effective and requires far less energy than recycling.

When Expo 74 came to Spokane, all kinds of buildings were constructed to house the exhibits. When it closed, the city recycled those buildings and their contents. A subcontractor friend of mine bought the hardwood flooring from one of the buildings and I helped him install it in his rec room. He now has flooring that was once part of a major historical event. People from all over the country — all over the world — walked on that very floor! Reusing materials like that creates a history for the home and makes living in it a more unique experience.

It can be the same for you and your customers. Just think of using beams that may have once been in a ship, an early industrial building, or even an 18th or 19th century warehouse. In many cases suppliers can actually provide documentation to go along with these historical products. Customers get excited about building with items that belonged to a different era. They're conversation pieces that add interest to their home, and you know how your customers like to show off their castles.

There are a few salvage companies in Seattle with huge warehouses filled with furniture, building components and materials. When I walked through the doors of one, I swore I had stepped back in time. You wouldn't believe the things they had on display! What especially caught my eye was a beautiful oak staircase, fully intact, including balusters and handrails, that had been salvaged from a turn-of-the-century mansion. It was absolutely stunning! The asking price plus shipping was less than it would cost to recreate and install such beautiful workmanship, even if you could find someone who could do it. As a matter of fact, it would probably be difficult (if not impossible) to duplicate it today at any cost. This piece, with its class and heritage, warranted installation in a very special home. Don't you agree?

My point is that if you can use salvaged materials, you're recycling not only material, but history as well. And installing an existing component rather than using virgin materials is a choice to use alternative methods of construction. Other materials you might want to consider recycling are bricks, fixtures, hardware, and metal and tile roofing. Speaking of roofing, you should have seen the brass and copper weathervanes on display at that salvage company!

Looking into the Future

As you can see, whether you use new materials made from recycled materials, new but longer-lasting products, or salvaged materials, you're still helping to conserve natural resources and the environment.

Are alternative materials right for every facet of every job? Not always. You'll want to carefully compare their costs against traditional construction materials and techniques. The consensus seems to be that alternative methods and materials may have higher initial costs than virgin wood products, but may offer other desirable advantages well worth considering. These include increased thermal efficiency, fire resistance, or durability.

Of course, some products could be more cost-competitive if their structural properties were fully used. Engineered wood products (EWP) are a good example of this. Better use of the longer spans now offered by some EWP manufacturers could eliminate the need for intermediate framing supports, and their related costs. Continuing lumber price increases and reduced availability of raw materials will eventually make alternative methods and materials more cost-competitive. However, this process will take time.

Where do you go from here? You only have to turn the page to begin . . .

2

Foundations

■■■

Where you live, houses may not normally have the luxury of a basement. In some areas, they're just not necessary. In the Northwest where I live, just about every home has a full basement — and often the basement is incorporated into a multi-level home. Because we have a required 3-foot-deep frost line here, most people simply say, "Why not go down a few more feet and have a basement?" Turning all those basements into "habitable environments" kept me busy for years! (That's quite a mouthful, but that's the terminology the building department seemed to prefer.)

It seemed odd to me at the time, but people usually only wanted their basement finished when they were preparing to sell their house. I guess they were hoping to increase the selling price. But that's too bad. Had they finished the basement sooner, they might not have needed to put the house on the market. A nicely-finished basement adds warm, comfortable living space to any home.

It's amazing how many homes built in my area at the beginning of the century have rock foundations. I guess the contractors were working with the materials at hand, but those rocks sure made finishing a basement difficult! Today, contractors have many more choices when it comes to building a foundation.

There are more than just a handful of manufacturers, methods, and systems out there. You don't have to rely on the conventional method of having a foundation crew come in to set up forms and then having concrete poured. But you do need to research the systems to learn which are most suitable for your area and which you'll be comfortable installing. Some of these systems can be used as both foundation and walls — referred to as "insulating concrete forms," or ICF. Others, such as structural panels, are installed after the footing and foundations are in place. Structural panels have a foam core sandwiched between oriented strand board (OSB) sheets. Still others include welded-wire sandwich panels (foam core sandwiched between steel mesh) or concrete wall systems.

These systems could include, but are not limited to:

■ Permanent wood foundation

■ Concrete blocks (mortared and mortarless)
  Preinsulated
  Structural lightweight
  Aerated concrete

■ Poured-in-place — conventional
  Panel (flat)
  Grid (post-and-beam)
  Shotcrete

■ Panelized (prefabricated)

In this chapter, and the next, I'll cover the most common manufacturers and systems currently on the market. Some of these you may want to research in more detail. I've included a list of recommended reading material as well as an Appendix with a list of manufacturers that produce alternative materials. You can never have too much information — so stock your library!

Permanent Wood Foundations

Because I'd never installed one, I never gave permanent wood foundations much thought until I learned that the concrete and rock foundation on my grandmother's house was deteriorating. Since she was on a limited budget, cost was a big factor. She had a partial basement that someone had dug by hand. It had about 5 feet of interior concrete and a ledge about 2 feet in depth. The rest of the area was crawl space. Under the circumstances, it was easier to remove one small section of the foundation at a time and build a new wooden foundation. Once the exterior stucco was installed, you couldn't tell the foundation was made of wood. Some contractors say they save 15 to 50 percent in foundation costs by using such a system. The only way you'll know for sure is to build a few and then compare costs.

Permanent wood foundations aren't new to the industry; they were introduced in 1969. But before you decide to go with this system, touch base with your local building department and the manufacturer of the specific materials you plan to use. Because this isn't conventional construction, the building code outlines specific requirements you must meet. The manufacturers will provide you with recommendations, specifications, and guidelines for the proper installation of their products. The most important factors in a permanent wood foundation are the dimensional lumber and the plywood — they have to be all-weather wood foundation material. You *can't* use standard pressure-treated wood!

Materials for all-weather wood foundations must be treated to 0.60 pcf retention. For those of you who aren't familiar with the phrase "retention levels," it means the amount of preservative that remains in the cell structure after processing is completed. These levels are expressed in pounds of preservative per cubic foot (pcf) of wood. The higher the number, the harsher the conditions the wood will endure. The material you normally use for decking, fencing, sill,

railings and joists — all above-ground uses — is treated to 0.25 retention. That's not enough for a foundation.

The pressure-preservative-treating process permanently protects the foundation from fungi, termites, and other causes of decay. Besides having a retention level of 0.60 pcf, wood used for permanent wood foundations *must also be kiln dried after treatment* (KDAT). Always check the quality mark for this. (We'll discuss quality marks shortly).

For complete design and construction recommendations, contact Hickson Corporation, American Forest & Paper Association, or the Southern Pine Council (their addresses are listed in the Appendix).

To learn more about the advantages of a permanent wood foundation, contact the Southern Pine Council and ask them to send you the three publications listed below. The first one you can give to your customers to read, and the other two are good references for your library.

- *A Home Buyer's Guide to the Permanent Wood Foundation* — This provides you and your customer with a quick overview of the permanent wood foundation and its advantage as a basement alternative.

- *Facts About Permanent Wood Foundations* — An outline, for both you and your customer, of the facts regarding permanent wood foundations and the wood industry's response to the concrete industry's claims that wood basement materials are treated with arsenic, creating a health risk.

- *Permanent Wood Foundation — Design & Construction Guide* — This will answer a lot of your questions on the elements of the system. The section on soil conditions and drainage is particularly interesting.

Treated Lumber

We've used wood products containing chemical preservatives for the last 60 years. These waterborne preservative chemicals are forced into the wood's cells under pressure inside a closed cylinder. A chemical reaction between the preservative and wood sugars forms an insoluble compound. It's chemically fixed during the drying process, permanently adhering the preservative to the wood's fibers. This process creates a product commonly called treated lumber.

While there are a number of wood preservatives on the market today, only one is used extensively in residential and commercial construction. It's known in the industry as CCA (chromated copper arsenate). It uses water as a solvent or carrier and copper as the primary fungicide to prevent rot and termite attack. The copper reacts with the sun's UV (ultraviolet) rays as the lumber dries. The arsenate in the preservative is primarily an insecticide to ward off termite attack, and the chromium facilitates its fixation.

The oxidizing process turns the material a greenish color, which fades to a natural gray color as it weathers in the sun. While this is an acceptable color for wood on the East Coast, many customers on the West Coast don't like it. To increase the marketing appeal, some manufacturers add a dye or color additive stain before and/or after the pressure-treating process to turn it brown instead of green.

Your customers may be concerned about these chemicals and their impact on the environment. According to the experts, the insoluble compound renders the CCA preservative largely nonleachable. Several decades of field testing have shown that only insignificant amounts of arsenic, copper, and chromium components have leached from pressure-preserved materials using CCA. Overall, treated lumber is a very stable product.

To learn more about the use of treated lumber, request a copy of *Answers to Often-Asked Questions About Treated Wood* from the American Wood Preservers Institute (AWPI). Your customers may enjoy reading it as well. It runs $19.95 (plus S&H). Order from:

American Wood Preservers Institute
2750 Prosperity Ave., Suite 550
Fairfax, VA 22031-4312
http://www.awpi.org (Web site)

Quality Marks

You should know exactly what you're getting in a treated wood product before you buy. Check the quality mark! Make sure that any preservative-treated lumber you get carries the quality mark of an approved independent testing agency. The stamp or

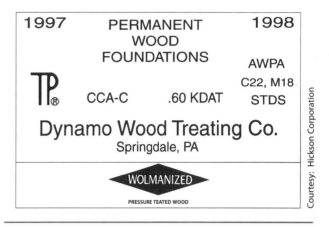

Figure 2-1
Typical quality mark for treated lumber

end tag must show that it complies with standards of the American Wood-Preservers' Association (AWPA) and the identity of the accredited agency. Figure 2-1 shows a sample quality mark. Notice the locations for the type of preservative treatment, the preservative retention pcf, and the AWPA standards. Look over this sample carefully, and use it to help you interpret what quality marks mean.

Construction Notes — Wood foundation material can be in direct contact with the ground. A footing isn't necessary. Instead, just dig a trench and fill it with gravel. But make sure the trench dimensions comply with your area's building codes. Slope the trench so any water can drain away from the building into a sump. This prevents the buildup of pressure against the foundation and prevents leaks. Tamp the gravel until it's 95 percent compacted. Then lay a 2 × 8 sill (or whatever meets your code requirements) around the perimeter in the center of the compacted gravel.

From this point on, it's like framing any walls in a new home. Prebuild the walls and add the plywood. Build the walls in sections small enough that you can handle, unless you're using a crane. Overhang the plywood by the dimensional width of your framing members on each end of those walls where the adjacent wall will tie at the corner (Figure 2-2). Remember to caulk all plywood joints before you install and fasten the walls. Then stand them up and nail through the bottom plate into the sill. Don't forget to nail the plywood at the corners to lock the entire system together (Figure 2-3).

Courtesy: Hickson Corporation

Figure 2-2
Prebuilt walls of Wolmanized wood
for permanent wood foundation

Courtesy: Hickson Corporation

Figure 2-3
Nail plywood at corners to lock system together

I prefer to install footings with concrete bolts. This provides a uniform and level surface to work from, as well as bolts to anchor the wall to the footings. Stainless steel hardware is required below ground level.

MAS Mudsill Anchors

Speaking of bolts, you may also want to try the MAS Mudsill Anchors by Simpson Strong-Tie. They're suitable for slab or stemwall construction. You can install them before (by nailing into the bottom plate) or insert them into the concrete after you pour. No more predrilling bottom plates; no more lifting walls up and over anchor bolts; no more bolts directly under framing members! You can probably guess that I like these conveniences. Figure 2-4 shows the Mudsill Anchors and SP Stud Plate Ties, which were used to secure the framing member to the bottom plate. This protects the wall from an uplift or seismic activity.

Waterproofing

You would think that the higher the retention rating of the preservative, the fewer the concerns about waterproofing — but that's not so! You still need to protect the wood from moisture.

There are many ways to waterproof a foundation, depending on the type of soil present at the construction site. The method I prefer with permanent wood foundations involves applying polyethylene membrane around the foundation. Install the membrane so it fastens above ground level and drapes over the footings. Overlap the corners and seal the laps with plastic sealer tape or a compound like clean butyl sealant that will stick but doesn't contain chemicals (solvents) that will break down the membrane. Then, instead of backfilling with earth up against the membrane, install gravel halfway up the foundation wall (the height of the wall that you'll be burying) and out as far as the footings. Install a second layer of the membrane back up the foundation wall and over to cap off the gravel. Backfill with earth after the basement floor and first floor joists are installed.

You may wonder how you keep the gravel in place while you try to backfill. Well, you can't entirely. But you may find it easier if you dig the hole for the foundation about 18 inches wider than the end of the footings. This will give you about 2 feet of space, enough room to tie the corners together and install the membrane. Of course, the wider the hole, the more gravel it takes to fill it. The type and amount of backfill you need for this will vary with the on-site soil conditions.

MiraDRI Moisture Protection Products

MiraDRI Moisture Protection Products and their technical services can help you meet all of your waterproofing challenges. Using a combination of these products, installed with their recommended techniques, can create an effective waterproofing system. And it's easier to install than the polyethylene membrane and gravel method I just described.

Figure 2-5 shows a MiraDRI system attached to a standard concrete foundation. Notice how the system wraps over the footings. The products shown include the following:

▌ MiraDRI All Weather Primer (AWP) — A latex-based primer that you can roll or spray onto concrete and masonry surfaces, gypsum, and wood surfaces.

▌ LM-800 — Designed to use at termination points, such as flashing to curbs and parapets on plaza decks, mechanical rooms, and other similar horizontal waterproofing applications. Cold-applied, it can create a fillet, as shown in Figure 2-5.

▌ MiraDRI 860/861 — These are self-adhering 3-foot-wide membranes suitable for both vertical and horizontal applications. They're composite materials consisting of specially-formulated rubberized asphalt that's laminated to high-impact-resistant plastic film. Once the wall surface is dry and free of dirt, loose aggregate, or other foreign materials, you can remove the release paper and apply the membrane starting from the bottom and working your way toward the top.

▌ MiraDRI M-800 Mastic — Rubberized asphalt sealant in a caulk tube that you use to seal off all exposed membrane edges.

▌ MiraDRAIN 6200 — These have a three-dimensional, high-impact polystyrene core, and a nonwoven filter fabric bonded to the dimples of the molded panels. The filter fabric prevents soil particles from passing into the core, permitting free drainage. Five types of filter fabrics are available to meet different soil conditions. You can hold the panels in place with MiraBond tape or MiraStick adhesive. Cover the top of the core's edge with the filter fabric flap by tucking it behind the panel.

MiraDRI products help reduce hydrostatic pressure against below-grade structures, and filter saturated soils by collecting and conveying groundwater to a drainpipe for discharge. You still need to tie the

Courtesy: Simpson Strong-Tie Company, Inc.

Figure 2-4
MAS Mudsill Anchors

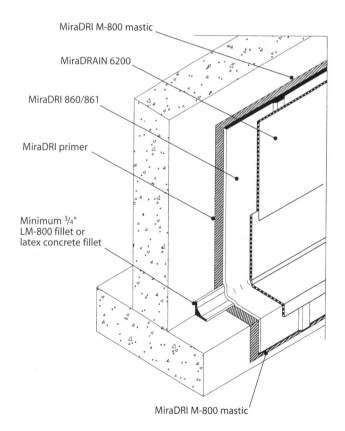

MiraDRI M-800 mastic

MiraDRAIN 6200

MiraDRI 860/861

MiraDRI primer

Minimum ³/₄"
LM-800 fillet or
latex concrete fillet

MiraDRI M-800 mastic

No Scale   Courtesy: MiraDRI Moisture Protection Products

Figure 2-5
MiraDRI moisture protection system

system into a perimeter drain system at and around the footings. Cover the top of the system with at least 6 inches of soil and backfill immediately once it's installed. To learn more about this system, contact MiraDRI Moisture Protection Products. Their address is in the appendix.

Don't limit waterproofing just to permanent wood foundations. Standard concrete foundations, ICF systems and any other system used below ground will require waterproofing as well.

Backfill

It's very important not to backfill against the foundation until you've completed these steps to provide support against the weight of the backfill:

1. Install floor joists and subflooring to tie the foundation walls together into one system.

2. Pour the interior concrete floor right up against the bottom plate of the foundation's framing. Consider pouring the concrete over gravel and a vapor barrier for a complete drainage system. You can also put in a wood floor system instead of concrete for maximum comfort for your customer.

Completing these two steps will prevent the top and bottom of the foundation from caving in when you backfill. It's also a good idea to check for radon and install the proper ventilation if needed before you finish the basement floor. Backfill in 6- to 8-inch layers and tamp to compact.

Manhours — Keep in mind that these manhours are general guidelines for your use. Always consider what you and your employees are capable of handling. You'll have to compare the manhours I've listed to your own bidding experience. You may want to bid a little higher if you're not experienced in the installation. Remember, these manhours are just starting points to help prevent you from underestimating your manhours.

Building a permanent wood foundation is really no different from framing a building with wooden studs. The key is to build in sections (prebuilt panels) that can be easily handled by a crew of two. Long walls, complete with sheathing, require more assistance or a crane to lift into place. In this case, you'll need to allow for the extra labor or for the rental equipment and operator.

Estimate 0.049 manhours (MH) per square foot (SF) for a crew of two to frame 2 × 6 walls, 8 feet high, including the installation of $^{23}/_{32}$-inch plywood sheathing. For example, if you're framing a 1,000 square foot basement, you multiply your manhour figure, 0.049, by 1,000 square feet to get an estimate of 49 manhours for the project. Adjust your estimate to include an additional 30 minutes (0.50 manhours) for each corner, because the inside and outside corners require caulk and must be secured. For a 1,000 square foot basement with four corners, add 2 hours to the labor figure to get a total of 51 manhours. Assuming an 8-hour day (if there is such a thing), it would take 6.375 days to frame this basement after the hole is dug and the gravel base is properly installed. The recommended crew of two will cut this time to approximately three days.

To install MiraDRI's complete moisture and drainage system, a crew of two can normally handle the self-adhering drainage system, but three or four would certainly speed up the process. For the purpose of calculating manhours, I'll break it out into two parts since you can use both systems independently or install them as one complete unit — moisture and drainage system.

1. To install primer, fillet, self-adhering membrane, and caulk seams, figure 8 hours for two men for 2,000 square feet, or 250 square feet in one hour, or 0.24 manhours per square foot.

2. To install the filter fabric drainage panels, again figure two men for 8 hours per 9,600 square feet, or 1,200 square feet per hour, or 0.05 manhours per square foot.

If you had a 4,000 square foot foundation to cover using both systems, how would you figure your total hours? There are a couple of ways to achieve total manhours for two. One way would be to divide 4,000 by 250 square feet for 16 hours, and 4,000 by 1,200 square feet to get 3.33 hours, for a total of 19.333 manhours for a crew of two. Or you could add 0.24 plus 0.05 to get 0.29 times 4,000 square feet, and get 1,160 minutes. Then divide 1,160 minutes by 60 to get 19.333 manhours. Divide this by 8 hours for a total of 2.416 days to complete this project. Don't forget that it takes at least two people to complete this project and that this is only an approximate figure.

Concrete Wall Systems

It seems no matter where you travel, you see commercial buildings with concrete block walls. Because of its relatively low insulating properties, concrete block isn't used much in the residential market. But in areas where the energy code isn't a big issue, like Florida, it's an economical way to go. Since the rapid inflation of the 1970s, the average price of concrete products has been moderately stable (compared to lumber). Still, while concrete is made of the most abundant materials around, many contractors (including me) didn't consider it the norm. It just hasn't been commonly used in houses. Maybe this is because of the extra work involved (like furring the interior) to raise the R-value, or create space to channel electrical wires, or give a solid backing to fasten the drywall. You have to consider all of these key elements.

One more thing to consider is the makeup of the concrete used in the block itself, and concrete in general. The weight of the *cement* used can be reduced by 15 to 30 percent by substituting fly ash for cement, depending on the type of fly ash used and the intended application. Fly ash is a by-product of coal-fired electric generating plants. Today, about 54 million tons of fly ash are produced each year, with about 9 million tons used in engineering applications and in concrete products. Besides reducing the weight, fly ash adds strength and durability to the final product. It's also easier to work with and provides a slightly smoother finish. On the other hand, concrete containing fly ash cures more slowly than traditional concrete.

To learn more about fly ash and whether this is the right product for your project, contact the American Coal Ash Association (ACAA). The ACAA works to advance the management and use of coal combustion by-products (CCBs) in ways that are technically sound, commercially competitive, and environmentally safe. Their library and newsletter, *Ash at Work*, are both packed full of useful information. For more details, contact them at:

American Coal Ash Association, Inc.
2760 Eisenhower Avenue, Ste. 304
Alexandria, VA 22314-4554
703-317-2400
ACAA-USA@msn.com (e-mail)
http://www.ACAA-USA.org (Web site)

Today, the combination of insulating and structural materials is coming on strong and is widely accepted in the construction industry. Products such as insulated, structural, lightweight, and aerated concrete blocks are both more user-friendly and more thermally efficient than conventional concrete block. You can use them below ground for foundation applications (providing you make provision for a moisture barrier), but they're really designed for a complete system above ground. The next chapter will cover above-ground applications more in depth (no pun intended!).

Stay-In-Place Systems

With the development of insulated wall systems, concrete offers new opportunities for the residential and commercial market. One of these systems is the insulating concrete forms (ICF). Basically, it consists of stackable interlocking polystyrene insulating blocks that are lightweight and easy to install. After installation, their cavities are filled with concrete and the forms are left in place, creating a wall with an R-20 to R-25 insulating value (depending on the product) ready to accept interior wallboard and exterior finish.

The beauty of this system is that you're not limited to the foundation. You can build the entire house this way. The framework of the ICF is made up of three basic systems: panel, plank, and block (which I'll discuss later on in the chapter). If you strip away the polystyrene to get to the concrete cavity, you would find three basic shapes: flat (similar to conventional concrete wall), grid (similar to that of a breakfast waffle), and post-and-beam (Figure 2-6).

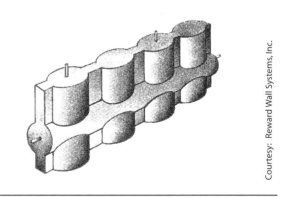

Figure 2-6
Interior concrete configuration of post-and-beam ICF wall

Courtesy: Reward Wall Systems, Inc.

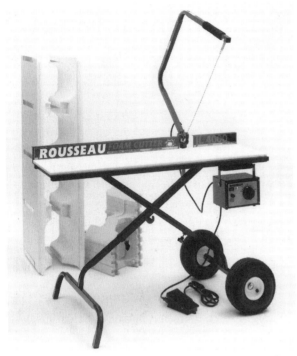

Courtesy: Rousseau Company

Figure 2-7
Hot wire cutting machine

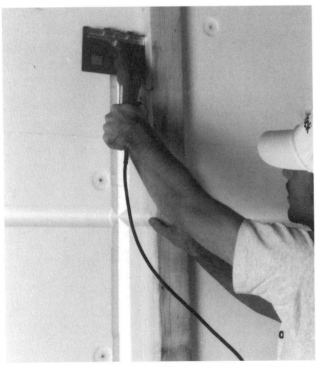

Courtesy: Avalon Concepts Corporation

Figure 2-8
Quick-Cut thermal hot knife

In this system, the posts and beams are spaced as close together or as far apart as you want. While the post is continuous (vertically), the beam (horizontally) is achieved when you install a special block called the *lintel block*.

The blocks are held together with plastic or metal ties. Some ties are designed carefully enough so the concrete easily flows horizontally and fills the vertical cavities, but others don't permit proper movement of the concrete. This is something you want to research before choosing a system. Most systems require concrete with smaller aggregate and a medium- to high-slump to ensure that the mix will flow thoroughly through the cavities within the system. (The *slump* indicates how thin or runny the concrete is.) Be sure to check with the manufacturer on the recommended concrete mix for their system. Some manufacturers also incorporate nailing strips. Some are exposed; some, like the ends of the tie, are embedded. Whatever the case, it's recommended that you use screws or hot-dipped galvanized nails to attach these nailing strips, as common nails don't hold.

This quick overview of some of the systems available certainly isn't the final word on the subject if you plan to enter this market. A little later I'll discuss some specific products. But first, we need to talk about the tools you'll need to work with the products we've covered.

Tools

Even the most efficient system still requires the proper tools to achieve precise component fit. The conventional handsaw, keyhole saw and utility knife may end up as some of your basic tools. But there are other tools on the market to help speed up the process. You may want to check into them, especially if you plan to install these systems on a regular basis. When dealing with expanded polystyrene (EPS), extruded polystyrene (XPS), exterior insulation and finish systems (EIFS) products, and insulating concrete forms (ICF), you may find a hot wire machine a little faster and cleaner than a handsaw. Figure 2-7 shows the Rousseau JL4000 with a power foot control that frees both of your hands. This tool is specially designed to give an accurate cut without the cutting debris you usually get with other cutting methods.

You can use a utility knife or router to cut chases for electrical wire and/or boxes, but I can suggest an easier way. Using a 2 × 4 as a guide and a thermal hot knife with a V blade (Figure 2-8), you'll glide through the foam for fast and accurate cuts. I suggest you consult EPS and XPS manufacturers before attempting to cut into any of the insulation currently on the market. Ask specifically for the material safety data sheet (MSDS) so you can learn about the material's flash point and toxicity. There's some concern about toxicity, and the product does give off an odor that isn't particularly pleasant. It goes without saying that you need to work with these products in well-ventilated areas.

If you've ever used a router, you know just how much of a mess it can create. In polystyrene the mess resembles . . . well, just think of a bean-bag chair falling off the back of a pickup at 55 mph! Polystyrene beads all over the road! Now imagine what that would be like in an enclosed area. A hot knife makes the job a whole lot simpler and cleaner. The nice thing about this tool is that the foam you cut out from the chase is in one continuous piece. That means you can glue it back into the chase for insulation before you install the wallboard. As you can see in Figure 2-9, there's no limit to what you can cut with the right accessory blade. For cutting or shaving high spots, a thermal skinner, a hand-held tool that works on the same principle as a hot wire machine, would be just the ticket.

Depending on the product, you may find yourself gluing joints. It's important to choose an adhesive that's compatible with the polystyrene. Consider using a polyurethane foam that serves as both an insulating sealant and an adhesive. It expands and assumes the shape of a cavity, forming an airtight seal. It's great for filling holes and gaps, especially around pipes. You may also find that an industrial applicator such as an NBS gun would be a wise investment (Figure 2-10). The long nose gets you into those hard-to-reach areas.

Because rebar is used in ICF systems, consider renting a rebar cutter-bender. If you're serious about these installations, consider purchasing one. Concrete has to be pumped into the forms, so look for a reliable subcontractor with a boom and/or grout pump. After you've put up a few systems, you'll find yourself picking up a few more tools for your collection.

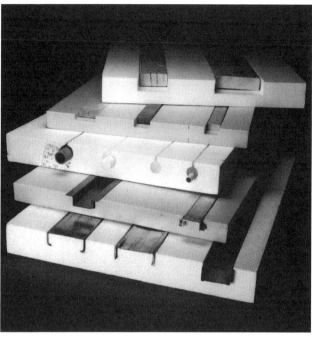

Courtesy: Avalon Concepts Corporation

Figure 2-9
Wind-lock's blades for a hot knife

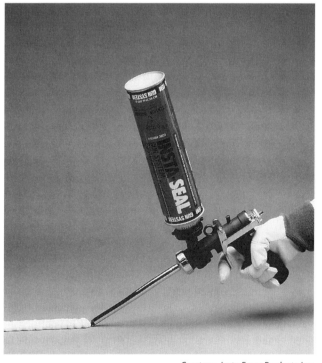

Courtesy: Insta-Foam Products, Inc.

Figure 2-10
NBS gun for gluing joints

Courtesy: American ConForm Industries, Inc.

Figure 2-11
SmartBlock wall

SmartBlock

Insulating concrete forms have been on the market worldwide for decades. About seven years ago, American ConForm Industries developed a refined product that uses interlocking teeth, called SmartBlock (Figure 2-11). This interlocking system of gray blocks was designed for speed and accuracy and to ensure straight and plumb walls with perfect right angles. Special end-caps for the corners prevent concrete spillage. With an R-factor between R-22 and R-24 (depending on the product), SmartBlock is made from expanded polystyrene (EPS) that weighs in at 2 pounds per unit. It's available in a standard fixed width size of 10 × 10 × 40 inches, with a concrete wall thickness of $6^{1}/_{2}$ inches. Uniquely, it's also available in variable widths with removable ties. (Figure 2-12 shows some of the end-caps and ties.) The ties are color coded to represent different sizes.

Any width interlocks with any other, creating a permanent system — providing you've installed the first course correctly. Besides using it for an entire wall system or foundation, you can also consider it for creating pier and grade-beam foundations, fences, retaining walls, landscaping, and swimming and lap pools.

With some special and specific detailing, you can create radii and curved foundations/walls (Figure 2-13). Consider the difficulties involved in creating this radius using conventional methods of building foundation walls! ConForm provides a working chart to help you figure the width of cuts required for an inside or outside radius. You make the cut through one side of the block and into the center of each interior cell (the open space between a solid bridge or plastic tie). The cell is the area that's filled with concrete.

Additional Considerations

Stucco, waterproofing, wallboard, siding, and other finishes can be applied directly to the foam surface, allowing a complete range of design options that are available with traditional building techniques. However, you need to consider the types of finishes you want to use and how you plan to install them before filling the forms. I suggest you check with your local building department for prior approval. Here are a few things to think about before you pour the concrete:

▌ The concrete has to be pumped into the forms. With that in mind — will a boom and/or grout pump be able to get into the area where you're building the system?

▌ Rebar must be installed vertically into the footings to tie the wall to the footings. The forms

Courtesy: American ConForm Industries, Inc.

Figure 2-12
Ties and end caps for SmartBlock

themselves, according to engineering specs, require rebar vertically as well as horizontally.

▌ Do you plan to face the exterior of this system with brick? Don't forget to install the galvanized anchors before the concrete is poured.

▌ What about installing first- or second-story floors? How do you plan to attach the floor joists? Ledger boards need to be installed by anchor bolts in the concrete. Don't forget those bolts!

▌ How will you install floor joists (low wall), trusses, or rafters? They should rest on a sill plate held in place by anchor bolts supported by the concrete.

▌ Have you considered doors, sliders, and windows? Again, install anchor bolts in the concrete for both the header and sill. Be sure to sink the anchor in the concrete so when the washer and nut have been recessed into the wood, the end of the anchor doesn't protrude from the wood. If it does, you can always cut off the bolt once the header and sill are firmly secured.

▌ Do your plans to create a chase for the electrical and plumbing affect the structure of the foundation? If so, then you need to take special measures. Consult an engineer to help plan ahead for this.

▌ How do you plan to install the electrical: conduit or Romex? If conduit, then it needs to be run before the concrete is poured, including the electrical boxes. If Romex, then you need to rout a chase $1\frac{1}{2}$ inches deep and use spray foam to hold the wire in place (Figure 2-14).

▌ Water pipes need to be buried $1\frac{1}{2}$ inches as well. Anything less than $1\frac{1}{4}$ inches requires you to protect the pipes from mechanical damage. That's to prevent you from accidentally penetrating the wire or pipe with a fastener, like a wallboard screw.

▌ The plastic ties you normally use to fasten wallboard may not meet local thermal barrier requirements (to keep the foam from melting within 15 minutes) in your area. If not, you'll

Courtesy: American ConForm Industries, Inc.

Figure 2-13
Curved wall using SmartBlock

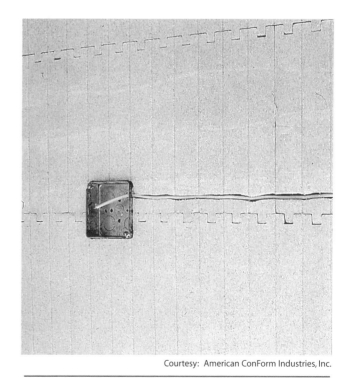
Courtesy: American ConForm Industries, Inc.

Figure 2-14
Romex embedded in SmartBlock wall

have to do special detailing to this area. One way is to apply adhesive directly to the back of the wallboard. You have two choices:

1. Apply a continuous bead $^3/_8$ inches in diameter, spaced not more than 12 inches on center, starting 6 inches in from the edge.

2. Spot the adhesive (not less than 2 inches in diameter by $^1/_2$ inch thick), spaced not more than 16 inches on center. Otherwise, secure galvanized steel strips (24 gauge) or square tabs (20 gauge) into the concrete before the concrete is poured.

Construction Notes — Here's a brief outline of how to install the SmartBlock system:

1. It's very important to get the first course of blocks level, straight, and square, so begin by making sure the footings go in just that way. You may want to do a dry run of the blocks so you know where to install the vertical rebar.

2. Once the footings have dried, snap chalk lines to guide block placement.

Courtesy: American ConForm Industries, Inc.

Figure 2-15
Bracing a SmartBlock wall

3. Start in one corner with an end block. Intersect this block with a second block and mark and cut the first block to allow concrete flow. Do the same to all inside and outside corners. Once this is done, lay the blocks, working your way toward the middle. You may need to cut block(s) to fit.

4. The second row is just like the first row but with one major exception: overlap the intersecting blocks to lock the corners together. Do this to every other row, just like building a log home. The third row is the same as the first, the fourth is like the second, and so on. It's also important to overhang the block joints by at least 10 inches to produce staggered joints. But check down inside the blocks to see that the cells line up with the ones underneath. (This is for the SF-10 block.)

5. Install the horizontal rebar as required, wire-tying it to the vertical rebar. Work your way up to the first level (unless this is the only level) — and don't forget to cut out for doors and windows.

6. Take the extra time to make sure foundations are properly braced before you pump in the concrete. You don't want a blowout in the middle of a pour! Brace all corners and tape the block at all cut joints. Plumb the walls as needed. Attach vertical bracing (both sides of the block) by fastening to the form with screws, or build a ladder bracing and place over the form according to the manufacturer's bracing schedule. Fasten a kicker, and stake it to hold the wall in the correct position. Brace all doors and windows. Be sure to leave an opening in the sill to pump the concrete. The manufacturer has detailed instructions on how to adequately brace their system (Figure 2-15).

7. When pumping concrete, make sure the slump is correct and use a 2-inch hose (3 inches maximum). Pump slowly and away from the corners, allowing the concrete to flow on its own into the corners. Pump 4 feet high all the way around the foundation. This will give the concrete time to set up before you start the second pour. Do the pours in multiple lifts until you reach the top. If a second story is planned, tape

the teeth before pouring the concrete below the last course of blocks. This will keep concrete out of the interlocking system so you can easily stack the next story once the tape has been removed.

8. Make sure the concrete flows throughout the system. Use a piece of rebar to poke the concrete to fill voids. You can also use a hammer and a piece of plywood to tap on the side of the form to get the concrete to flow. Always check for adequate consolidation (no air bubbles or voids) immediately after pouring.

9. Keep some plywood and extra bracing handy for quick repairs.

Manhours — SmartBlock's standard fixed block (10 × 10 × 40 inches) covers 2.75 square feet, so it would take 269 blocks to cover 742 square feet (after adjusting wall height from 48 to 50 inches for 10-inch standard forms). Assuming "worst case" reinforcing with #4 rebar 10 inches on center vertically and horizontally, this would equal 85 inches of steel placed through one block (269 blocks × 85" ÷ 12"), which equals 1,905 feet of rebar.

For labor time, assume a three-man crew to lay out forms, tie rebar, and brace. (This doesn't include footings or pouring of concrete.) Calculate 269 forms times 3 minutes per form, or 807 minutes. Divide by 60 minutes to find that it'll take about 13.5 hours. That's an average of one block per minute per three-man crew. Productivity is slightly slower with a two-man crew and may drop dramatically with four or more. With a three-man crew, one can build the blocks, another can stack, and the third can install and tie rebar. Three create a good rhythm.

The first time around, I'd bid higher than 3 minutes per block. Another consideration to keep in mind is that radii and curved walls require extra time — possibly 5 minutes per block. Add an extra minute per block for the variable-sized forms.

Remember, these manhour figures are based on familiarity with the system. American ConForm Industries provides free on-the-job services and training to contractors and their crews. To learn more about their system and the services they offer, give them a call at the number given in the Appendix.

Reward Wall Systems

Reward Wall Systems, Inc., introduced the Reward ICF (insulating concrete forms) system in 1994. The interior of this block is a monolithic post, beam, and web design — four vertical posts that are 12 inches on center and a horizontal beam that's 16 inches on center. It weighs in at 5 pounds each for a unit that's $9^1/_4 \times 16 \times 48$ inches.

While looking over this product, I noticed some qualities about the system that are worth mentioning:

■ The plastic ties have two advantages over the usual metal ties. They reduce thermal transfer (heat loss) by over 20 percent, and allow you to use larger aggregate while minimizing the possibility of voids.

■ The edges (including the ends of the blocks) feature a lap joint design which helps to lock the entire system together. That design is 50 percent stronger than tongue and groove and you don't have to remove concrete from the grooves.

■ There are small vertical lines on the outsides of the blocks where the ties are located. These marks help you line up the cells so you don't have look down into the system, and they help you locate the ties when fastening wallboard and external finishes. The recessed ties don't get in the way of stucco application.

■ The blocks have a left and right, so they're designed to be installed in one direction. In Figure 2-16, you can see the plastic ties with a built-in rebar chair (V-shaped) where the rebar rests.

■ You don't have to make cuts for every corner of every course, because 90-degree and 45-degree corners are available.

Construction Notes — Reward's installation is similar to that of other ICF systems. Of course, their installation manual gives special details for their particular system.

1. I mentioned earlier the importance of good footings, but don't forget to install the vertical precut rebar (5 to 6 feet in length with a 6-inch hook) into the concrete. If you look closely down the row in Figure 2-17, you'll notice rebar already in position.

Courtesy: Reward Wall Systems, Inc.

Figure 2-16
Reward Wall System

2. If you plan to make unique corners by cutting unusual angles, then these joints (as well as any other blocks that you've cut) need to be taped with duct tape before you pour the concrete. The premolded corners require less bracing and will save up to 30 percent labor.

3. How about an intersecting wall? It's important to ensure the wall being intersected is cut to accept the entire width and height of the incoming form wall (Figure 2-18).

4. If you plan to face the exterior wall with brick, you'll need galvanized anchors and a brick ledge constructed before the concrete is poured (Figure 2-19).

5. A recessed furring strip (spanning the entire height of the form on 12-inch centers) allows wallboard and paneling hangers to attach wall coverings more easily.

6. Earlier I reminded you to install the anchors for your ledger. While this is true, you still need to go one step farther. Figure 2-20 shows how to achieve a secure ledger or rim joist.

Courtesy: Reward Wall Systems, Inc.

Figure 2-17
Starter course being installed

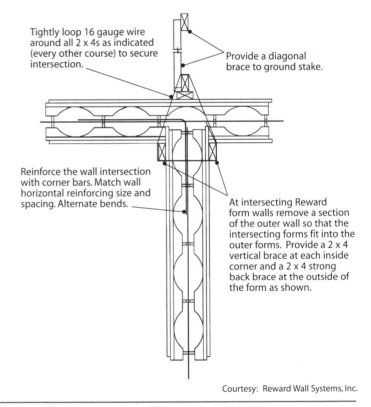

Tightly loop 16 gauge wire around all 2 x 4s as indicated (every other course) to secure intersection.

Provide a diagonal brace to ground stake.

Reinforce the wall intersection with corner bars. Match wall horizontal reinforcing size and spacing. Alternate bends.

At intersecting Reward form walls remove a section of the outer wall so that the intersecting forms fit into the outer forms. Provide a 2 x 4 vertical brace at each inside corner and a 2 x 4 strong back brace at the outside of the form as shown.

Courtesy: Reward Wall Systems, Inc.

Figure 2-18
Intersecting wall detail

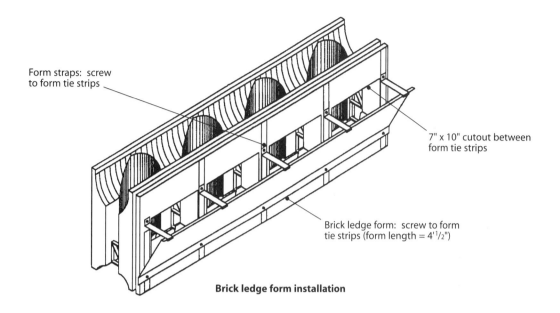

Form straps: screw to form tie strips

7" x 10" cutout between form tie strips

Brick ledge form: screw to form tie strips (form length = 4'1/2")

Brick ledge form installation

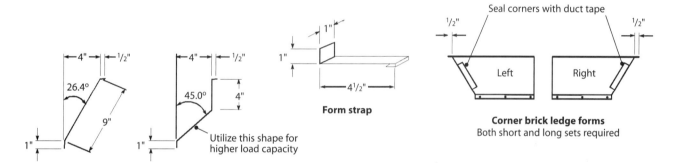

26.4°

4"   1/2"

9"

1"

45.0°

4"   1/2"

4"

1"

Utilize this shape for higher load capacity

Brick ledge form shapes
Fabricate forms from 16 gauge galvanized steel sheet material

1"

1"

4 1/2"

Form strap

Seal corners with duct tape

1/2"   1/2"

Left   Right

Corner brick ledge forms
Both short and long sets required

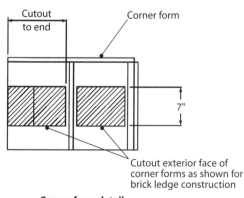

Cutout to end

Corner form

7"

Cutout exterior face of corner forms as shown for brick ledge construction

Corner form detail
Both short and long corner forms required

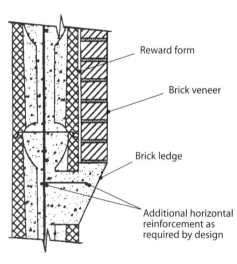

Reward form

Brick veneer

Brick ledge

Additional horizontal reinforcement as required by design

Wall section at brick ledge

Courtesy: Reward Wall Systems, Inc.

Figure 2-19
Brick ledge form details

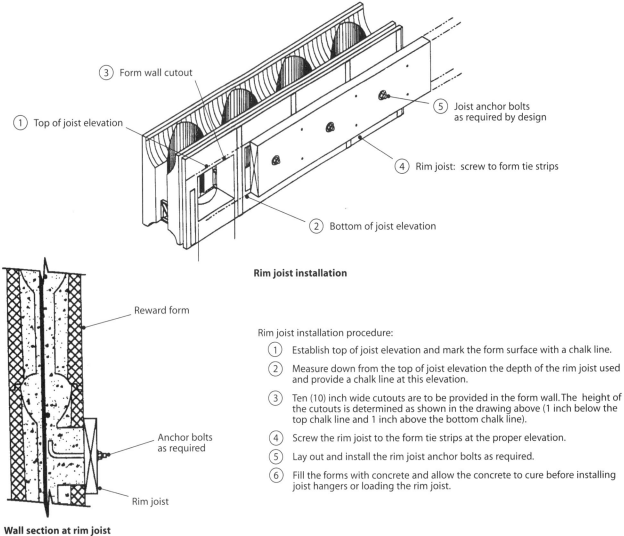

Courtesy: Reward Wall Systems, Inc.

Rim joist installation

Wall section at rim joist

Rim joist installation procedure:

(1) Establish top of joist elevation and mark the form surface with a chalk line.

(2) Measure down from the top of joist elevation the depth of the rim joist used and provide a chalk line at this elevation.

(3) Ten (10) inch wide cutouts are to be provided in the form wall. The height of the cutouts is determined as shown in the drawing above (1 inch below the top chalk line and 1 inch above the bottom chalk line).

(4) Screw the rim joist to the form tie strips at the proper elevation.

(5) Lay out and install the rim joist anchor bolts as required.

(6) Fill the forms with concrete and allow the concrete to cure before installing joist hangers or loading the rim joist.

Figure 2-20
Rim joist installation

	Crew's experience with Reward Forms		
Complexity of project	**No previous experience**	**Some experience**	**Completed 4+ projects**
Simple design (four corners with few openings)	6 forms/MH	8 forms/MH	10 forms/MH
Average design	6 forms/MH	7 forms/MH	8 forms/MH
Complex design (many corners and many cutouts)	3 forms/MH	5 forms/MH	7 forms/MH

Figure 2-21
Manhour estimates for installing Reward Wall Systems

These ICF systems appear easy to install — and they are — but there are certain steps you can't overlook. Be sure to read the manufacturer's installation manual. Don't assume that because you know how to swing a hammer, you automatically know how to install them. Take the time to read that manual, and if you still don't understand, ask! To learn more about this system, contact Reward Wall Systems, Inc. The address is in the Appendix, along with the 800 number for technical assistance.

Manhours — To help calculate manhours on this product, the manufacturer has provided a table to guide you in estimating labor costs (Figure 2-21). The table provides the number of forms one man can install in one hour, but keep in mind that these numbers are only *approximations*. Your actual manhour requirements will vary, depending on your crew's ability and efficiency. The manhours in the table include layout, stacking the forms, installing and tying rebar, bracing, pouring the forms, and removing the bracing after the concrete has set.

Suppose you're bidding on a one-story house that requires 356 forms. How many manhours would it take to install this system? Assuming the house is an average design (without excessive corners and windows) and the crew has never used this system, look at the table under "No previous experience" and "Average design." You'll find 6 forms per manhour. Divide 356 forms by 6 to get an estimated 60 manhours to install this system for a crew of one. A crew of three may be able to cut this time to 20 hours.

Diamond Snap-Form

If you're still stuck on conventional foundation materials and methods, then perhaps this next product will convince you to make a change. AFM Corporation, established in 1980, has come on strong in this market with alternative products and systems to make the job go smoother and faster in the field. We'll discuss some of these products in the next chapter, but let me introduce you to the Diamond Snap-Form EPS ICF system.

This system uses 1 × 8-foot panels, 2 inches thick, and green Diamond Snap-Ties (Figure 2-22). And that 8-foot length can sure speed up the process of putting up the foundation. The ties are available for solid concrete wall thicknesses of 4, 6, 8 and 10 inches. The diamond face, approximately 4 inches

Courtesy: AFM Corporation

Figure 2-22
Components of Diamond Snap-Tie system

square, gives plenty of support along with a large fastening surface. Figure 2-23 shows a 4-foot-high wall with the diamond-shaped ties set 12 inches on center. The cradle of the tie can hold rebar up to #6 in size. It's symmetrical in design, so there's no top or bottom to the tie. Their half-tie is used to start and cap the forms.

To speed up installation, the manufacturer has designed preformed corners for both 90- and 45-degree corners. New to their family of products is Perform Guard — an insect-resistant EPS insulation. If you work in an area where carpenter ants or

Courtesy: AFM Corporation

Figure 2-23
Wall formed with Diamond Snap-Tie system

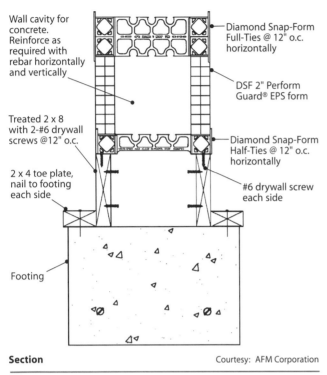

Wall cavity for concrete. Reinforce as required with rebar horizontally and vertically

Diamond Snap-Form Full-Ties @ 12" o.c. horizontally

DSF 2" Perform Guard® EPS form

Treated 2 x 8 with 2-#6 drywall screws @12" o.c.

Diamond Snap-Form Half-Ties @ 12" o.c. horizontally

2 x 4 toe plate, nail to footing each side

#6 drywall screw each side

Footing

Section

Courtesy: AFM Corporation

Figure 2-24
Baseboard attachment with Diamond Snap-Form

termites are a major concern, this product may be the answer for below- or above-ground installation. The insulation has a built-in formula using a natural mineral to resist these wood-boring insects. The Environmental Protection Agency (EPA) has approved this natural mineral additive and the product has been thoroughly tested. It's safe, noncorrosive, and nontoxic, claims the manufacturer. Ask for documentation or a copy of the test results concerning these claims.

Construction Notes — The Diamond Snap-Form system doesn't require staggered joints because the snap-tie design supports the EPS regardless of joint location. As long as the maximum dimension of unsupported EPS is 7 inches or less (that's 7 inches measured from point to point of the diamond-shaped ties), the Diamond Snap-Tie will support the EPS. Other areas to consider during the installation or before the concrete is poured include:

1. If you have concerns about attaching base moldings, Figure 2-24 shows how to achieve a guaranteed wood surface for attachment.

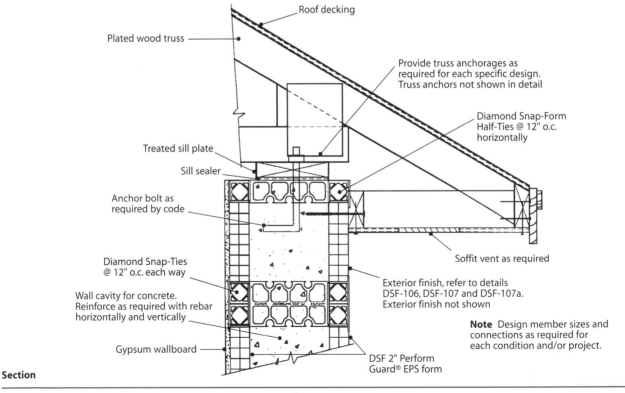

Roof decking

Plated wood truss

Provide truss anchorages as required for each specific design. Truss anchors not shown in detail

Diamond Snap-Form Half-Ties @ 12" o.c. horizontally

Treated sill plate

Sill sealer

Anchor bolt as required by code

Diamond Snap-Ties @ 12" o.c. each way

Wall cavity for concrete. Reinforce as required with rebar horizontally and vertically

Gypsum wallboard

Soffit vent as required

Exterior finish, refer to details DSF-106, DSF-107 and DSF-107a. Exterior finish not shown

Note Design member sizes and connections as required for each condition and/or project.

DSF 2" Perform Guard® EPS form

Section

Courtesy: AFM Corporation

Figure 2-25
Soffit and truss details in Diamond Snap-Form wall

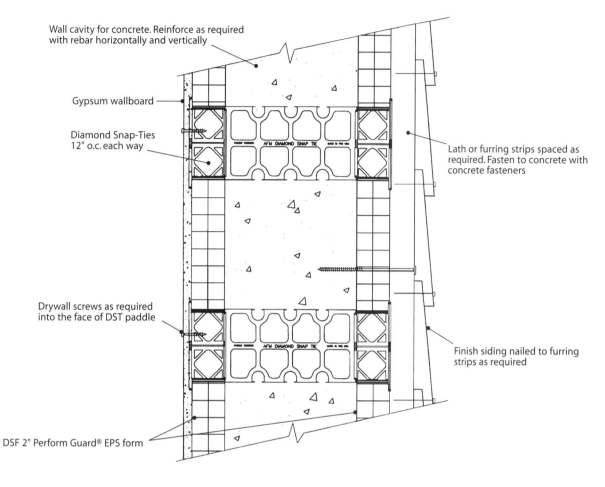

Wall cavity for concrete. Reinforce as required with rebar horizontally and vertically

Gypsum wallboard

Diamond Snap-Ties 12" o.c. each way

Drywall screws as required into the face of DST paddle

DSF 2" Perform Guard® EPS form

Lath or furring strips spaced as required. Fasten to concrete with concrete fasteners

Finish siding nailed to furring strips as required

Section

Figure 2-26
Wallboard and wood siding application details

2. Figure 2-25 shows how to attach a sill plate for the trusses, and provides details for a soffit application. Notice the fastener that holds the ledger board has been installed into the concrete after curing.

3. What about traditional wood siding? You'll need furring strips, and it's best to attach these strips to the concrete after the concrete has set. Figure 2-26 gives examples for both interior and exterior applications.

4. Consider installing the waterproof membrane after the first 4-foot wall has been poured and before you continue building the form (Figure 2-27). The membrane is self-adhering. Just peel off the backing and stick.

Figure 2-27
Installing a waterproof membrane

Courtesy: AFM Corporation

Figure 2-28
Filling Diamond Snap-Tie forms with concrete

Courtesy: Lite-Form Incorporated

Figure 2-29
Lite-Form sections folded for storage

Courtesy: Lite-Form Incorporated

Figure 2-30
Lowering a Lite-Form section into place

5. When using a boom pump, try not to let the concrete fall from the full height of the boom into the forms. For best results, use two 90-degree elbows or an "S" at the end, or the discharge hose, to control the concrete's fall (Figure 2-28). If you have a 6-inch discharge hose, reduce it to 4 inches.

Manhours — With this system you can figure around 50 to 100 square feet per manhour, depending on the complexity of the walls and the number of window cutouts. That includes installing and tying rebar, building scaffolding once the wall reaches the height where you can't see the bottom when pouring the concrete, and bracing the system. This manhour estimate doesn't include footings or pouring concrete. Normally you'd hire a subcontractor to pour the concrete, and that sub can give you the labor charge.

For those interested in manhours for concrete, allow 7 minutes per yard to place concrete (for example, 41 yards × 7 minutes = 287 minutes ÷ 60 = 4.78 hours). It's more efficient when two handle the end of the pump hose during the pour. The manufacturer recommends a crew of three to four for installing their system. For more in-depth information on the Diamond Snap-Form system and how you can incorporate it into your next project, contact AFM Corporation. The address is in the appendix.

Lite-Form

Lite-Form International was founded in 1986 by Pat Boeshart, a custom home builder who was first introduced to the concept of stay-in-place concrete forms while attending a home builders' convention in the mid '80s. Even though he was skeptical, he gave the system a try. Shortly thereafter, he patented his own concrete foam form system. This product is known in the industry as Lite-Form. His family of other forming systems includes Folding Lite-Form and Fold-Form.

The Lite-Form system uses common brands of XPS or EPS foam 2 inches thick. The insulation is cut into 8-inch × 8-foot planks with slots every 8 inches along the plank. Lite-Form plastic spacer ties fit in the slots to assemble form walls of almost any shape. They can be assembled at the job site or pre-assembled in major sections and delivered to the job.

The forms are available in concrete wall thicknesses of 4 to 24 inches and have an R-factor of 20 plus. Spacer ties for corners and T-intersections are available as well.

When XPS insulation is used in the form walls, it can be stripped (removed) and salvaged. Once the form is stripped, the exposed concrete wall resembles a standard foundation wall. The manufacturer claims that 90 percent of the XPS insulation can be reclaimed for use on other projects.

Folding Lite-Form is basically the same system except that the ties fold. It's preassembled by a Lite-Form representative into 2 × 8-foot sections and the corners are precut and packaged as a kit. The spacer ties fold so the 16-square-foot form sections fold flat for easy transport and storage (Figure 2-29).

The Fold-Form system, on the other hand, is available in wall thicknesses of 4 to 16 inches and features preassembled 1 × 4-foot EPS blocks that also fold flat for shipping and handling. What's unusual is the perimeter design. Essentially, it's a tongue-and-groove system, but the design is castellated. Its pattern of indention is similar in appearance to a battlement, that low protective wall with openings normally found up on the outer wall of a castle. The form walls are assembled course by course (row by row) by sliding the tongue-and-groove ends of the blocks together and pressing the castellations together. Each row of sections alternates, keeping the entire wall form locked together. The ties are roughly 12 inches on center and the form has an overall R-value of 18-plus. Unlike the Lite-Form systems, Fold-Form can't be stripped because the form uses EPS insulation which bonds to the concrete surface.

While you're reading about system installation in this chapter, remember that I'm only giving the basics — the key factors to consider as you assemble a system. Each manufacturer has specific guidelines to follow, which means you need to get with the manufacturer and read their manual. Here are some key points to remember about Lite-Form:

1. The joints are staggered and secured with nylon twine wrapped around the ties on both sides of the form. In Figure 2-30, you can see the staggered joints on the block being lowered into place. Figure 2-31 shows how to wrap and tighten the twine. This acts as a clamp to secure the system and to force two form sections into proper alignment.

Wrapped and tightened with a nail

Use a 6 foot length of twine. Wrap around 2 tie pads as shown.

Ends of twine are forced behind tie pads on either side of common seam

Common seam between corner and wall section

Twine is tightened with a nail. When tight, point of nail is pressed into insulation

Courtesy: Lite-Form Incorporated

Figure 2-31
Securing Lite-Form joints with twine

2. Have you made provision for utility lines? Be sure to insert a sturdy object that can be removed one hour after concrete placement for any lines (water, gas, etc.) that will run through the forms.

3. Don't forget to support the walls prior to pouring the concrete (Figure 2-32). Here's a good way to do it. Insert baling wire through the form, around the form's tie, and back out through the form. Then twist the wire around a steel stud containing a wood insert.

Courtesy: Lite-Form Incorporated

Figure 2-32
Supporting walls for concrete pour

Courtesy: Lite-Form Incorporated

Figure 2-33
Tying conventional forms into Lite-Form walls

4. You can still tie in conventional formwork like the one shown in Figure 2-33 to create a main support.

Manhours — For all their products mentioned, figure one manhour for every 100 square feet. This includes stacking, installing and tying rebar, and bracing/scaffolding. This doesn't include footings or the concrete pour. For special walls, radii, and brick ledges, add 20 to 25 percent to your manhours. This figure is based on having installed their system at least three times. The manufacturer recommends a crew of three.

Whether I've convinced you or confused you about ICF systems, I hope I've steered you in the right direction. This isn't the final word on the subject. In Chapter 3 I'll discuss other systems that you can use with the systems in this chapter. Of course, it depends where you're building and whether you're building above or below grade level. I suggest you not make any decisions until you've read the entire book and had an opportunity to request sample products.

3

Wall and Roof Systems

■■■

By now, you probably realize that there are a lot of alternative materials systems out there, offering you plenty of construction options. For instance, an all-wood foundation gives you the flexibility to replace an existing foundation one section at a time. It also lets you install the foundation without having to hire a subcontractor.

You can install ICF systems yourself, and they also feature a couple of bonuses. First, they're an all-in-one system — both form and insulation. Second, you have the option to use the forms to build a complete wall system for an entire structure.

Pour-in-place systems using XPS foam panels are also available. They have the same characteristics as an ICF system, plus they give you the option to cut the ties and salvage up to 90 percent of the foam board. You can use the foam board on your next project, so there's a nice savings for you. You can give your customer the option of a solid concrete wall or not.

However, it doesn't end there. We've only scratched the surface of the different systems that are available. In this chapter I'll cover:

■ Lightweight concrete blocks that allow you to use standard construction tools, like a handsaw, to cut the blocks

■ Mortarless block systems that lock together with rebar while the cavities are filled with grout and insulation

■ Structural insulated panels (sheets sandwiching rigid foam) that can be used for building floors, roofs, and walls

■ EPS foam panels flanged with steel wire mesh that's later sprayed with concrete

■ EPS molded into steel frame panels

■ Concrete panels reinforced with steel

Now the question becomes, "What system do you want to use?" Keep in mind that the system you choose may be decided by where you're building and/or whether you're building above or below grade. Once you've zeroed in on a couple of systems, call or write the manufacturer (addresses in the appendix) to get the complete information you'll need to make an informed decision.

Aerated Concrete Wall Units

I thought I had seen just about every product on the market, but somehow this one slipped past me. It's an autoclaved aerated concrete (AAC) wall unit by Hebel that's been on the market since 1943 when its founder, Josef Hebel, starting producing them in Munich, Germany. Today, more than 8 million cubic yards of AAC have been produced in 51 licensed plants worldwide. Simply put, AAC is a lightweight, precast concrete wall unit — weighing about *one-fifth* of a standard concrete block — made from sand, cement, lime, water, gypsum, and a proprietary expansion agent. Once mixed, the slurry is poured into a mold and allowed to "rise." During the expansion process, millions of tiny bubbles are dispersed throughout the AAC. After it's removed from the mold and sectioned into predetermined sizes, it's sent to an autoclave (pressurized chamber) where it's steam-cured under pressure to achieve its structural strength. Hebel offers AAC in a variety of strength classes, from exterior to interior and from load-bearing to non-bearing walls.

What's unique about the product is its uniform cellular structure. It's *so easy* to cut with a handsaw, and nailing into it — well, you might think you were nailing into wood. The characteristics of the product are unbelievable. You can even cut utility chases with an electric router (Figure 3-1). And it's easy to make architectural designs. You've got to finish the AAC with some type of finish material — stucco, brick or paint.

The standard block is 24 inches long by 8 inches high by stock thicknesses of 4, 6, 8 and 10 inches. An 8 × 8 × 24-inch block weighs only 28 pounds, compared to a regular block that weighs from 37 to 42 pounds. An exterior wall of the same thickness achieves a fire rating of a maximum of four hours and has outperformed an R-30 stud wall. This is only the beginning, as the company produces complete systems and products to use to build an entire structure. These include:

▌ wall units

▌ jumbo wall units

▌ floor, roof and wall panels

▌ lintels

▌ solid staircases

▌ thin-bed mortars

▌ interior plasters and exterior stuccos

The material is energy efficient and fire resistant. Exposed to fire, it emits no toxic fumes. It's durable (solid walls make it difficult for uninvited guests to enter the home) and versatile (the product can be used for both loadbearing and nonbearing walls).

Products

As mentioned earlier, Hebel offers a whole family of products (Figure 3-2) for use in both residential and commercial markets. You may want to contact the manufacturer to learn more about these products and how they could benefit you and your customer on your next project. I'll only discuss the installation on the wall units themselves in detail.

▌ *Wall panels:* Prefabricated steel reinforced structural panels whose primary application is for solid wall construction for load-bearing exterior or interior partitions. Unreinforced wall panels are available for nonbearing applications. Both products are available in 1-inch increments from 3 to 12 inches thick, up to 20 feet in length, 2 feet in width, and have a tongue-and-groove (T&G) profile.

Courtesy: Hebel Building Systems

Figure 3-1
Cutting a utility chase
in an aerated concrete wall

Courtesy: Hebel Building Systems

Figure 3-2
Hebel Integrated Construction System components

▌ *Floor and roof panels:* Very similar to wall panels, i.e., in geometry, physical characteristics, and erection, floor and roof panels are steel reinforced with a T&G profile. They're available in various strength categories and sizes to suit a range of spans and design loads.

▌ *Arches and staircases:* These design elements come in a variety of thicknesses and lengths for both residential and commercial markets.

▌ *Lintels and wall lintels:* These reinforced supports (headers) are used above doors and windows. Wall lintels are the same as standard lintels except they're longer and designed for use in larger spans with higher loads.

Construction Notes — Aerated concrete wall units can be installed quickly and efficiently, but there are a couple of key factors to keep in mind. First, you can install this system below grade, but you must use a waterproof membrane to protect the wall's surface from moisture. Second, make sure the footings are level and square. This ensures that the first level of the wall unit will be precisely aligned, and that the following courses will follow along, both level and aligned. Use a leveling bed mortar mix on the first course to help with leveling. (You can see it in Figure 3-3, where the builder is laying an 8-inch wall unit as part of the first course for an exterior wall.) Then apply Hebel's thin-bed mortar adhesive to all vertical and horizontal surfaces. Install the wall units with staggered joints. For corners or load-bearing partitions, lay the blocks with alternate joints (interwoven) to lock the system together securely.

1. When tie-down rods are required, drill the blocks using a flat wood bit.

2. If your area calls for tension tie-down rods, tie each rod into the foundation and into the roof's sill plate (Figure 3-4).

3. How do you install floor joists using this wall system? One way is to create a pocket for the joist to fit into (Figure 3-5). Important here — and with tension tie-down rods generally — is that you must install U-section units anyplace there'll be downward pressure. Then fill the channel with concrete so it can support the design load after it's cured. It also provides one way to install anchor bolts horizontally or vertically.

Courtesy: Hebel Building Systems

Figure 3-3
Leveling the aerated concrete wall units

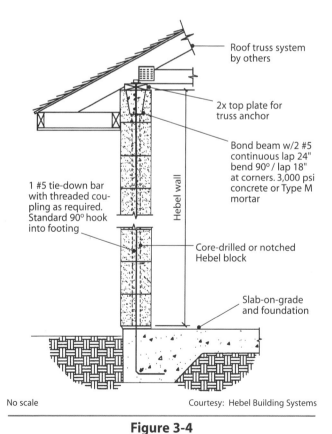

No scale   Courtesy: Hebel Building Systems

Figure 3-4
Hebel exterior wall section
with roof system tension tie-down rod

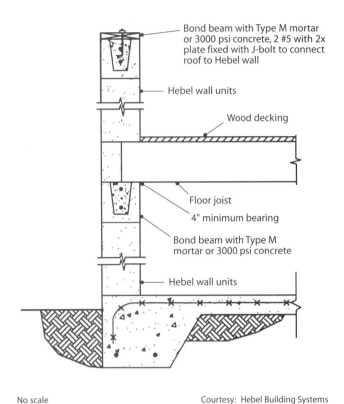

Bond beam with Type M mortar or 3000 psi concrete, 2 #5 with 2x plate fixed with J-bolt to connect roof to Hebel wall

Hebel wall units

Wood decking

Floor joist

4" minimum bearing

Bond beam with Type M mortar or 3000 psi concrete

Hebel wall units

No scale   Courtesy: Hebel Building Systems

Figure 3-5
Hebel wall — floor joist detail

4. Create door and window openings by installing a factory-made lintel or using a U-section and filling the channel with concrete and steel bar reinforcement top and bottom (Figure 3-6).

5. If your plans call for exterior wood siding, the exterior walls will require wood furring strips attached with an approved fastener. Properly-spaced deck screws are sufficient and no pilot holes are required. If you're facing the wall with brick, don't forget the wall ties (Figure 3-7).

Fastening and anchoring into an aerated cell structure requires this very simple thought: *You're not working with wood.* You can't use common nails for fasteners, but you *can* use tapered or zinc plate cut nails. For most applications you should use smaller fasteners, more closely spaced. There are approved fasteners and methods for any situation you meet during construction. To find out what they are, check with the manufacturer.

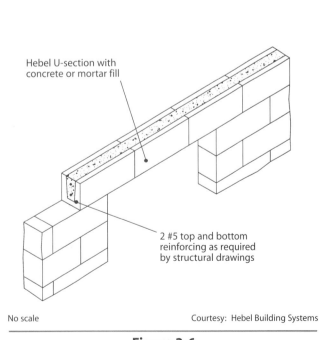

Hebel U-section with concrete or mortar fill

2 #5 top and bottom reinforcing as required by structural drawings

No scale   Courtesy: Hebel Building Systems

Figure 3-6
Hebel lintel section

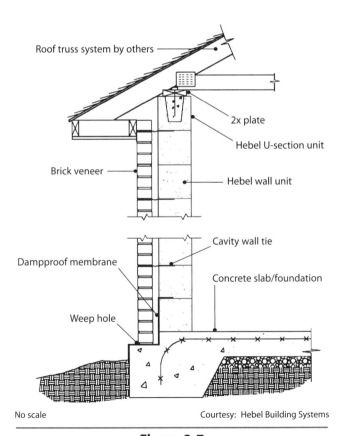

Roof truss system by others

2x plate

Hebel U-section unit

Brick veneer

Hebel wall unit

Cavity wall tie

Dampproof membrane

Concrete slab/foundation

Weep hole

No scale   Courtesy: Hebel Building Systems

Figure 3-7
Hebel wall — brick connection detail

Take, for example, wallboard. In some cases it can be fastened directly to the surface of the wall, but the preferred method is to install furring strips. In either case, the screw has to penetrate at least 1 inch into the wall system. While a wallboard screw isn't an approved fastener, a deck screw *is* (because of its coarser threads). Adhesive plus screws creates a better fastening system. Exterior plaster and acrylic-based finishes provide a good wall finish; just be sure the products you choose are approved for AAC systems.

Hebel manufactures jumbo wall units (40 inches long × 24 inches high × stock thickness of 4 to 12 inches) in addition to the standard wall units. In Figure 3-8, they're using a hydraulic hoist to make the job go a little smoother.

Manhours — Because Hebel offers so many types of products for different applications, you may find an excerpt from their product description, production and equipment chart (Figure 3-9) helpful to pinpoint

Courtesy: Hebel Building Systems

Figure 3-8
Installing Hebel jumbo wall units

Product & Application	Product Size			Coverage Sq.Ft.	Labor Crew	Units/Day Per Crew	Production Sq.Ft./Day	Regular bandsaw	Jumbo bandsaw	Mini crane	Standard scaffold	Lifting gear	Lifting crane
	T	H	L										
Residential Construction													
Precision wall units	12	8	24	1.33	A	160	213	***			***		
Precision wall units	10	8	24	1.33	A	184	245	***			***		
Precision wall units	8	8	24	1.33	A	225	300	***			***		
Precision wall units	6	8	24	1.33	A	250	333	***			***		
Precision wall units	4	8	24	1.33	A	260	347	***			***		
Commercial Construction													
Jumbo wall units	12	24	40	6.67	B	58	387		***	***	***		
Jumbo wall units	10	24	40	6.67	B	60	400		***	***	***		
Jumbo wall units	8	24	40	6.67	B	64	427		***	***	***		
Jumbo wall units	6	24	40	6.67	B	68	453		***	***	***		
Jumbo wall units	4	24	40	6.67	B	70	467		***	***	***		
Residential & Commercial Construction													
Mini jumbo wall units	12	12	20	1.67	A	100	167				***		
Mini jumbo wall units	10	12	20	1.67	A	120	200				***		
Mini jumbo wall units	8	12	20	1.67	A	128	213				***		
Mini jumbo wall units	6	12	20	1.67	A	138	230				***		
Mini jumbo wall units	4	12	20	1.67	A	140	233				***		
Residential & Commercial Construction													
10" floor panels	10	24	240	1.00	C	60	2400					***	***
8" floor panels	8	24	192	1.00	C	60	1920					***	***
6" floor panels	6	24	144	1.00	C	60	1440					***	***

Notes:
Crew A (4 crew) Lead installer, installer, helper, laborer
Crew B (3 crew) Lead installer, installer, helper
Crew C (3 crew) Lead installer, installer, helper

Figure 3-9
Hebel product descriptions, production rates and equipment needed

what products to consider. Keep in mind that lifting equipment will be required when you're working with jumbo wall units and floor panels. The production in square feet per day in the chart is based on 8 hours for the size of crew indicated, and assumes the use of this equipment. Don't forget to add in the cost for this equipment in your estimate. Hebel can train or recommend subcontractors in your area that are familiar with the system. They also can supply installation services along with material.

As mentioned before, the company provides its own exterior plaster and acrylic-based finishes for both interior and exterior wall applications. The wall finishes include:

- Interior unpainted (residential): Two coat application consists of a $\frac{1}{4}$-inch base coat and $\frac{1}{8}$ inch of smooth veneer plaster. After the first coat has been applied, wallpaper can be installed.

- Exterior finished (residential): Two coat application consists of a $\frac{3}{8}$-inch base coat that's normally hand applied directly to the AAC blocks and a $\frac{1}{8}$-inch thick finish coat hand troweled to the desired texture. Color is applied with an acrylic-based paint.

- Interior unpainted (commercial, three- to six-story hotel or apartment): First, $\frac{1}{8}$ to $\frac{3}{16}$ inch of base smoothing plaster, followed by a sprayed knock-down texture. The knock-down texture is typically wallboard joint compound.

- Exterior finished (commercial, three- to six-story hotel or apartment): A two-part operation where the base coat is sprayed on and then smoothed with a darby trowel (long, narrow trowel). This allows more production at about the same manhours. The finish coat is sprayed on and then hand troweled to the desired textured finish.

To calculate manhours for residential finishes for both interior and exterior applications, figure a crew of four can apply finish to 700 square feet of wall surface in 8 hours. This is based on hand work; if the finish is applied with a spray gun, you could increase production by 20 percent. For interior and exterior applications for commercial projects, figure a four-man crew can do 1,000 square feet of wall surface in

8 hours. This is again based on hand work, but a spray finish could increase production by 50 percent.

Both figures include installation of corner beads. However, for exterior detail work such as decorative banding, trim around windows, and quoining (adding dimensional outside corners), figure one man can install and finish 10 linear feet in 8 hours. Normally, that would be 2 inches thick by 6 to 8 inches wide for banding and trim and up to 1 foot 6 inches square in design for quoin application. Also, quoins can be made with different thicknesses of scrap materials, such as 2-inch-thick pieces of Hebel blocks attached to the wall with adhesive and using fasteners ready for exterior finish.

Mortarless Block System

The patented IMSI Insulated Reinforced Masonry System features a specially-designed concrete block with uniquely-shaped cores filled with insulation inserts. They're stacked without mortar, and you need to reinforce the walls with rebar (every 4 feet in most applications) and grout. Once stacked, blocks are finished with surface bonding cement, and troweled like stucco or plaster on both sides of the wall. That locks the entire wall system together.

Then the system requires a second coat as a finish coat. You can apply any facing to the surface of the block: elastomeric paint, regular or synthetic stucco, full or thin brick, artificial stone, or siding. The design reduces assembly time while creating a cost-competitive, high-strength, highly thermal-efficient, airtight, water-resistant, vermin-proof and virtually maintenance-free wall system. What's more, the design performs equally well in all applications — residential, commercial, industrial and institutional buildings.

The actual cost of building with the IMSI System is generally competitive with any insulated wall material. In fact, the manufacturer claims that savings in heating and cooling costs can repay any extra cost of installing this system in as little as three years. Maintenance costs are almost nonexistent. And you can't put a dollar value on the comfort and security it's adding to the building.

In the mid 1980s, the heavyweight industry struggled under new Federal guidelines as they tried to achieve high thermal resistance R-values and low-cost construction. In 1985, IMSI designed an engineered, field-ready, insulated heavyweight construction product to satisfy the required thermal resistance values. Unlike typical concrete blocks, the IMSI block has an interior core with an unusual design. The blocks' cavities not only allow for insulation to be inserted during construction, but they also allow reinforcement with rebar and grout to provide a structurally-sound, thermally-efficient system (Figure 3-10). The insulated insert is formed from fire-resistant EPS foam that has interlocking features.

The natural effects of thermal lag are increased because the IMSI block design creates an extended indirect thermal path from the interior to the exterior of the wall. (*Thermal lag* refers to the additional time it takes for heat energy to be conducted through heavyweight systems like block, brick, stone, or concrete.) Greater thermal lag results in substantially less heat gain, fewer cold and hot spots, and improves overall thermal efficiency. The result? Lower utility bills with greater comfort. Even though the IMSI blocks are 8 or 12 inches wide, the thermal path is 12 or 20 inches long due to staggered cross-webs between a central longitudinal web. The right-hand block in Figure 3-11 illustrates the elongated thermal path. The achieved thermal-dynamic values for an 8-inch unit are R-19 and R-30 for the 12-inch unit.

Products

Figure 3-12 shows the components that make up IMSI concrete masonry units (CMUs). All block-producing companies can make IMSI block using molds provided by IMSI. Blocks are available in two main sizes: 8 × 8 × 16 inches and 12 × 8 ×16 inches (full dimensions) as well as metric equivalents required in some foreign countries. Two additional widths will be available soon — 6 inches and 10 inches. The 8-inch wide CMU system consists of four different block sizes:

▐ Left end (for corners)

▐ Right end (for corners)

▐ Square half (for use at edges of openings)

▐ Stretcher (main block)

▐ Half stretcher

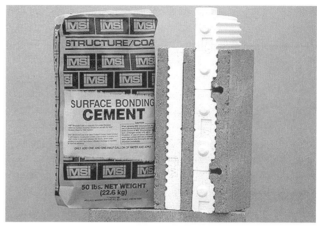

Figure 3-10
Cross section of IMSI block

The 12-inch-wide block system consists of three block shapes:

▐ Right or left end (block is symmetrical)

▐ Half

▐ Stretcher

The 8 × 8 × 16-inch stretcher shown in Figure 3-11 shows two primary insulation openings and two secondary (half) openings. The outside opening, the

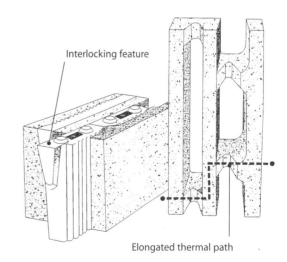

Interlocking feature

Elongated thermal path

Figure 3-11
IMSI stretcher (8" x 8" x 16")
showing elongated thermal path

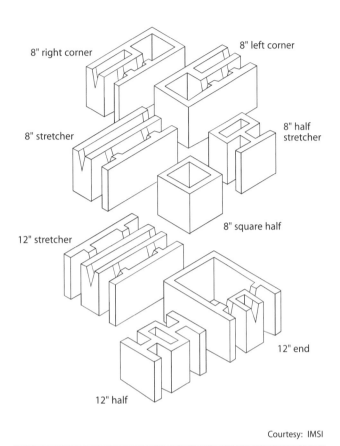

Courtesy: IMSI

Figure 3-12
IMSI concrete masonry units

Courtesy: IMSI

Figure 3-13
Applying the surface bonding cement over IMSI units

long thin one, is always used with inserted insulation. The inner (shorter) opening can accept either insulation or rebar and grout when they're required by the building code. The open half ends of these inner cells are used as bridge insulation cells. They're designed to break air transmission between blocks, aid in alignment, and help interlock the blocks. When you use rebar and grout, the reinforcing cells are still isolated from most outside influences by the outer long insulated cell. Insulation inserts are expanded polystyrene (zero toxicity, flame retardant) and available in sizes to fit both kinds of cells.

Surface bonding cement consists of a mixture of portland cement, fine silica sand, chopped fiberglass fibers, and other chemical plasticizers and anhydrators. (IMSI's brand is called IMSI Structure Coat.) Once the 50-pound bag is mixed with $1^1/_2$ gallons of water, it can be applied to both sides of the wall, with a minimum thickness of $^1/_8$ inch. While it's still damp, you can apply different finishes. Figure 3-13 shows the surface bonding cement being applied. Notice the anchors installed to hold the ledger board in place. It's common to use skip troweling for exterior or smooth troweling for interior applications. No additional furring or other building materials need be applied to either side of the wall. It can be painted or left a natural white.

Construction Notes — The blocks can be used below grade, and exterior bonding surface materials are sufficient for below-grade use. But as a cautious contractor, I'd rely on conventional waterproofing materials as well. A good installation requires accurate footings; lay the blocks in a leveling bed mortar to help assure a true and level first course. Then lay the rest of the blocks with joints staggered to the previous course (a running bond). For a straight and true course, it may be necessary to shim an inaccurately produced block. Shims must be corrosion-resistant metal or plastic with a minimum compression strength of 2000 psi. Here are some other considerations to keep in mind before and during the installation of the system:

1. Make provision for water and utility lines to pass through.

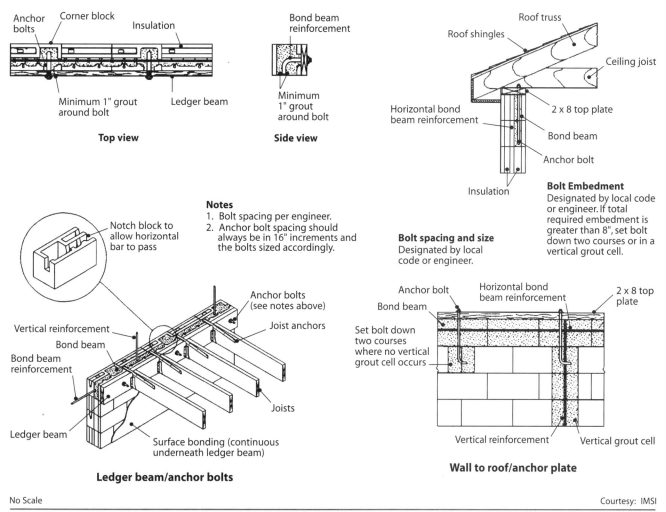

Figure 3-14
Anchoring requirements with IMSI walls

2. Don't forget about contraction and expansion. All building materials are subject to movement, and concrete masonry walls are no exception. Build control joints into the system to help eliminate possible cracking caused by movement in the wall from expansion and contraction. Any movement will occur at the control joint and be inconspicuous.

3. Rebar, used both horizontally and vertically along with grout installation, is important to lock the system together. Rebar and grout are a must around all openings.

4. Anchor bolts play an important role in securing sill plates and ledgers. Use corner blocks (notch block at both ends to allow horizontal rebar to pass) so the anchor can be placed in the larger cell of the block. Depending on application, a full-length vertical grout cell along with a horizontal bond beam may be required by your local building department. The horizontal bond beam is the last course where the complete row of cells is reinforced with steel and grout (Figure 3-14).

5. Correct positioning of rebar plays an important role in this system. Figure 3-15 shows the rebar dead center of a corner block, and other pieces of rebar placed in the smaller (interior) cell of the block. Make sure you meet your building code requirements when installing rebar and grout.

6. Insulation fills the exterior cell of the blocks (Figure 3-15).

7. Masons install electrical boxes and conduit as walls are stacked, providing the raceway for the wiring later. Conventional masonry saws cut blocks or, when only a small amount of cutting is required, you can use an electric hand saw with a masonry blade.

8. Good design will limit the amount of plumbing in the exterior walls of any building, but when it's unavoidable, water or gas pipes may run through the inner cores. Larger sewer pipe can run through corner block placed within the wall, forming a large vertical cell.

9. Standard masonry fasteners are used to connect floor joists, ceiling joists, etc.

Manhours — A crew of five (two masons and three helpers) can create a good working rhythm for installing this product. Using standard block (8 × 8 × 16 inches, or 0.88 square foot), a crew of five can install at least 700 blocks in 8 hours. This includes installing and tying rebar and pouring grout. Since the interior and exterior bonding surface materials have to be applied to both sides of the block's surface, a crew of three can apply 1,500 square feet per side (3,000 square feet total) of wall surface material in 8 hours. Keep in mind that while these figures are based on experience with the system, they're only approximations. Use them as a guideline until you have figures of your own.

Structural Insulated Panels

One unique alternative building technology for building envelope construction is the Structural Insulated Panel (SIP). According to my source, the Structural Insulated Panel Association (SIPA), these panels can provide efficient solutions to concerns such as energy efficiency and dwindling natural resources. The technology for SIPs is not new. It emerged as early as 1952, when Alden B. Dow, founder of the Dow Chemical Company, began designing homes using SIPs. The first one, built in Midland, Michigan, featured foam-core SIPs for exterior walls, interior partitions, and roofs. Many of these homes are still occupied today.

SIP construction is an engineered system. The technology presents a viable alternative to conventional stick-built construction methods, offering strength and structural performance as well as superior energy efficiency. It's also far more environmentally-friendly. These panels are primarily used as the exterior walls and roofs of low-rise residential and commercial buildings. Figure 3-16 shows an SIP being lifted into place at a Ryan Homes townhouse project.

Courtesy: IMSI

Figure 3-15
Correct rebar placement in IMSI units

Photo: Russel Roeding   Courtesy: Structural Insulated Panel Association

Figure 3-16
Structural Insulated Panel wall
installed in a townhouse project

Product types vary in the industry, but all SIPs share one basic design: two exterior skins adhered to a rigid plastic foam core. Panels are available in sizes from the standard 4 × 8 feet on up to 8 × 24 feet, and in thicknesses from 2 to 12 inches. These options allow builders to select the optimum size for a specific application.

The two panel skins on an SIP can be made of the same or of different materials. The most commonly-used faces are OSB, waferboard, plywood, sheet metal, and gypsum board. The rigid foam cores are composed of EPS, XPS, polyurethanes or polyiso-cyanurates. EPS is the most common because of its low cost, but the cores are thicker to equal the higher insulating properties of other foams.

It's widely recognized by energy performance specialists that some foam plastics, over time, are subject to "thermal drift" — the out-gassing of blowing agents. In some plastics, a higher R-value is attributed to the blowing agents, like HCFCs. Over time, some of the blowing agent, but not all, dissipates from the cell, reducing the R-value of the foam. EPS foam isn't subject to thermal drift because all of the blowing agent is dissipated in the manufacturing process and the R-value is based on the air trapped in the cell. With R-values of 5 per inch, XPS cores are listed widely for design values, indicating that this is the long-term constant after all thermal drift adjustments. Most manufacturers of other foams also quote R-values at the fully-aged rate, but this needs to be confirmed by designers.

The process of assembling skins and cores into an SIP varies among the manufacturers. Most use either adhesive bonding or foam-in-place.

▌ *Adhesive bonding* involves applying a structural-grade adhesive to both sides of an unfaced, preformed rigid foam core. Then the core is placed on top of a clean sheet of facing material, the second panel is positioned on the opposite side, and pressure is applied. The panels are then set aside for approximately 12 hours until the adhesive has completely cured.

▌ With the *foam-in-place* method, the facing boards are held apart by panel framing or specially-made spacers. The chemical components of the foam core (and a blowing agent) are combined and then forced between the braced skins. The expanded insulation material bonds to the facing material and no adhesives are needed.

Unlike fiberglass batts, SIPs are resistant to moisture absorption. While you should make every attempt to keep the panels dry, SIPs will retain their R-value even if they do absorb some moisture. Wood-frame walls need vapor barriers installed on the warm side of fiberglass or mineral wool to prevent water vapor penetration, which may condense and degrade insulation performance. By contrast, SIPs don't need vapor barriers because moisture doesn't materially affect performance.

SIPs form structural envelopes that are effective against air infiltration, a major source of energy loss. This advantage over conventional construction is primarily due to the large uninterrupted areas of insulation in the panels. Frame walls have frequent joints between sheathing at studs which break the continuity of the envelope. They also have nail or screw penetrations at every stud on both sides of the wall. Other common points of leakage such as electrical outlets, vents, and other penetrations are more difficult to seal in frame structures. Even if these penetrations are poorly sealed in an SIP structure, insulation performance isn't compromised by air circulation in the insulation cavity. The bottom line is that SIP houses are exceptionally tight when compared to conventional framed structures, allowing very low levels of air infiltration. This makes the building more energy efficient.

Innovations in plastics and wood products industries are largely responsible for the rapid growth of new products now used in SIPs: first plywood, then OSB. These products have a common goal: to provide improved, economical products while conserving scarce resources. Panel manufacturers can remove the strength-reducing characteristics of wood (like knots and splits). They produce superior engineered products from moderate-cost, low-quality hardwoods and plantation thinnings. As a result, trees are more fully utilized and fewer wood fibers are used to produce a more consistent product than materials used in conventional framing.

Designers can optimize building design using SIPs, resulting in more efficient use of construction materials. They can significantly reduce process and construction site waste, requiring less landfill disposal. SIP openings for windows and doors are often precut at the factory, making the job easier for the contractor and again, reducing debris on the job site and its accompanying disposal costs.

During panel manufacture, foam core materials are customized for a particular application. Creative design and resource management limit waste. Unused foam generated during the manufacturing process can be returned to the foam manufacturer for reprocessing or recycling. Recycling is the preferred method for handling waste. But if recycling isn't available due to a job site's location, foam plastic can be safely put in a landfill. SIP foams are stable and won't biodegrade or create leachate or methane gas, the two major problems with all landfills. Construction materials are often used in stable landfills where the ground is later reclaimed for parks, stadiums, and similar applications.

SIP foams can also be safely incinerated at regulated waste-to-energy facilities. Its energy value is greater than some soft coals. EPS burns cleanly and produces nontoxic ash which doesn't require hazardous landfill disposal.

Air quality is another ongoing concern to the public, regulating agencies, SIP producers, and foam manufacturers. EPS foam cores are produced from materials which have never had any adverse effect on the earth's protective ozone layer. Many, if not all, polyurethane and extruded polystyrene foam core producers have switched from chlorofluorocarbons (CFCs) to hydrochlorofluorocarbons (HCFCs) for blowing agents with substantially reduced ozone depletion potential. Plastic industry members are working to exceed air quality standards, both current and future, through improvements in materials, processing and control equipment.

Noise pollution (unwanted sound) is another matter of concern. SIPs provide excellent barriers to airborne sound penetration because of their closed construction (no air movement in the panel wall) and their extremely tight joint connections.

Despite recessionary times in the construction industry, SIPA members have experienced significant market growth since the late 1980s. This growth is fueled by lumber shortages, concerns about energy costs and environmental sustainability, and demand for construction quality. There's also a greater understanding and acceptance of this technology by builders, designers, homeowners, and other decision makers.

Products

There are many Structural Insulated Panel Systems manufacturers on the market, but I want to zero in on R-Control Structural Building Panel by AFM Corporation. It features rigid EPS insulation laminated between engineered wood, and is designed for both the residential and commercial markets. Engineered wood (made of wood strands bonded together with structural adhesive) yields a product *that's stronger than regular wood*. The wood is produced from fast-growing common trees, like aspen, poplar, and southern yellow pine. These underused species are less expensive, and manufacturers can use a greater proportion of each tree.

The panels are available in sizes from 4 × 8 feet up to 8 × 24 feet for use as exterior walls, roofs/ceilings and floors. You can install the SIP on top of exposed beams (Figure 3-17), or use the beam or joist as your spline for a flat ceiling. Factory spline joints are available (Figure 3-18). Standard panels come in the following thicknesses and R-values:

Courtesy: AFM Corporation

Figure 3-17
R-Control roof panels

Panel thickness (in)	R-value at 75°F	R-value at 40°F
4½	14.88	16.00
6½	22.58	24.33
8¼	29.31	31.63
10¼	37.00	39.88
12¼	44.71	48.31

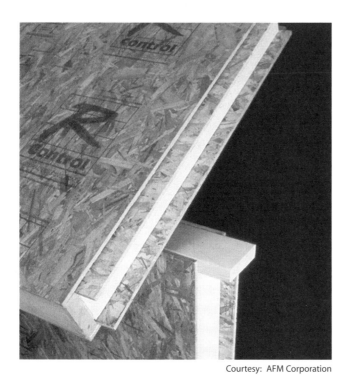

Courtesy: AFM Corporation

Figure 3-18
Factory spline joints for R-Control panels

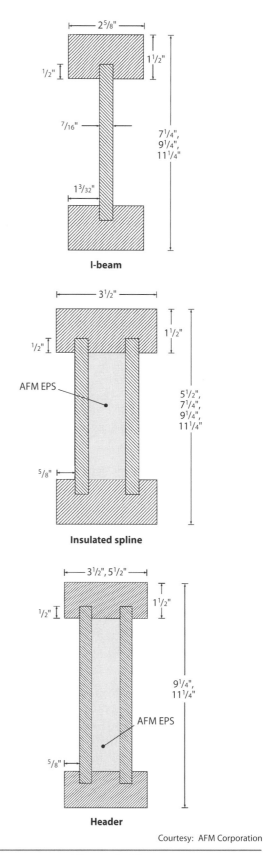

Courtesy: AFM Corporation

Figure 3-19
AFM I-beam, insulated spline and header

R-Control SpecLam is a nailable roof panel designed for application over a structural roof deck. The panel, which consists of one OSB skin laminated to EPS, is available with an optional spline joint. Their I-beam, insulated spline and header systems are single and/or double OSB webbed beams with 2 × 4 or 2 × 6 top and bottom flanges. Figure 3-19 shows all three. The I-beam has a single OSB web with 2⁵/₈-inch flanges, available in depths of 7¹/₄, 9¹/₄ and 11¹/₄ inches. The insulated spline has 3¹/₂-inch flanges with depths the same as the I-beam. You can get the header with either 3¹/₂- or 5¹/₂-inch flanges with depths of 9¹/₄ and 11¹/₄ inches or more. The cavity of a double-webbed beam is laminated with OSB. The company also supplies premium grade adhesive/sealant, fasteners from 4 up to 14 inches, and specialty tools for foam cutting.

Construction Notes — To give you a simplified outline of SIP installation, I'll zero in on the wall panel. To better understand the entire process, contact the manufacturer to get their Construction Details Manual for R-Control Panel Building Systems. You have two options for using these systems. First, you

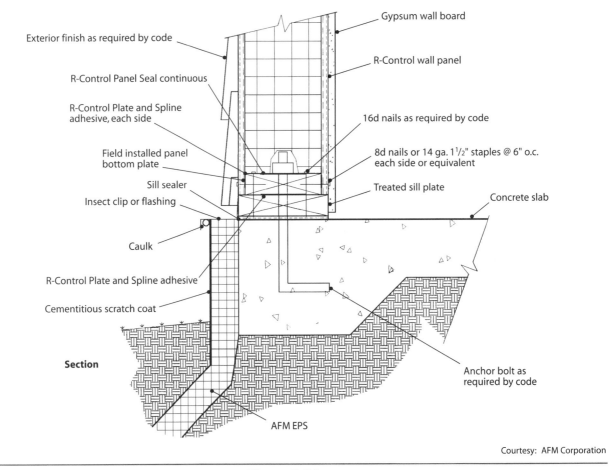

Exterior finish as required by code

R-Control Panel Seal continuous

R-Control Plate and Spline
adhesive, each side

Field installed panel
bottom plate

Sill sealer

Insect clip or flashing

Caulk

R-Control Plate and Spline adhesive

Cementitious scratch coat

Section

Gypsum wall board

R-Control wall panel

16d nails as required by code

8d nails or 14 ga. 1$\frac{1}{2}$" staples @ 6" o.c.
each side or equivalent

Treated sill plate

Concrete slab

Anchor bolt as
required by code

AFM EPS

No Scale

Courtesy: AFM Corporation

Figure 3-20
Slab foundation framing for the R-Control panel

can let the manufacturer fabricate a complete structural system according to your plan specifications. Or, second, you can bring in stock panels to install using the manufacturer's specifications. When putting up a panel wall system, consider the following:

1. This panel system isn't recommended for installation below grade level. When installing a wall on a concrete foundation or slab, be sure to install anchor bolts according to building code requirements. Consider using a treated sill plate (Figure 3-20) to provide a good barrier and an extra 1$\frac{1}{2}$ inch of nailing surface for base moldings once the bottom plate has been installed to the sill plate. Remember that the sill plate has to be the same size as the wall's thickness and set so that the outside skin of the R-Control is supported by the treated sill.

2. Avoid running plumbing through any of the exterior walls because it reduces insulation value and may cause frozen pipes. Try to bring plumbing up through an interior wall, build a plumbing case or false wall, or bring it out of the floor and under a cabinet toe kick — for instance, a kitchen cabinet located on an exterior wall.

3. The manufacturer can install 1$\frac{1}{2}$-inch electrical chases that are large enough for you to snake a wire horizontally, providing you're snaking from the electrical outlet's rough opening. If you can't get a particular horizontal run in, then consider installing the wire vertically from the top down (on a slab) or from the bottom up (when there's a basement or crawl space). Remember that when you install wiring from the top down, it takes more wire.

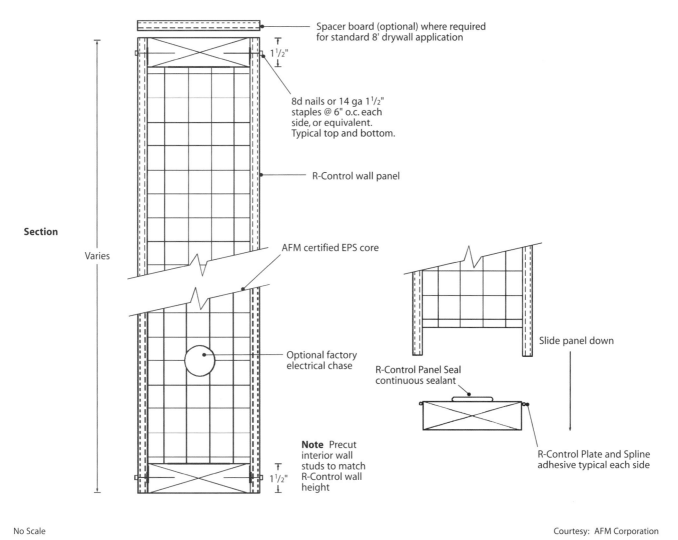

Section

Varies

Spacer board (optional) where required for standard 8' drywall application

1½"

8d nails or 14 ga 1½" staples @ 6" o.c. each side, or equivalent. Typical top and bottom.

R-Control wall panel

AFM certified EPS core

Optional factory electrical chase

Note Precut interior wall studs to match R-Control wall height

1½"

Slide panel down

R-Control Panel Seal continuous sealant

R-Control Plate and Spline adhesive typical each side

No Scale

Courtesy: AFM Corporation

Figure 3-21
Plate connections for the R-Control panel

When coming up through the floor, be sure to mark the floor where the chase is located in the wall and drill your hole before standing the wall up. If you plan to use a double top plate, don't forget to drill your hole for the wire. Use spray foam to seal all wire entrances and exits. If a commercial project requires conduit, check it out with your building department. You may need a licensed electrician on hand to install the conduit.

4. When installing top and bottom plates, be sure to follow the manufacturer's specifications and use the correct adhesive to secure and seal the entire system (Figure 3-21).

5. Bottom plates are always installed first and held back $^7/_{16}$ inch from the edge of the platform you're working on. They're also held back by $1^{15}/_{16}$ inches at the end where a corner will be established so the vertical end plate can slide past the bottom plate. The adjacent plate (corners or partitions) also needs to be held back so the first panel can slide past (Figure 3-22).

6. Another area of concern involves floor joist installation. Choices include joist hanger/wall panel, joist hanger and ledger beam, or the floor truss bearing on the wall panel, the same as conventional framing (Figure 3-23).

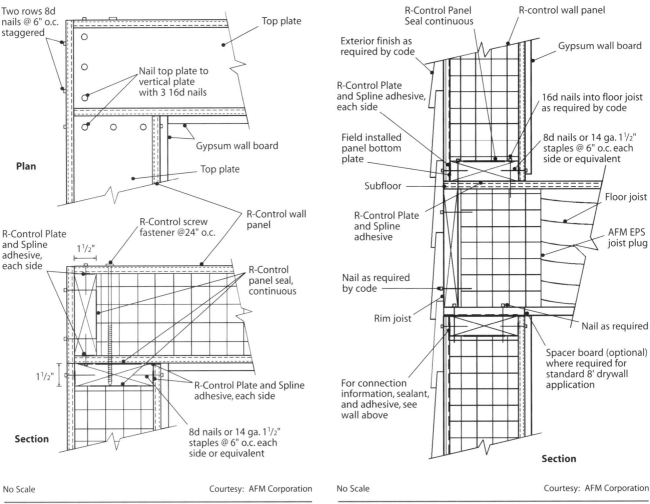

Figure 3-22
Corner connections for the R-Control panels

Figure 3-23
Floor joist bearing on wall panel

7. What about door and window openings? They can be cut out at the factory or you can cut them out in the field after the walls are installed. Once the openings have been cut, don't forget to frame inside the rough openings. You'll need to use the special tool (hot knife) to cut back the EPS foam by 1½ inches so the framework can be recessed. If you cut into the 12-inch header area or extend beyond the 8-foot span, you'll be required to install a header in this area. Consult the manufacturer before ordering materials.

Manhours — Before you can estimate manhours, it's important to know what size panels you're going to use. The larger the panel, obviously, the heavier it gets. A 4 × 8 foot with a 3½-inch EPS core and OSB skins at ½ inch weighs about 122 pounds (3.8 pounds per square foot). While a single person can install a 4 × 8 foot panel in a wall, it takes a second set of hands to stabilize the panel while it's being secured. A crew of two can handle panels up to 8 × 12 feet before it gets too awkward. If the panels have been constructed in a full-length wall or if you need longer lengths for roof/ceiling and floor applications, I'd seriously consider using mechanical equipment such as a crane.

I suggest that you contact the manufacturer about weight sizes for the project you're bidding and design considerations to determine what crew and mechanical help to use.

Based on 4 × 8 foot panels, figure about 50 to 100 square feet per manhour. This includes plating, cutouts, caulking, installing splines, headers, fastening, and bracing. Keep in mind that in order to be efficient with this system, a three-man crew may work better than a two-man crew. The first time around you may want to figure 100 to 150 square feet for a crew of two; that's roughly three to five panels an hour for wall application.

Insteel 3-D Panel System

When the Insteel 3-D Panel System sample showed up in my office, I wasn't quite sure what to make of it. It wasn't until I watched the video from ICS 3-D Panel Works, Inc. and studied their specs that I realized what this product was all about. What really got my attention was the materials used in the first step in achieving a concrete wall — insulation sandwiched between two layers of welded wire and then sprayed with shotcrete. Insteel has been marketing their product since 1990, but it wasn't until the last few years that their 3-D Panel System gained international recognition. In particular, it was when Hurricane Andrew (Mother Nature at her worst) blew through south Florida in 1992. During this storm, 80,000 to 85,000 homes were destroyed, but the 15 concrete homes built by *Habitat for Humanity* using the Insteel system survived with minimum damage. These homes withstood winds in excess of 140 miles per hour.

This type of system was first introduced to the United States back in the early 1960s. Unfortunately, at that time our technology didn't allow the system to be mass produced economically and efficiently. After seven years of research and development, an overseas company designed and patented the computerized machine that mass produces the 3-D panels today. By changing the materials and fasteners used to construct roof systems, incorporating this concrete wall system, and by using improved windows and doors, people living in disaster-prone areas can build structures that can withstand most weather.

Products

The components of the 3-D System include an EPS core flanked by wire mesh, connected with diagonal galvanized truss wires, and field-coated with concrete. The wire used is from recycled steel. This insulated welded wire space frame provides astonishing strength along with overall system design flexibility (Figure 3-24).

Panels are available in any length, in 8-inch increments. They're lightweight and easy to cut, handle and put in place. A 4 × 8 foot panel weighs only 38 pounds. The insulation core ranges from $1^1/_2$ to 4 inches thick. Applying $1^1/_2$ inches of concrete to each side of the panel gives a fire rating of $1^1/_2$ hours; 2 inches yield a 2-hour rating. Ratings increase as concrete or other coatings and finishes are added.

Depending on the thickness of the core and the thickness of the concrete, the R-value is between 7 and 33. For example, $2^1/_2$ inches of insulation and $1^1/_2$ inches of concrete on both sides (for a total wall thickness of $5^1/_2$ inches) create an R-value of 11.

The product can also be used in any radiant floor system. First apply flowable concrete fill over the ground surface. Then lay the panels on top of the concrete (Figure 3-25). Panels will automatically

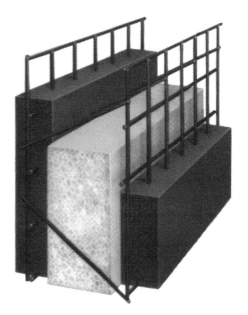

Courtesy: ICS 3-D Panel Works, Inc.

Figure 3-24
Insteel 3-D Panel System

Courtesy: ICS 3-D Panel Works, Inc.

Figure 3-25
Installing the 3-D panel in wet concrete

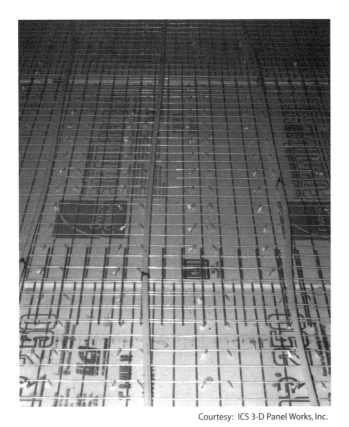

Courtesy: ICS 3-D Panel Works, Inc.

Figure 3-26
Radiant heat installed on 3-D panels

self-level in wet fill and can be immediately walked on, allowing radiant system installation to proceed (Figure 3-26). There's also the option of installing the panels in a roof application.

Construction Notes — This system appears to be fairly simple and doesn't require many tools. You use tie wire to connect panels together and install cover mesh to the panels. However, specialized tools can speed up the process. If you plan to make this system part of your regular building materials, consider investing in these tools. You'll find them described in the company's installation manual. Buying all of the recommended tools would be a major investment, but using subs that specialize in spraying concrete, for example, would reduce the number of tools you need to buy.

It's also important to think ahead and do some careful planning, especially when it comes to hanging a floor system and cutting out for windows. The manufacturer can outline the specialized tools required as well as provide technical information — just be sure to consult them.

Here's a brief summary of how to install the system:

1. Whether you're on a slab or foundation, it's important to install #3 rebar 12 to 18 inches above the slab, 24 inches on center, and according to building code requirements. If it's an existing slab, drill and epoxy the steel in place. Be sure to take precise measurements for rebar placement. When you lift the wall up and over the rebar, the steel has to slide behind the mesh and in front of the insulation core. Then tie off the steel with baling wire to the back of the mesh (Figure 3-27). The panels can be installed below grade.

2. Wall panels butt together with a lap-mesh on both sides of the panels to tie the two together. Corners and partitions butt up against the next panel. Corner mesh, both interior and exterior, ties the system together. Again, you can use baling wire or a specialized tool to clamp the mesh. Figure 3-28 shows a typical corner profile.

3. After the walls are in place, cut out for windows and doors. Cut back each rough opening by 1¹/₂ inches. Then, using a foam cutter, cut

the foam back an additional $1^{1}/_{2}$ inches and install 2-by treated material between the mesh. Fasten it with staples. Then determine the thickness of the finished concrete wall. Finally, measure and cut 2-by material to build your rough opening frame and install it into the rough opening. Fasten it with screws to the first framework. Be sure you have equal measurements on the overhang of the rough opening frame on both sides of the unfinished wall. Pay special attention at the header area. It's possible that the core will have to be removed in this area and rebar installed to provide additional strength. With the plywood attached to one side of the mesh, you can fill the cavity with concrete. Once the concrete has set a little, remove the plywood, then shotcrete the other side. The process is the same for truss pockets, ledgers, beams and columns.

4. Install electrical and plumbing between the insulation and mesh. For the electrical, use conduit and masonry outlet boxes, set out far enough to allow for the thickness of the concrete (and don't forget to install mud rings). Depending on winter temperatures in your area, avoid installing plumbing in an exterior wall. Instead, stay on the interior side with a false wall or come up out of the floor.

5. Figure 3-29 shows how to tie in a roof system. Notice that the hurricane strap was installed before the concrete.

6. Bracing is required on walls before installation of concrete.

7. Depending on the size of the structure, it may be possible to install wooden trusses before the concrete.

Manhours — Erecting the 3-D Panel System doesn't require a sub or a factory installer. But applying the concrete (the second step in the process) can be done by several distinctly different means. It's usually (but not always) performed by subs. I'd recommend using a sub instead of investing in the tools. Times vary widely, depending on how complicated the project is. Consider the number of windows, straight walls vs. curved, single- vs. multistory, the quality of the finish required, and so on. Here's an estimate of

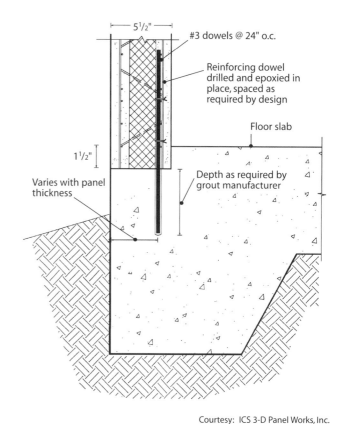

Courtesy: ICS 3-D Panel Works, Inc.

Figure 3-27
Footing to wall connection for 3-D panel

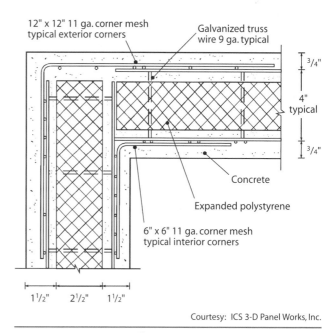

Courtesy: ICS 3-D Panel Works, Inc.

Figure 3-28
Typical corner for 3-D panels

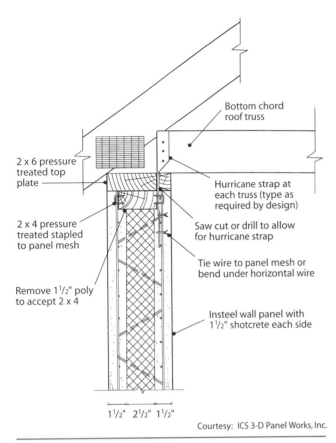

2 x 6 pressure treated top plate

2 x 4 pressure treated stapled to panel mesh

Remove 1¹/₂" poly to accept 2 x 4

Bottom chord roof truss

Hurricane strap at each truss (type as required by design)

Saw cut or drill to allow for hurricane strap

Tie wire to panel mesh or bend under horizontal wire

Insteel wall panel with 1¹/₂" shotcrete each side

1¹/₂" 2¹/₂" 1¹/₂"

Courtesy: ICS 3-D Panel Works, Inc.

Figure 3-29
Tying down a truss to a 3-D panel

the manhours to construct a structurally-sound basic one-story structure:

■ Erection: 2,000 square feet per three-man crew in an 8-hour shift = 24 manhours (3 × 8); 2,000 square feet ÷ 24 manhours = 83.333 square feet per manhour.

■ Shotcreting: A cubic yard of concrete will cover both sides of 100 square feet of panel at 1¹/₂ inches thick. So 20 cubic yards = 2,000 square feet per four-man crew in a 9-hour shift. That's 36 manhours (4 × 9). And 2,000 square feet ÷ 36 manhours = 55 square feet per manhour.

■ Total process: 24 manhours + 36 manhours = 60 total manhours; 2,000 ÷ 60 square feet = 33.33 square feet per manhour and 2,000 ÷ 33.33 square feet = 60 manhours. The 60 total manhours represent two days to complete this project: one 8-hour shift (three-man crew) and one 9-hour shift (four-man crew).

ThermaSteel System

Thermastructure XT, Corp. has been producing and marketing a lightweight structural insulated system (panels) for over 20 years. Since 1976, the panels have been marketed under various names, like RADVA Panels, Wallframe, and Thermastructure steel-framed EPS panels. You can use them for foundation walls, floors, exterior and interior walls, as well as ceilings and roofs. The panel is designed to perform four functions in one high-tech step: structural framing, insulation, sheathing, and vapor barrier, which decreases labor costs and increases the speed of construction. Since then, the product has been refined and is marketed as ThermaSteel, but it's also referred to as Thermastructure Panel.

In 1995, Anguilla, an island of the British West Indies, took a direct hit from Hurricane Luis. A million dollar transmitter was housed in a commercial building constructed with this system. During the storm, in which winds peaked at over 175 mph, some water leaks occurred around the door (which is to be expected), and one of the walls suffered impact damage when a 40-foot container sitting on the ground some 15 to 20 feet away was picked up and thrown into the side of the building. After the container was removed, the only damage was to the corner of the building, and the two corner panels were later replaced. A thorough inspection yielded no other signs of structural damage.

Products

The panels use composite technology, so the whole is stronger than the sum of the parts. The manufacturer uses three components: G-90 galvanized steel, special heat-activated glue and Carbon Enhanced Modified (CEM) EPS in a patented molding process (Figure 3-30). The studs are 16 and 14 inches on center. The standard panel is 3¹/₂ inches × 4 feet × 8 feet. You can also get heights up to 12 feet, and standard thicknesses of 4, 5¹/₂, and 7¹/₂ inches. A panel 3¹/₂ inches × 4 feet × 10 feet weighs roughly 48 to 50 pounds. But as Figure 3-31 shows, you still need enough bodies to get a full wall off the decking, even though it's lightweight.

Both sides of the panel have a shiplap joint which reduces air infiltration by 34 percent. The effective R-value is R-16 for a 3¹/₂-inch, R-23 for 5¹/₂-inch,

Figure 3-30
Typical ThermaSteel panels

and R-33 for a $7\frac{1}{2}$-inch panel. The manufacturer indicates a minimum of 33 to 50 percent savings in utility bills by using this system. Using CEM EPS as a major material component gives the panel the following advantages:

- It's rigid and lightweight (Figure 3-31)

- It has stable R-values due to the closed cellular structure which contains stabilized air

- Insulation values are consistent over the life of the structure

- Contains no CFCs or HCFCs and there's no off-gassing

- Resists fungus, decay, and moisture gain

- Will not rot; highly resistant to mildew

- Environmentally safe

- Eliminates risks of cancer and respiratory and skin irritations you can get from some other types of insulation

- No food for termites and other common wood eaters

ThermaSteel can be ordered as a predesigned package or you can incorporate stock panels into virtually any house or building design (Figure 3-32). The manufacturer can provide premolded electrical

Figure 3-31
Raising a ThermaSteel panel

Figure 3-32
ThermaSteel panels in a house design

chases in the CEM EPS core of each panel, or you can use a hot knife to cut wiring and plumbing chases. And you aren't limited to just walls; you can use this system in roof, floor, or foundation assemblies. Door and window rough openings are premolded to plan specifications.

Construction Notes — It's essential that the slab or foundation you're working off be level and square. Because this system attaches to a sill plate or steel channel, the sill plate also has to be as level as possible. If it's not, you'll have to install it on a bed of leveling mortar. Don't try to level the panel with shims. Here are highlights to consider when out in the field:

1. The two most important tools you'll use are a screwgun (if it's cordless, be sure you have extra backup batteries) and a caulking gun for the adhesive. You might use an industrial applicator, as discussed in Chapter 2 under "Tools."

2. Throughout this system you'll be using *shear plates* to secure the panels. Inside and outside angle shear plates are used on corners and partitions and flat shear plates are used to attach wood top plates to the wall panel and when attaching the wall panel to a wood sill plate (Figure 3-33). This is when the screwgun will come in handy.

3. When adjusting for wall length, you can leave gaps up to 1/4 inch at each joint. Then if there are still gaps after you've adjusted and fastened the panels, seal them with an insulating spray foam.

4. To secure the panels, you'll need top plates. Either use a plate that's long enough to cover the distance of the wall or install a plate that reaches halfway into the next panel. As with conventional framing when installing a double plate, always overlap the top plates at the corner. One last thing: Be sure to predrill your plates for electrical and plumbing before installing.

5. When cutting in the rough openings for doors and windows, the framing can be done with a "C" channel or wood. If you plan to double up on your framework, be sure you allow enough on the cut for the double thickness. There's nothing more discouraging than discovering that the rough opening is too large or too small when you're ready to install a window, for instance. Measure twice, cut once!

6. When using the panel for a roof application, sheath the surface with 1/2-inch plywood before installing the roofing material.

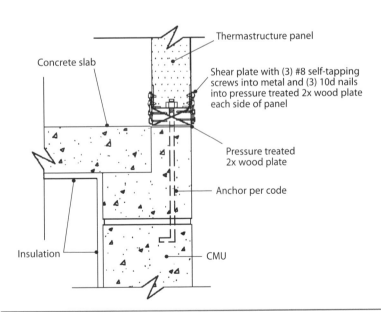

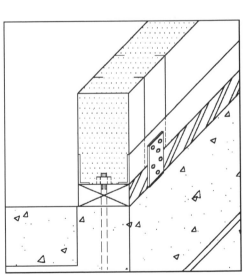

Courtesy: Thermastructure XT, Corp.

Figure 3-33

Attachment details — ThermaSteel panel to wood plate

7. The metal used in the panels could be sharp; make sure you and your workers wear gloves when moving them.

8. Because you're dealing with a lightweight panel, it's not a good idea to work in high wind conditions — you may find yourself airborne!

9. Interior sides of panels must be sheathed with some form of cladding to give a minimum 15-minute fire rating. It's usually ½-inch wallboard.

Manhours — A 3½ inch × 4 foot × 8 foot panel weighs roughly 45 to 48 pounds, so it's comfortable for one person to handle. But for installation, it's better to have a two- or three-man crew. Since window and rough openings are premolded to plan specifications at the plant, there's no labor to figure for cutouts. A 40 foot × 24 foot (128 linear foot or 960 square foot) building can be installed in about 5 manhours with a crew of two — and that includes bracing. That's 192 square feet per manhour. This manhour figure is based on having experience with the system. The first time out, I'd figure around 162 square feet per manhour with a crew of two.

Courtesy: Royall Wall Systems, Inc.

Figure 3-34
Royall wall panels on site

Panelized System

Have you ever wondered how an entire shopping mall can appear in just a matter of weeks? Most likely the contractor used a panelized system. They're common on industrial and commercial job sites to speed up the construction process. Basically, these systems involve prefabricated wall panels composed of concrete and rebar, manufactured either in a controlled environment or out on the job site. Today's new technology and materials make these precast units exceptionally strong. Perhaps that explains why these systems are used more and more frequently in disaster-prone areas.

Products

One manufacturer of these strong and energy-efficient panelized wall systems is Royall Wall Systems, Inc. Working out of Florida since 1990, Royall is working to license other manufacturers to fabricate these patented systems elsewhere in the United States. The panels themselves are composed of an outside layer of steel-reinforced concrete with EPS foam on the inside (Figure 3-34). Steel studs ("C" profile) run vertically every 2 feet on center, plus horizontal studs at the top and bottom. In Figure 3-35, notice that the studs are flush with the EPS. The "C" channel of each stud is filled with concrete, which anchors it to the wall.

Most wall panels come 5½ inches thick, although panels 4 to 10 inches thick are also available. The R-value varies from 8 to 30, depending, of course, on the panel's thickness. They come in heights up to 22 feet and lengths up to 30 feet, the maximum that can be trucked to a job site. The manufacturer handles design, production, and installation and offers a ten-year warranty against major structural defects — but not against Mother Nature herself. However, the wind speed resistance of Royall's panels varies from 110 mph to over 260 mph, so these panels are extremely hurricane resistant. No wonder they're moving into the residential market!

Courtesy: Royall Wall Systems, Inc.

Figure 3-35
Installing Royall wall panels

Construction Notes — The panels themselves sit on a shelf 1 1/2 inches deep by 6 inches wide built into the foundation or slab (Figure 3-35). This shelf design helps prevent water from penetrating the joint between the panel and the slab. Panels are lifted into position with a crane (tilted-up) and held there with adjustable screw jacks while additional panels are lifted into place. To permanently secure the panels, remove the foam to expose the steel U-shaped "buckets" fabricated into the panels at the factory. Bucket positions are predetermined by the uplift load calculated at the factory (never greater than 6 feet on center). Once you've drilled into the concrete, insert expansion anchors through the predrilled holes in the buckets. Then reinstall the foam.

On a straight wall, join panels at their mating lap joints and hold them in place with bolts placed through preformed holes in the panel edges. Corners, on the other hand, use one flat end that fits into a recessed area of the adjacent panel. They're also held in place with bolts. The entire system is allowed to have gaps in the joints up to 1/2 inch. Later in the construction process, you'll seal these joints with spray expanding foam.

There's room to feed the electrical through sleeves in the concrete beam at the top of the panel. The factory places these, and also cuts electrical raceways into the foam along the steel studs. All the electrician has to do is remove the foam, feed the wire from the top down, and install the outlet box attached to the steel stud. The foam can then be friction-fitted back

into the panel following inspection. The factory can also install sleeves from the bottom if a crawl space or basement is planned.

Fasten wallboard to the steel studs on the interior side of the panel. Leave the exterior surface rough to provide good adhesion for stucco application. If you're planning a brick exterior, incorporate a brick ledge into the foundation or slab. This construction system allows the walls of a typical single-family residence go up in less than four hours! Try to beat *that* with stick framing.

Manhours — To install this product, you must either have a factory-trained crew, or let the manufacturer's crew install the panels. This system is custom-built to plan specifications, then delivered to the job site, where the panels are unloaded and set in place by a crane. A crew of three can install a 2,500-square-foot residential structure (with 8-foot walls) in about 3 hours, ready for trusses. This assumes the crew is experienced, having built at least three to four homes.

What Now?

While Chapters 2 and 3 are packed full of ideas and products, remember that this book isn't the final word on the subject. There are many more products on the market! What I've tried to do in these chapters is to give you a new outlook and some insights into what you can expect throughout this book. They're also a starting point to consider alternative materials and methods. If there's a new product out there that could let you build better structures for less, and put more profit in your pocket, it's not going to do you any good if you don't know about it. Not every product will be right for you, but I hope the information in this book will help you zero in on the ones that are. The key is to start somewhere — with trial and error — until you find that "just right" product that you're comfortable with. That's the product you should be selling and installing.

The next chapter deals with more conventional framing materials. These are products that can help to save on labor cost, provide some design flexibility, and that you can incorporate with products described in these first two chapters. Most of them are also more environmentally friendly than the older products and methods, a good selling point for your customers and the right thing to do for the planet.

Framing Materials

A few years ago I framed a house that was a six-hour drive from my home. I can't remember exactly why I took this job, but I think hunger may have played a part. It was close to winter, and I was using unfamiliar materials. Although that job was some 300 miles from my home, even this short distance determined whether I was using fir and larch or hemlock. It seemed that every nail I drove split the hemlock studs. It didn't even help if I blunted the nail's point (as if I had time for that)! The lumber itself simply wasn't up to the standards I was used to. In fact, it seemed that the quality of the lumber overall was falling —almost as fast as the price was increasing.

Unfortunately, building with wood was all I knew how to do — and it's something I thoroughly enjoy. Perhaps you feel the same way. Nothing is more pleasurable than working with high-quality lumber, and the aroma when you cut into it creates a real natural "high." But lumber quality has gone downhill. You struggle and fight just to get a straight, uniform piece that isn't missing half the stick. Generally, you order lumber over the phone so you never know what to expect until you cut the bundle strap. Then stand back. Who knows what will fly out of that bundle!

But today we have choices that weren't available (and weren't necessary) when lumber was cheap, plentiful, and good. Now we have alternative products like *engineered lumber* or *steel studs*, products that consume fewer trees in the manufacturing process. Understanding how to use these alternative

products offers a wealth of new opportunities for our customers, our profession, and our creative minds. Education is the key to success, and perhaps survival.

Log Homes

Living in the Northwest allows me to live as close to nature as you can get. Camping in nearby national forests is part of everyday life. I grew up in this area, and I remember when my family (including my grandparents) would all head up to the forest to camp, fish, hunt, and just have fun. Even though my grandparents have passed on, nothing has changed much after all these years. Trees are just as big now as they were then. What sticks in my memory of those old days is an old abandoned log home. What a feeling it was to stand in the middle of that building, knowing that someone had once built it by hand. Today many of these structures still stand, now taken over by the forest and its creatures. Perhaps somewhere up in the hills one of those old log homes is occupied by Bigfoot himself!

The era of log homes doesn't have to end. We can build log homes today that meet all the building code requirements. The real beauty of it is that we can construct log homes without harvesting old-growth trees. One company that specializes in these homes is Anthony Forest Products Co. They've been in the industry since 1916 and they actually own and manage their own timberlands. The company operates an engineered wood laminating plant that takes advantage of new-growth trees. They believe in sound management and the utilization of forest resources through environmentally-sound practices.

PowerLog

Because of concern for the environment and the surprisingly-large demand for log homes, Anthony Forest Products Co. developed the *PowerLog* in 1988. Since then, the product has been refined and incorporated into a kit of designer log homes sold under the name of Anthony Log Homes. Aside from the environmental issues, the company realized that even with proven drying methods, conventional logs could never be dried to the point that they wouldn't settle, crack, split, twist or warp. Instead, they developed engineered laminated beams milled into various log shapes. Glued under high pressure, the new log is stronger than the wood it's made from. And with an average moisture content of 12 percent, it's drier than conventional framing materials.

Anthony Log Homes has taken a different approach to the market than other log home companies. They offer only the components necessary to build the log house itself — no doors, windows or roofing. This way you can work with your customers to select finishes that match their personal tastes. What Anthony delivers are the logs and roof system, plus other engineered laminated products, such as the Power Beam that you can use for headers, girders, joists, posts or ridge beams. They also offer decking, log siding and paneling as well as edge-grain flooring. Figure 4-1 shows a three-story home using laminated logs in the D profile.

Products

Finished logs are available up to 30 feet in length, and precision end-trimmed pieces run from 8 to 16 feet. Popular profiles include D-shaped (8×6 and 8×8) and round-on-round (8×8) with finish sizes of $6^{1}/_{2} \times 5^{1}/_{4}$ inches and $6^{1}/_{2} \times 7^{13}/_{16}$ inches. Figure 4-2 shows a D-profile log in a butt and pass corner style. You can clearly see the lamination on the ends of the log. Both interior and exterior surfaces are made of Southern Yellow Pine. Each log is composed of various lengths of finger-jointed materials. The joints (depending on the profile) may be more visible than normal — both for interior and exterior surfaces.

An 8×8 log has an R factor of around 9. Unfortunately, R-9 may not meet the energy code in your area. But consider that energy efficiency can be measured several different ways. R-value measures resistance to heat transfer, but it doesn't measure

Courtesy: Anthony Log Homes

Figure 4-1

Log home under construction

Courtesy: Anthony Log Homes

Figure 4-2

D-shaped laminated logs

how much energy it will take to heat and cool a home. Nor does it take into account the effects of *thermal mass* — a characteristic of log homes. I'm told that log homes perform as if the walls were insulated to a much higher R-value. The problem is that most code officials won't take into account that a log home costs no more to heat and cool than a conventional home. Their concern is "by the book." They look at R-value, not performance.

There may be a way to get the log home to meet local requirements through an alternate method, however. The Federal Energy Policy Act of 1992 directed states to establish energy standards based on the Council of American Building Officials (CABO), creators of the CABO Model Energy Code (MEC). Unfortunately, the MEC isn't a national code, but it may have been adopted in your state. Check and see if it has. It defines various approaches for compliance, including an *envelope report*. That's an alternative method of measuring energy efficiency by evaluating the overall insulated structural envelope. For example, even though the building's log walls don't meet required R-values, you can increase other areas such as the roof and crawl space to compensate. High-energy glass (Low E) also contributes to the entire envelope. The envelope report assigns a value to these areas of increased R-values. Those values may make the log home equal in efficiency to the required R-value, and thus in compliance with the MEC.

The envelope report doesn't acknowledge the value of *thermal mass* in log homes — it's just an alternative measure. Perhaps your local utility company can do an envelope report. If not, you may be able to hire an engineer to produce the report. I suggest you consult with your local building department to learn the requirements before you're asked by a potential customer.

Construction Notes

As mentioned earlier, the success of a well-built structure is, of course, in the foundation. Log homes are no exception. You have to plan ahead when installing a solid log, anticipating the placement of electrical, plumbing, doors and windows. The key word is "preplanning." Additionally, you have to direct a lot of thought toward how to install that first log course. It can go on the foundation or on the decking and floor joist. Here are some points to consider:

1. Consider installing logs on the decking. Remember that the entire outside perimeter of the decking area must be doubled up, including the first and last joist and band or rim joist, to carry the weight of the logs. Overhang the first course of logs just far enough to match up the log siding that will cover the joist and rim joist. Figure 4-3 is a detail showing the

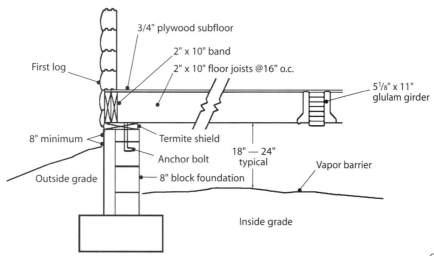

Courtesy: Anthony Log Homes

Figure 4-3
Laminated logs installed directly on the decking

prescribed distance from grade to the framing members. Be sure to cover the crawl space with a plastic vapor barrier to contain the natural ground moisture.

2. It's important to embed the first course in caulking material to prevent water from penetrating under the log or onto the finish floor or subfloor. Logs are held in place by $^3/_8 \times$ 10-inch lag screws countersunk $^1/_2$ inch with a washer in the center of the log. Hold them back from corners, doors, and windows by 3 to 6 inches. Every course thereafter is the same. Place caulk on both tongues.

3. Hold the butt joints in place by lag screws 4 inches in from the joint. Once the joint has been secured, drill a 1×6-inch-deep hole in the center of the joint. Install glue in the hole and drive in a 1-inch oak peg.

4. Exercise care on cuts made at all butt joints and other exposed joints. This could include areas around all doors and window openings.

5. Plumbing is installed in the interior walls and under toe kicks of cabinets. If plumbing has to go on an exterior wall, you'll have to build a fake wall.

6. Electrical should be installed from the crawl space if one is available. If this is the case, predrill into the double rim joist and joists on an angle in order to get to open space. This is easiest to understand by viewing the cross section of the front wall, as shown in Figure 4-3. Notice how the logs sit directly over the double rim joist. You'll also have to drill holes into the logs before each course is installed. Outlet boxes are installed directly in the logs themselves. And there's no room for error. Be very careful in cutting the rough opening for an outlet box. Try to locate the hole so it falls at the log's joint (Figure 4-4.) If you're lucky, the joint will fall within the code's limits of 12 to 18 inches above the floor. Whether it's at the joint or in the center of the log, consider using a template and router for a cleaner cut. And it's always best to do it before installing the log.

Manhours

While you don't need to be factory trained, you do need good cabinetry skills to install this system. A four-man crew can put up an average 2,000 linear foot home in 16 hours. That's 64 manhours, or two 8-hour days. A home this size (2,000 linear feet) works out to be roughly 28×38 feet or 1,064 square feet. Using their 8-inch logs (actual height, $6^1/_2$ inches), it works out to 15 logs for a wall 8 feet $1^1/_2$ inches high. The manhour estimate doesn't include gables, roof framing, interior walls or installation of doors and windows. It does, however, include door and window cutouts.

Steel Framing

Steel frame construction has been an industry-proven building method in high-rise and low-rise commercial and industrial markets. Steel has become the dominant building material for the following reasons:

▌ Steel has a higher strength-to-weight ratio than other common building materials.

▌ It's produced in strict accordance with national standards and isn't subject to regional inconsistencies as are other building materials.

▌ Steel won't warp, creep, rot, or be damaged by termites.

▌ It won't contribute fuel to the spread of a fire.

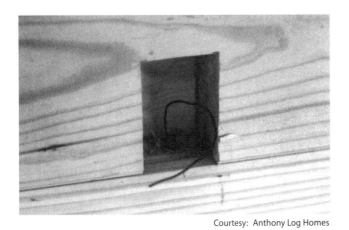

Courtesy: Anthony Log Homes

Figure 4-4
Electrical outlet box will be installed directly in the log

Yes, I've used steel studs, but not in the residential market. I used them when I built a retail outlet inside a shopping mall under construction. I wasn't an expert on steel studs, but I basically had to use them to meet the fire code requirements for interior construction in the mall. However, even though it was my first time, it really didn't take long to understand the system.

I do remember several of the difficulties I encountered with steel framing on that job site. First, because the ceilings were so high in the mall, I had a hard time securing the top plate for one of the partition walls. Then, right smack in the middle of the wall, there was a metal-framed door. Even though we installed ⅝-inch wallboard on both sides of the wall, it didn't have a solid feel when I closed the door. In the bathroom, I struggled to hang the upper cabinet. The metal screws I used just wouldn't hold in those metal studs. It seemed as though there just wasn't enough meat to the stud.

I can't really say I had a bad experience. Rather, it was that I didn't have enough time to learn how to work with these unfamiliar materials. I think that if I could have used wood around the door openings, for a double plate, and in the bathroom where the cabinet was to be installed, I would have had fewer problems. I know now that some experience handling the product would have really helped. Since that time, techniques and products have changed and there are new fasteners on the market especially designed to work with light-gauge steel.

One fastener I'd like to bring to your attention is the EverTite screw from Metaltite, distributed by GRABBER Construction Products. This unique self-driller, self-tapper is designed to resist the forces of bending, vibration, expansion, and contraction and installs easily with a standard 0-2,500 rpm screw gun. Its patented features ensure easy installation and superior holding power. It's designed for use in light-gauge metal, plastics, wood, fiberglass, and combinations of all of these materials. Figure 4-5, showing an enlarged cross section of a Metaltite screw joining two pieces of light-gauge steel, demonstrates how the screw pulls metal into the recess under the head to provide superior holding power. The company claims the EverTite screw is ". . . so stress-resistant it is unlikely to 'back out' under any circumstance."

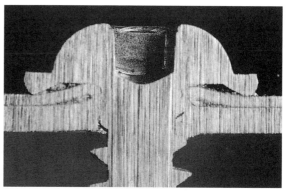

Courtesy: Metaltite Corporation, Inc.

Figure 4-5
EverTite™ screw

Would I use steel again? Yes I would, mainly because many of the steel products currently on the market are so user-friendly. Steel is durable (no shrinking, warping, or swelling), lightweight, insect resistant, uniform in design, noncombustible, stable in price, and — for the most part — less expensive than wood. Because it's lightweight, however, it damages more easily than wood during handling and installation.

In the steel vs. wood debate, how does using steel affect our environment? Even though the steel contains a large amount of recycled material, the raw materials aren't renewable resources like wood. And because it takes more energy to produce steel initially, the environmental benefits aren't as high as I'd like. But a typical 2000-square foot wood-framed house takes 40 to 50 trees, about an acre's worth. With steel framing, it takes about the equivalent of six scrapped cars.

According to the Steel Recycling Institute, the overall steel recycling rate for 1995 was 68.5 percent. This saves on valuable energy and natural resources. Each year, steel recycling saves enough energy to electrically power about one-fifth of the households in the United States (or about 18 million homes) for one year. Every ton of steel recycled saves 2,500 pounds of iron ore, 1,000 pounds of coal, and 40 pounds of limestone. For more information about steel recycling, call the Steel Recycling Institute at 800-876-7274.

There's also another factor to consider: Steel increases thermal bridging through exterior walls, which actually makes heating and cooling less efficient. (Look back to "Mortarless Block Systems" in Chapter 3.) But installing insulated exterior sheathing to create a thermal barrier can reduce heat transmission through studs and improve the thermal performance of walls. And the American Iron & Steel Institute (AISI) points out that steel framing results in less air loss around windows and doors as well as foundations and roofing connections.

I also wonder if this increased thermal bridging has anything to do with the dark vertical lines (ghost marks) that sometimes appear on walls where the steel studs are located. This can be a problem if you don't properly insulate the wall to create a thermal break. The AISI publishes the "Thermal Design Guide for Exterior Walls" that addresses this issue and provides solutions.

Don't get me wrong — steel has been used successfully for years in the commercial and industrial market where construction methods are different than they are in the residential market. It's more economical to achieve strength, noncombustibility, and design versatility with steel and concrete. But in the residential market, steel is a real newcomer.

My experience using steel made me wonder whether or not we can continue to use the carpentry and woodworking tools we already have if we expand into using this material. I looked into it and learned the answer — we need to equip ourselves with metal tools such as electric shears and/or nibblers, chop saws, hole saws, screw guns, and C-clamps and seamers, just for starters. However, I don't consider these minor additions to the tools we use a serious barrier to using steel.

The bottom line is that if you plan to use steel framing systems in your projects, you need to educate yourself to understand them. That's the only way you can use them efficiently. And that also holds true for building officials, engineers, and architects.

The History and the Future of Steel

While steel framing is a familiar sight on commercial construction projects, light gauge steel framing in residential construction is generally regarded as something new, although it has a long past. The 1932 Chicago World Fair featured a steel-framed home. During the 1930s, companies developing new welding technologies began building steel frame homes in an effort to create a large new market for those new technologies. During the 1940s, Lustron Homes factory-built literally thousands of steel framed homes. In more recent times, interest in the use of steel framing in residential construction has grown whenever lumber prices have risen, only to abate when they subside.

But steel framing has never been used on a larger scale in residential construction than during the 1990s. Around 55,000 homes were framed using galvanized steel in the early 1990s. It's projected to be 325,000 new homes by the year 2000. What's driving the current expansion in the use of steel framing in residential construction? Most attribute its foothold in the market to two changes in the lumber market. First, lumber prices are rising and highly volatile. Second, lumber quality has declined because lumber companies are harvesting younger trees (most old growth forests have already been harvested or are protected on public lands).

Steel, as a construction material, offers extremely stable prices as well as some superior performance characteristics. It's dimensionally stable and uniform in quality. Steel doesn't rot, warp, split, crack or creep. It can't be consumed by termites and doesn't burn. With all of these advantages, and with lumber problems expected to continue long-term, the domestic steel industry is strengthening its commitment to this new market.

In 1998, the member companies of the AISI formed the Residential Steel Partnership (RSP). This group's mission is to create an environment that enables and encourages the practical, economical, and widespread use of steel in residential construction. RSP's activities will include creating standards, training framers and other tradespeople, defining efficient distribution channels, and promoting steel framing to consumers in the new-home buying market.

RSP's name may be changing, but you can always get in touch with them through the American Iron & Steel Institute at:

American Iron & Steel Institute
1101 17th Street, NW, Suite 1300
Washington, DC 20036-4700
800-797-8335
www.steel.org

AISI is collaborating with the NAHB to increase awareness of the advantages of residential steel framing. They built an interactive steel-framed house on-site at the 1997 NAHB Builders' Show in Houston (Figure 4-6). The house, approximately 900 square feet, combines the use of steel framing with traditional wood framing. This house was created to highlight the fact that working with steel doesn't have to be an "all or nothing" decision. You can combine many other framing applications with steel.

The Residential Construction Department of AISI also offers several publications, from technical to promotional, related to residential steel framing. One is the *Builders' Guide to Steel Studs*, designed to provide professionals with instructions and details (industry practice) for constructing homes with cold-formed steel studs. It's just a basic guide, but it makes some interesting points, especially for those considering entering this market.

Thermal Design Guide for Exterior Walls outlines test results of thermal resistance in steel-framed walls. Thermal resistance or R-value is the resistance to heat flow through a wall. Because walls typically are comprised of various materials (studs, tracks, cavity insulation, sheathing, etc.), an effective R-value must be determined for the entire system.

Published in 1993, *Fasteners for Residential Steel Framing* still provides insight into the development of an efficient fastening system and provides guidance on the use and design of fasteners for cold-formed steel framing.

Carpenter's Steel System

HL Stud Corporation was formed in 1994 by a couple of people experienced in the residential and commercial market. Understanding the structural and economic advantages of steel in the commercial market, they saw an opportunity for steel to cross over into the residential market. But they understood that it had to be more user-friendly if it was going to succeed. They created a family of products that offer consistent pricing and quality and a simple application for site builders and manufactured housing builders. Their U.S. patented design combines wood and steel in the framing system. Figure 4-7 shows a premanufactured house builder attaching Carpenter's Steel Joist to a rim joist with a pneumatic nailer. The built-in nailing flange and bent shear tab eliminate the need for joist hangers.

Courtesy: American Iron and Steel Institute

Figure 4-6
A demonstration building showing
a combination of steel and wood framing

Courtesy: HL Stud Corporation

Figure 4-7
Carpenter's Steel System

Wood framing material is used only in plates. Notice the wood top, bottom and double plates, rough opening frames, cripples, headers, and the backer/nailer of corners and partitions in Figure 4-8. Those will overcome most of the problems I had with my first steel framing experience in the mall.

HL Stud successfully combines wood and steel by using a nailing flange and bent shear tab. You can see it in the photo of the joist in Figure 4-9. The top nailing flange simplifies the on-center alignment and holds the member in place for nailing through the bent shear tab. This overall design makes the entire procedure simple and quick. Each stud and joist is

Figure 4-8

Steel studs with wood plates

Figure 4-9

Carpenter's Steel Joist showing
nailing flange and shear tab

Figure 4-10

Shear tab bent for installation

prepunched for easy electrical and plumbing. Floor joists can be manufactured to specific span requirements. Their Carpenter's Steel System includes the following:

- 2 × 4 (true dimensions $1^1/_4$ × $3^1/_2$ inches) non-bearing Carpenter's Steel Stud — 25-gauge studs are used for nonbearing interior walls and are available with one nailing flange and one shear tab. The shear tab is bent on the job site. In Figure 4-10, you can see the bent shear tab, which was attached to the bottom plate with a pneumatic staple gun. Lengths are available in sizes comparable to wooden framing members.

- 2 × 4 ($1^5/_8$ × $3^1/_2$ inches) loadbearing Carpenter's Steel Stud — 20-gauge studs are used for loadbearing interior and exterior walls. They're available with two nailing flanges and one shear tab. Shear tabs are bent out in the field.

- 2 × 3 ($1^1/_4$ x $2^1/_2$ inches) Carpenter's Stud "HUD" Stud — 25-gauge studs are used for non-loadbearing interior walls in premanufactured homes.

- Carpenter's Steel Floor Joist — Available in 20-, 18-, 16-, and 14-gauge thicknesses. Its sizes are $1^5/_8$ × $5^1/_2$ inches, $7^1/_4$ inches, and $9^1/_4$ inches. Shear tabs are bent at the factory for the 18, 16, and 14 gauge.

These products are made from recycled steel (70 percent auto salvage). One thing I found surprising is that they sell directly to the builder, in truckload quantities, eliminating the dealer's markup. You can pass a savings like that on to the customer. However, HL Stud wants to develop dealer relationships in an effort to provide the best service for its product line. This is a great combination: wood and recycled steel working to conserve our natural resources, not to mention the builder's bottom line.

Manhours

Since their wall studs use a combination of both wood (top and bottom plates) and steel (studs), it makes it more difficult to determine manhours. For the most part, it's the same as stick framing, after the initial learning curve with this system. Based on 2 × 4 walls 8 feet high, the first time around I would figure 0.023 manhour per square foot. After you learn the system, it could drop down to 0.016

manhour. This wouldn't include framing the door or window openings. With a crew of two, the initial manhour time is 0.0115 (0.023 MH ÷ 2).

For door openings framed in wood (up to 3 feet, including header, double vertical studs, cripples, blocking, and nailing), I'd use 0.830 per opening, assuming a crew of two. If you had 20 openings, how long would it take to frame each opening?

0.830 MH × 20 doors =
16.6 MH ÷ 8 hours = 2.075 days ÷2 (crew) =
1.0375 days to frame 20 openings

1.0375 × 8 hours =
8.3 ÷ 20 doors = 0.415 × 60 =
24.9 minutes per opening for a two-man crew

Or you can simplify the calculation:

0.830 MH × 60 =
49.8 ÷ 2 (crew) = 24.9 minutes

Figure window openings the same way because they include the same elements as the door opening plus the subsill plate, and top and bottom cripples. For window openings over 4 to 5 feet wide (average), use 1.73 manhours per opening. You can find more manhours and costs for door and window size openings in *National Construction Estimator*. It's listed in the order form in the back of this book.

If you plan to use HL Stud's Carpenter's Steel Joists, be aware that they're also attached to wooden rim joists. Their weight can vary from 1.355 to 3.3489 pounds per linear foot depending on the width and gauge thickness of joist. For installation, figure on 0.020 manhour per square foot based on a wood 2 × 10, which is close to the measurement of steel studs ($1^5/_8$ × $9^1/_4$ inches). This doesn't include beams, blocking, or bridging but does include the band or rim joist. Keep in mind that you'll probably need a crew of two for production and training for both wall and joist installation, depending on the application.

Electrical Fittings in Steel Studs

Steel studs come prepunched ready to accept electrical and plumbing. That's the good news. The bad is that you can't run wire and plumbing pipes through the studs the way you can with lumber studs. The inside edges of the prepunched holes are sharp, so the electrical and plumbing codes require that you protect wires and pipes from the sharp edges. Don't

let this fact get past you. You *don't* want to discover after you have the entire home wired that you're in code violation. Even worse — what if the cable is pulled and it gets cut at one or more locations? Typically you wouldn't discover this until the electrical inspector visited the site, and then you'll get to start tearing out the walls to find it! The best way to avoid these problems is to use plastic stud bushings.

Bushings are very simple. Just pass one end of the joined connector through the hole in the stud and snap the bushing and washer ends together. This little precaution will save you time and money. Figure 4-11 shows the SB13 stud bushing by Arlington. It fits all hole configurations in prepunched steel studs. And just think how easily the cable will slide through the bushing when the wire is being pulled!

There is one other matter to bring to your attention, and that's how to tie a cable down to the face of a steel stud. The code says that whenever you enter or leave an electrical box, you need to secure the wire within 12 inches of a plastic box or 8 inches for a metal box. You also have to fasten the wire every $4^1/_2$ feet thereafter. How do you do this with steel studs? Simple! Install a cable support. In Figure 4-12 you can see an adjustable cable support that will hold three cables securely in the center of the stud, meeting code requirements. It keeps cables back and $1^1/_4$ inch away from the front edge of the stud.

Courtesy: Arlington Industries, Inc.

Figure 4-11
Plastic stud bushing

Courtesy: Arlington Industries, Inc.

Figure 4-12
Cable support on steel stud

Of course, the prepunched holes won't always be just where the electrician needs them to use the cable support. You may want to loan the electrician your stud punch tool. If you don't already have one, it would be a good idea to add it to your collection of metal tools. Arlington, the maker of the stud bushings and cable supports, has a stud punch tool available. Check out their model #MSP2. Stud and cable supports are normally installed by the electrician or his helper. It generally takes less than a minute per each if the holes are prepunched.

Prefabricated Homes

According to Lorin Sorensen's book, *Sears, Roebuck and Co. 100th Anniversary, 1886-1986* (a Silverado Publishing Company Book), prefabricated homes were big business for Sears Catalog mail-order sales. Between 1909 and 1937, Sears distributed more than 100,000 homes to various locations around the country. Customers who ordered these homes wouldn't receive the entire project at once; instead, they would receive it in sections. The floor and the framing would show up first, and then the next section would arrive just in time for the next construction phase, and that would continue until the house was completed. That's quite an accomplishment, especially for the time period. This same tradition is now being carried on by Tri-Steel Structures, Inc., a company that sells *Lifetime* Homes.

What makes this package different from the Sears Catalog homes is that Tri-Steel homes are computer designed and constructed on-site using steel instead of wood, bolts instead of nails, and screws instead of staples. All parts are prepunched, prewelded, and fabricated in their factory. Once the package arrives, you simply bolt the framing pieces together like a giant steel erector set (Figure 4-13).

Tri-Steel has devoted over 20 years of technical research and development to this system. Each package contains:

▌ Heavy gauge bolt-together steel columns and rafters

▌ Metal roof purlins

▌ An 8-inch steel stud system for exterior walls (allowing for R-30 insulation)

▌ Metal framing components for all dormers, roof and porch saddles

▌ Metal sub-fascia material for roof overhangs

▌ $3^5/_8$-inch metal studs and track for all interior walls

▌ Metal furring channels for all ceiling surfaces

▌ 8-inch metal second-floor joists for all second-story applications

▌ All bolts and fasteners for:
    Structural columns, rafters, joists and purlins
    Metal studs and furring
    Metal sub-fascia framing

▌ Complete set of working drawings, including anchor bolt layout and erection instructions as shown in Figure 4-14.

The framing system can be erected with a crew of three to five people. For multilevel designs, they recommend lifting equipment. One interesting fact is

A Each home package is loaded and inspected for delivery

B Steel framing is bolted together on the ground

C Purlins are attached to frames with self-drilling fasteners

D A little landscaping and your Lifetime Home is complete

Courtesy: Tri-Steel Structures, Inc.

Figure 4-13
A Lifetime Home goes together

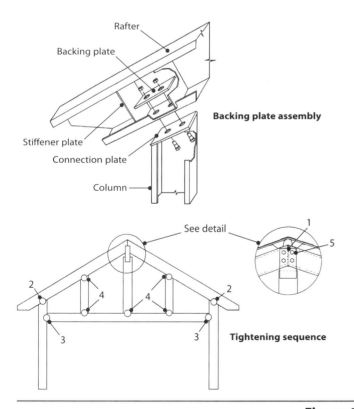

When assembling the frames, remember:

A The slots on the nuts identify it as a high-strength nut and should face away from the Red-iron so inspectors can identify them.

B Almost all frame connections require a backing-plate. The backing-plate is a steel plate that acts as a washer to distribute bolt forces and is critical to the connection. *Do not leave them out.* Refer to the assembly drawing for correct (color-coded) backing-plate and location.

C For frames using a steel truss design (i.e. Hillcrest, Heritage, Williamsburg, etc.), assemble each frame leaving bolts finger tight until all pieces are together. Once rafters, columns, stub columns, ceiling beams, and second floor beams are bolted together, tighten in the sequence shown. A good practice is to tighten all model frames from the top down.

D Assemble all frames with open "C" up. Once in the air, you can turn to face the appropriate direction.

E If a jackbeam is included in the structure, change the assembly sequence so that the supporting frames are raised before the supported frame(s). Example: If frame line 5 is supported by a jackbeam, assemble in order 4, 6, 5.

Courtesy: Tri-Steel Structures, Inc.

Figure 4-14
Assembling Tri-Steel frames

that you can build and cover the exterior shell first, allowing you do the interior framing in a protected environment if weather conditions are an issue.

Figure 4-15 shows some of the recommended manhour estimates per square foot for two of Tri-Steel's popular home styles.

I think anyone considering entering the home building business might want to consider such a system. Tri-Steel offers over 150 standard home designs that can be customized to meet your customers' wants and needs. Their engineers can also design homes to meet even higher local live-load or wind-load requirements — 155 mph +.

If you become a distributor (and it's not uncommon for a contractor today to become a distributor of building materials), you can get involved in their monthly hands-on construction training workshop. This will allow you to gain first-hand experience in proper tool and equipment usage for building with optimum safety. You'll also learn about slab and frame layout, primary and secondary frame erection, and installation of anchor bolts, purlins, soffits, and fascias. This is a great opportunity to sharpen your skills in this industry and get a better understanding of building with steel — so you can be a well-informed construction professional.

Sheathing

Have you ever wondered what it would be like to sheath a home with 1-bys in a diagonal application? I've seen lots of that during remodeling projects, but never had the urge to copy the craftsmanship. We all know that the fastest way to button up any project is with 4 × 8-foot sheathing. Covering the roof, walls, and floors — interior or exterior, it doesn't matter — is a snap. And for years plywood was the number one sheathing product, the standard for the industry.

However, the product itself has slowly diminished in quality. Besides having larger voids randomly located throughout the panels, it seems to delaminate all on its own. And one day I discovered, after ordering the material from the same supplier for years, that plywood is no longer the same thickness — it's now undersized. I'm not so worried about the product's structural integrity, but it's hard to match up to existing subflooring, underlayment, exterior wall, and roof sheathing. This "scanting" of the material is causing headaches out on the job site.

If you look around, everything is being downsized (except the cost). As professionals, we have to accept these changes over which we have no control and adapt to them. Even the new sheathings today, such as OSB, come in nominal thicknesses, not a full $1/2$, $5/8$, or $3/4$ inch. They are, however, designed to appropriate structural ratings. And they help conserve our

Red Iron System (Lifetime Homes) — Ridge built, bolted and frames raised into place		
Style	**Approximate square feet**	**Framing (days)**
Hillcrest	1,500	2-4
Alicante	2,100	2-4
Light Gauge System (Quick Silver Homes) — Platform built (conventional/stick frame)		
Style	**Approximate square feet**	**Framing (days)**
Shiloh	1,500	3-5
Glenview	1,500	3-5
The time estimates vary with the complexity of the project and familiarity with the project. The figures here are based on a crew of three, experienced in working with steel framing. ▌ Framing includes all structural steel interior and exterior. ▌ Does not include sheathing, felt paper, roof, windows, and doors.		

Figure 4-15
Manhours for installing Tri-Steel homes

natural resources because they're manufactured from fast-growing, underutilized trees such as aspen.

As for the differences in thickness, you just have to be a creative contractor, a problem solver. But for the record, I suggest you keep 15- and 30-pound asphalt felt on hand to use as shims. You can still find plywood at the full thickness if you search out a supplier and pay the premium price tag. In the meantime, take a look at some of the products we'll be discussing. There are many products on the market in this category. I'll just touch on a few that you may want to consider for your next project. When considering any new sheathing product, *be sure to check* that it's an approved, rated product. If you don't know, ask. Don't get caught out here unless you enjoy removing, replacing and reinstalling materials at your own cost! Be a safe and conscientious builder.

Solid-Core

What do you have when a panel product looks like plywood and feels like plywood, but it's not? And it's not waferboard or strandboard? Advanced Wood Resources took the best blend of available technology and natural resources to create a solid-core product called COMPLY Sturd-I-Floor. Just think — a solid-core plywood with no voids!

COMPLY is a five-layer composite panel of Douglas fir veneers and reconstituted wood fibers. The composite material is post-industrial resource, planer shavings, from the manufacture of Douglas fir lumber. The raw material goes through many steps before fabrication. The shavings are first dried and hammermilled (broken down into long strands or fibers). Then they sift the fibers in large shaker screens to remove strength-compromising dust particles. Finally, the fibrous strands are blended with an exterior rated phenolic resin. The dust-like fines are used as a heat source for the manufacturing process.

Under extreme heat and pressure, two interior layers of reconstituted wood fiber are joined with face, center and back veneers to create a solid-core panel that's stronger and stiffer than comparable plywood panels and highly resistant to moisture. It's the center veneer that gives this panel excellent dimensional stability in the cross panel (4-foot) direction. This center veneer is positioned with grain running in the 4-foot direction while the grain on the face and back

run in the 8-foot direction. Advanced Wood Resources helps our environment by efficiently using planer shavings for a larger proportion of the panel.

COMPLY is available in 4 × 8-foot sheets in four thicknesses: $^{19}/_{32}$, $^{23}/_{32}$, $^{7}/_{8}$, and $1^{3}/_{32}$ inches. Edges are square or tongue and groove. Certified by APA, the panels meet the specification of the major building codes. It's moisture resistant and guaranteed not to delaminate. The $^{7}/_{8}$-inch Sturd-I-Floor was specially designed to work with I-joists and has a 32-inch on center span rating. That could save on material cost because of the larger joist span. Glue-nailing is recommended. For tile, vinyl or glue-down carpet, use an approved underlayment.

Manhours

When considering the COMPLY $^{19}/_{32}$-inch square edge for a roof application, use 0.013 manhour per square foot for 24-inch on center. That's for any slope from 3:12 to 12:12, with a total height not to exceed 12 feet from the bottom chord to highest point on truss. Consider boosting your manhour figure above 6:12. Add 0.007 manhour per square for hip roof and 0.015 manhour for a steep pitch or cut-up roof.

When installing their $^{7}/_{8}$-inch T&G Sturd-I-Floor for subfloor/underlayment application 32 inches on center, estimate 0.015 manhour per square foot. For $1^{3}/_{32}$-inch, use 0.020 manhour per square foot. Both figures are based on hand nailing. Use a two-man crew.

Fire-Retardant

I never gave it much thought until Hickson Corporation brought to my attention the value of fire retardant treated wood (FRTW). It's available in both dimensional lumber and plywood. As I look around the office and think of all the valuable irreplaceable resources I've collected (like negatives) — they're absolutely right! I wish my office ceiling and walls were lined with FRTW; I'd sure hate to see my business go up in smoke. The purpose of this section is to show how FRTW can benefit your customer. But before we discuss benefits, we need to talk about the product itself. The specific product I'll be talking about is *Dricon* fire retardant treated wood licensed by Hickson, though there are other manufacturers of similar materials.

Performance

Dricon FRTW (lumber and plywood) is pressure-impregnated with a proprietary fire-retardant solution. Because the chemical is somewhat water soluble, you can only use it in interior applications, and areas protected from precipitation and direct wetting. In a fire, combustible materials generate gases that explode at a certain temperature, spreading flames in all directions. This effect, known as *flashover*, is one of the most common ways in which a fire spreads. An important feature of Dricon FRTW is that it automatically reacts when exposed to fire. The chemicals convert the combustible tars to carbon char and dilute the combustible gases with harmless carbon dioxide and water vapors. The carbon char insulates the wood and lowers the rate at which the cross section of exposed wood is reduced. The carbon dioxide and water vapor released dilute the combustible gases to help retard flamespread, allowing the wood structure to maintain its structural integrity longer, as shown in Figure 4-16A. Photo B shows a panel built of ordinary lumber and plywood which suffered severe fire damage (right). On the left is a treated wood panel, shown with the paneling stripped off. There's some charring, as intended, but the fire didn't spread.

FRTW also exhibits excellent resistance to fungal decay and termites when used in above-ground, weather-sheltered applications. Dricon FRTW contains the only fire-retardant chemical for wood treating registered for use as a wood preservative with the Environmental Protection Agency. In single-family detached homes, using FRTW is a personal choice of the homeowner. For many multifamily, commercial and public structures, however, building codes require the use of a noncombustible material or accepted substitute like Dricon FRTW. In these applications, it's best to check local codes for specific requirements.

Hygroscopicity and Corrosion

A material which gains or loses moisture from the atmosphere as the relative humidity changes is said to be "hygroscopic." The more hygroscopic a material is, the more moisture it will absorb during periods of high humidity. Wood is naturally hygroscopic, and fire retardants can significantly increase this attribute.

Corrosion is the tendency of a material to oxidize by chemical reaction. This process occurs more rapidly at higher moisture and temperature levels. Historically, FRTW has been considered corrosive because it picks up too much moisture in very damp locations. Conventional fire retardants can cause corrosion problems and failure in various types of metal hardware, fasteners and structural components. However, in some tests, Dricon treatment actually reduced corrosion of the protective zinc layer on galvanized steel truss plates.

A Charring retards flamespread

B Fire retardant treated wood vs. ordinary wood and plywood

Courtesy: Hickson Corporation

Figure 4-16
Dricon fire retardant treated wood vs. ordinary wood

Workability

If extensive cutting is anticipated, you should use a carbide-tipped saw blade. You can cut treated plywood in either direction without loss of fire protection. It doesn't change the surface-burning characteristics of the plywood. Cutting to length, drilling holes, joining cuts, and light sanding are OK, and it's not necessary to treat cut ends to maintain the flame spread rating. The recommended fastener is made of hot-dipped galvanized steel.

FRTW shouldn't be installed where it will be exposed to precipitation, direct wetting or regular condensation, or be used in contact with the ground. When using it as a roof sheathing, promptly cover it with felt for temporary protection until the roofing material is installed. Store the material off the ground and covered to protect it from precipitation. Take precautions when handling any treated material; wear dust masks and eye protection to avoid possible irritation from sawdust. Gloves will help avoid splinters. Hands should be thoroughly washed after handling any wood material. If you have any concerns about disposal of *any* treated materials you use, contact your local County Health District — Environmental Health Division and/or Solid Waste Management (Refuse) — Hazardous Waste Management Division.

Benefits

You'll probably use FRTW as an alternative to construction materials classified as noncombustible, not as an alternative to untreated wood. It's your responsibility to contact your building department to get approval when considering the use of FRTW in a project.

Here are some of the benefits of FRTW to your customer:

- Several major insurance companies provide rate reductions in homeowner's fire insurance.

- FRTW materials can prevent fire spreading to and through a roof system or interior non-bearing stud wall plates and fire stops with metal lath and plaster, or wallboard construction and partitions where noncombustible construction is required.

- FRTW may appeal to customers living in rural areas or in other locations where there may be inadequate water supply or fire protection.

- FRTW is worth considering in places where you can't readily install sprinkler systems, such as framing under raised platforms or theater stages, floors, walls, stud areas enclosed under roofs, and all types of remodeling work. Also consider it for studs, joists and sheathing in sensitive areas housing computer and electronic systems. Using FRTW may eliminate the need for sprinklers in these areas, which would prevent the damage that sprinkling systems can cause electronic equipment if they should happen to go off. Be sure to check with your local building department for approved application.

- FRTW would be very appropriate for roofs and walls in health care facilities where you need a versatile and economical construction system that will not compromise life safety.

Manhours

When using FRTW, you'll find only one significant difference from stick framing with conventional materials — safety considerations in handling the product could slow down production time. You must wear a dust mask and eye protection when using power tools.

To calculate manhours to frame 2 × 6 walls 8 feet high, use 0.029 manhour per square foot and 0.016 manhour per square foot for $1/2$-inch plywood or a total 0.045 manhour per square foot for the combination of both materials. For roof applications using $5/8$-inch plywood fastened to metal frame, use 0.026 manhour per square foot. Remember: to speed production, use at least a two-man crew.

Fiberboard

Can a durable, lightweight, weather-resistant structural sheeting be fabricated from 100 percent recycled newspaper? You bet it can! In fact, Homasote Company is the nation's oldest manufacturer of recycled building products. Most likely you've used their product but didn't realize what it was made of. When I contacted them concerning this book project, they replied, "While we are doing our part in helping to relieve the solid waste problems by manufacturing such products, the recycling job is not complete until these products are put back into use." So, to make the recycling circle complete, we all have to do our part by specifying and using recycled building materials in our projects.

Homasote also had some interesting facts about the use of recycled newspaper in their own products:

▌ Recycling not only decreases air pollution emissions by up to 73 percent but uses 40 to 70 percent less water than converting virgin wood pulp.

▌ Recycling uses up to 70 percent less energy.

▌ Each ton of newspaper recycled conserves about three cubic yards of landfill space. And we all know our landfills are already burdened to overcapacity.

▌ Homasote building products help in the conservation of more than 1,370,000 timber trees each year. They also eliminate more than 160,000,000 pounds of solid waste annually.

With that in mind, let's talk about two products that are environmentally safe, free of asbestos or formaldehyde additives, and have a long track record. I'm talking about Easy-ply Roof Decking and 4-way Floor Decking.

Easy-ply Roof Decking

You can use Easy-ply Roof Decking as a structural, loadbearing decking in residential and nonresidential buildings. Consider it for room additions, A-frames and log homes for exposed beam construction, primarily because the exposed application is eye-appealing. Easy-ply is fabricated from multiple plies for structural strength and comes with a 6-mil prefinished surface of decorative (flitter pattern) white vinyl film so you don't have to paint it. It's also available in a natural gray finish for those who enjoy being up on scaffolding and doing a little one-on-one with a paintbrush or roller.

The decorative vinyl also serves as vapor retarder, but Easy-ply isn't recommended for use over swimming pools, hot tubs, or in a sauna. Because the product comes with a finished surface, it requires more careful handling and installation than an unfinished material.

Easy-ply is available in one size: 2 × 8 feet (actual size $23^1/_8$ x $95^7/_8$ inches) with a tongue-and-groove profile on the 8-foot edge. It's also available in the thicknesses shown in Figure 4-17.

Here are some things to consider when working with this product:

▌ Prime the edges of all cuts made for fitting, clearance holes, or other openings with a quality primer-sealer and caulk them.

▌ Also caulk the T&G groove.

▌ Provide a gap of $^1/_8$ inch between square end butt joints and caulk joints.

▌ If an exposed beam will be stained, consider applying a polyethylene film over the beam before installing Easy-ply. After the beam has been finished, you can remove the film by slitting each side of the beam with a utility knife.

Panel thickness	Weight lb/SF	R-value	Maximum rafter spacing for live load (inches)*		
			40 PSF	50 PSF	A-Frame
1″	2.4	2.4	24″	16″	32″
$1^3/_8$″	3	3.27	32″	24″	48″
$1^7/_8$″	4	4.46	48″	32″	48″
$2^1/_{16}$″	5	5.0	48″	48″	48″

*Spans are limited by $^1/_{180}$ deflection expressed in lb. per square foot of roof area projected on a horizontal plane and over two or more spans; 10 PSF dead load assumed.

Figure 4-17
Easy-ply thicknesses available

▮ Nail from the panel center out to the ends, using five nails per rafter (or beam) and keeping nails back ³/₄ inch from edges.

Manhours —You need to take care when installing this product. Remember that it has a finished vinyl surface (one side), an 8-foot T&G edge, and a 2¹/₁₆-inch thick panel weighs 80 pounds. All cut edges need to be primed and caulked. When installing their thickest roof panel (2¹/₁₆ inches) for applications on flat, shed, or roofs to 5:12, estimate 0.047 manhour per square foot 48 inches on center, and an additional 0.019 manhour per square foot for steeper pitch roofs. The first time around, bid slightly higher and use at least a two-man crew. If you hope to achieve a higher R-value, you can install a second insulating sheet over Easy-ply. Also, you can apply most types of finish roofing materials directly over Easy-ply. Contact the manufacturer for technical support, installation instructions, or to get detailed application specifications.

4-Way Floor Decking

Similar in design to Easy-ply, its main purposes are structural, noise-deadening subflooring and insulating, and carpet underlayment. It's available in two panel sizes, 2 × 8 and 4 × 8 (actual size 47¹/₈ × 95⁷/₈ inches), and two thicknesses, 1¹¹/₃₂ and 1³/₄ inches. You can use them to span joists of 16 or 24 inches on center. Consider using this subflooring in low-rise condominiums and apartments, motels, nursing homes, professional buildings, and private homes where noise control is an important consideration. It's chemically treated for termite, rot, and fungus protection, and can support uniform live loads up to 100 lb./square foot based on a maximum deflection of ¹/₃₆₀ of the span. Here are some key points to know when installing this product:

▮ Use cross bridging to secure the floor joists, not solid bridging.

▮ Wall and partition plates can be applied directly over the material, across floor joists. If a partition is parallel with floor joists, install adequate support framing under the partition plate.

▮ Expansion and contraction are very real issues. Provide a ¹/₂-inch gap for every 50 feet of length and ¹/₂-inch gap at all abutting masonry

Figure 4-18
Apply continuous ³/₈″ beads of
APA-approved decking adhesive to install panels

walls. Leave a ¹/₁₆ to ¹/₈-inch gap at all butt joints. Snug all T&G edges together tightly.

▮ Adhesive is recommended with ring shank nails. In Figure 4-18, there's a continuous ³/₈-inch bead of APA-approved decking adhesive applied to the tops of the floor joists to install one or two panels at a time. Also apply two parallel beads to joists supporting butt joints and a continuous bead into the groove of the tongue and groove. Keep nails back ³/₄ inch from the ends and dimple heads below the surface by ¹/₁₆ inch.

▮ When installing nails, always begin in the center of the panel and work your way out to the ends. (Request the manufacturer's nail pattern and sequence.)

▮ For resilient tile, an approved plywood underlayment is required.

▮ Carpet should never be direct glued. Once tackless strips have been installed, add additional ring shank nails 8 inches on center.

Manhours — When installing the 4 × 8 T&G panels, either 1¹¹/₃₂ or 1³/₄ inch thick, at 16 or 24 inches on center in a hand-nailed subfloor application, use 0.020 manhour per square foot. Again, use a crew of two to speed production. When using adhesive, consider the following:

1. Applying a $^3/_8$-inch bead 16 inches on center takes 18 quart tubes for 1,000 square feet. Use 0.056 manhour per CSF (100 square feet). If you had a 3,000 square foot area, what would be your total manhours for the project? Here's the calculation:

   3,000 SF ÷ 100 CSF = 30 × 0.056 = 1.68 total manhours for one person to do 3,000 SF

2. Using a $^3/_8$-inch bead 24 inches on center requires 16 one-quart tubes for 1,000 square feet. Use 0.042 manhour per CSF.

Framing Members

Some of us aren't ready to make the plunge into using steel studs, but realize that if we rely only on wood, we're at the mercy of massive price fluctuations, shortages, poor quality, and losing out to competitors in an increasingly environmentally-conscious marketplace. The products in this section provide some choices to enable you to continue to work with wood, while avoiding the detriments just listed. These products come under the title of *engineered lumber*. Engineered lumber is stiffer, stronger, and more stable than conventional lumber, so it can span greater distances and support heavier loads. There are no natural defects such as knots or pitch pockets to be concerned about. There are four different manufacturing processes for engineered lumber:

- *Laminated Strand Lumber (LSL)* — An engineered lumber product made by reducing a log to thin strands up to 12 inches long that are bonded with adhesive to create a billet (large block) from which smaller dimensional pieces can be cut. Use it as a rim board, window and door headers, and millwork core material.

- *Laminated Veneer Lumber (LVL)* — The generic name for wood manufactured by peeling logs into thin veneers, which are then placed with grains parallel to each other, covered with a surface adhesive, and bonded under heat pressure. This process is used in creating micro-laminated

beams that you can use for window and door headers and structural beams. It's also used in the top and bottom flange of I-joists.

- *Parallel Strand Lumber (PSL)* — This is manufactured by peeling logs into veneers that are then cut into thin strands. The strands are laminated together with their grains parallel to one another. Parallel-laminated veneer is used in furniture and cabinetry to provide flexibility. Structural-grade PSL can be used as structural posts, columns, beams, and window and door headers.

- *Oriented Strandboard (OSB)* — A composite wood panel made of narrow strands of fiber oriented lengthwise and across each other in layers bonded with a resin adhesive under heat and pressure. This process manufactures sheathing for interior and exterior application and webs for I-joists.

Manufacturers use trademark names to identify their particular product, which may use any combination of these four different lamination processes. When contacting these companies for information, be sure to ask for a complete product binder that includes all the framing products they produce and offer. You may find the following products appealing.

Premium Studs

Studs without surprises — what a concept! Can you get a stud that won't bow, twist, or shrink once its been installed? Trus Joist MacMillan makes a product like that — TimberStrand LSL Premium Studs. This is one product where you'll never find a twisted stud. It's superior in quality and performance. This stud is manufactured from fast-growing trees where most of the log is converted to strands. The strands are bonded together with a polyurethane resin in a steam injection press to form a uniform billet. The billets are then sawn to the dimensions of nominal 2 × 4 and 2 × 6 lumber. The results are straighter and longer studs — up to 22 feet in length.

Look at Figure 4-19. You can see how straight and uniform the TimberStrand LSL studs are. If you like these, but your customer is on a budget, consider using TimberStrand premium studs only where high performance counts — at door and window openings

Figure 4-19
TimberStrand premium studs

Figure 4-20
Block placement in tall studs

and for straight kitchen and bathroom walls. Straight walls make cabinetry installation easier and faster.

TimberStrand LSL studs may be slightly more expensive than traditional lumber, but their benefits more than pay for themselves. Since they hold their shape and resist splitting, they'll reduce and possibly eliminate those expensive and annoying customer service callbacks.

Construction Notes

Some facts to keep in mind when working with TimberStrand LSL wall studs:

▌ They've got to be designed according to the manufacturer's current product literature. You can't just substitute them for solid-sawn studs.

▌ Walls that are 10 to 18 feet require one row of blocking at midspan. Figure 4-20 shows a good example of block placement. Walls over 18 to 22 feet need one row of blocking at two points, found by dividing the wall into thirds.

▌ You can use let-in corner bracing provided the maximum depth of cut is $^7/_8$ inch. Studs that are let-in within the middle third of the stud length must be doubled.

▌ Trusses (or rafters) and joists must be installed within 3 inches of stud locations when stud spacing is 19.2 inches or 24 inches on center.

▌ 2 × 4 studs-to-plates attachment requires two 16d end nails.

▌ Holes may be cut anywhere along the length of the stud but they must be a minimum of $^5/_8$ inch from the stud's edge. The maximum hole is $1^3/_8$ inches for a 2 × 4 and $2^3/_{16}$ inches for a 2 × 6.

▌ You can cut notches anywhere except the middle third of the length of the stud. The maximum notch is $^7/_8$ inch for a 2 × 4 and $1^3/_8$ inch for a 2 × 6.

▌ You can't have notches and holes in the same cross section.

Courtesy: Fiber Technologies Incorporated

Figure 4-21
FiRP glulam (left) is as strong
as the conventional glulam (right)

Manhours

Calculating manhours per square foot might be difficult because the quality and consistency of the studs mean fewer cuts during production. This alone could cut down on labor costs. While there are no standard studies or surveys that detail manhour experience, it's possible to save significant manhours when using these studs instead of traditional materials when framing a home. For the purpose of bidding, especially the first time around, base your calculations on 2 × 6 walls at 8 feet high using 0.029 manhour per square foot. This doesn't include door or window openings. Use a crew of two or more.

Fiber-Reinforced Glulam

I'd venture a guess that you've used a glue-laminated beam once or twice. If you haven't, try it. This is one framing member that can do a world of good in design and function. Once you start using them, you'll soon discover their convenience and start looking for ways to incorporate them into your work. Laminated beams can span a long distance and carry loads with minimum (or sometimes no) support. That gives you greater design flexibility. For example, using a laminated beam instead of 2 bys nailed together for a garage door header noticeably reduces sag.

Fiber Technologies, Inc., has gone one step further to create a product called FiRP (fiber reinforced product) glulam. This major breakthrough (1994) can significantly improve the strength, reliability, and performance of laminated structural wood beams. FiRP glulam can span the same distance and carry the same load as a conventional glue-laminated beam, but it's a lot smaller dimensionally. Figure 4-21 shows the difference. The FiRP glulam on the left is as strong as the larger conventional glulam on the right. What makes this possible? A mix of synthetic fibers such as carbon, aramid, and glass are aligned parallel to one another and encased in resins to create a thin, flexible, high-strength panel. The panel is bonded between laminated wood sections during the manufacturing process. Typically, a single layer is laminated to the bottom (tension) section of the beam. Another wood section often sandwiches the reinforcement for aesthetic and practical purposes.

From a natural resources perspective, how does this product help? The biggest impact is that up to 25 percent less wood is used compared to a conventional glue-laminated beam equivalent in strength. In addition, lower-grade, less-expensive western woods are used. FiRP glulams can carry 30 percent more load than a traditional glulam. Their lighter design makes them easier to handle, which could reduce labor and equipment costs. But how do you determine manhour costs when the product is custom-designed to plan specifications? And don't forget that it may take equipment to set the glulam in place, depending on its size and weight.

I recommend you use 0.133 manhour per linear foot as a guideline. For drilling holes and attaching connectors while at ground level, estimate 0.236 manhour per hole. Depending on the size of the glulam, figure at least a crew of four and one equipment operator.

LVL Beam

How many times have you purchased solid sawn timber for use as a beam or header and by the time you get it installed, it's already twisted? Or you begin to hang wallboard only to discover a huge bulge above the window in the middle of the header? I've seen it time and time again. It takes a lot of time to stop your rhythm in order to fix these trouble spots. Today you have choices, and with all the new engineered lumbers on the market, there's no excuse. Louisiana-Pacific has a really nice-looking Gang-Lam LVL beam (Figure 4-22). It actually looks so good you almost hate to install it.

This laminated veneer process can be used to make thicker, wider and longer beams than solid sawn lumber. You don't have to worry about twisting, splitting, and checking, common problems with solid lumber. And the carrying capacity per pound is also a lot greater. There's no need to worry about strength-reducing defects.

They're available in longer lengths and a variety of depths for any job. Because the product is uniform, it's possible to cut down on labor costs. I've included a few of the average sizes used in a home. For costs and manhours for other sizes, consult the *National Construction Estimator*.

- $3^1/_2 \times 9^1/_2$ — .094 manhours per linear foot
- $3^1/_2 \times 14$ — .115 manhours per linear foot
- $5^1/_4 \times 7^1/_4$ — .080 manhours per linear foot
- $5^1/_4 \times 14$ — .120 manhours per linear foot

It's always a good idea to use a crew of two or more when lifting beams. In some cases, depending on where the beam will be installed, lifting equipment may be necessary. If so, remember to factor these equipment costs into your bid. The next time you go to order header material, consider the LVL material.

I-Joist

One product I've never actually tried is the I-joist. I knew about them from the beginning, when they were first introduced to one of my regular suppliers, but I wasn't ready to try them. One reason is because they just didn't look good back then — I distinctly remember the dark adhesive smeared all over the product!

Today there are many variations of the product on the market, all with the same "I" design. The top and bottom flanges may be made from solid wood or laminated veneers. The web can be plywood or OSB panels. The web is glued into a groove located in the center of the flanges. Sometimes the web has knockouts for electrical and plumbing.

Courtesy: Louisiana-Pacific Corporation

Figure 4-22
LVL Beam

Figure 4-23 shows an I-joist by Boise Cascade with an OSB web (trade name BCI Joist). They're available in three series: 400, 450, and 600, with lengths up to 66 feet. The 400 Series comes in depths of $9^1/_2$, $11^7/_8$, and 14 inches; the 450 Series comes in depths of $9^1/_2$, $11^7/_8$, 14, and 16 inches; and the 600 Series has depths of $11^7/_8$, 14, 16, 18, and 20 inches.

You can use the joists for both floor and roof applications. Figure 4-24 shows a floor I-joist with a flange of solid wood. The hanger is an ITT by Simpson Strong-Tie, specially designed for use with I-joists when a web stiffener won't be used. However, while web stiffeners aren't always required, consult the I-joist manufacturer for specifications for their product.

Special laminated rim boards are designed to work with the I-joist. Perforated knockouts ($1^1/_2$ inch diameter) are located in the web 12 inches on center. This product has come a long way from the first time I was introduced to it!

Boise Cascade has a Specifier Guide that goes into more depth on proper use of their product. I've included four pages from this guide so you can get an idea of how to use this product in floor and roof

Courtesy: Boise Cascade

Figure 4-23
BCI Joist with OSB web

Courtesy: Simpson Strong-Tie Company, Inc.

Figure 4-24
I-joist in floor application

applications — of course, this is certainly not the final word on the subject (Figure 4-25, A-D).

With an experienced framing crew, you can install an I-joist floor system in approximately one-half the time of a conventionally-framed floor system based on 2 × 10s. For joist sizes $9^1/_2$ and $11^7/_8$ inches, estimate 0.017 manhour per square foot and 0.018 manhour per square foot for 14 and 16 inches (based on

16-inch centers). For scheduling purposes, estimate that a two-man crew can install 900 to 950 square feet of joist in an 8-hour day:

8 hours ÷ 900 SF = 0.0089 × 2 (crew) = 0.0177 MH per SF

This doesn't include beams, blocking, or bridging, but does include the band or rim joist.

Steel Web Joist

The first time I saw a steel web joist, I was struck by its unique design. And I was surprised to learn it's been on the market since 1976.

SpaceJoist by Truswal can be used in both floor and roof system applications and for A-frame construction. SpaceJoist has an open web design with top and bottom cores of solid 2 × 3 and 2 × 4 wood. The webs are galvanized recycled steel in a "V" design. They have two advantages. First, the electrical, plumbing, and heating can be contained within the depth of the truss. Second, the clear span capabilities can often eliminate the need for beams and columns. Several depths are available: $9^1/_4$, $11^1/_4$, $14^1/_4$, and $15^3/_4$ inches in lengths up to 38 feet.

Figure 4-26 shows SpaceJoist installed in floor and roof applications. Pay close attention to the details in this drawing, especially the different types of attachments suggested.

On-site "in place" costs are competitive with conventional joist or truss systems. In many cases, because you need less material with wider on-center spacing, you'll use less material. That can make the system less expensive than conventional framing. You should be able to reduce labor costs by one-third to one-half because its design allows for fast span placement. Support beams can often be eliminated through clear spanning. Based on wooden 2 × 10s on 24-inch centers, use 0.014 manhour per square foot. The first time around, bid higher than this figure (possibly 0.025 manhour per square foot based on 12-inch centers). Depending on the joist length, a crew of two may not be enough, and lifting equipment could be required.

BCI Joists

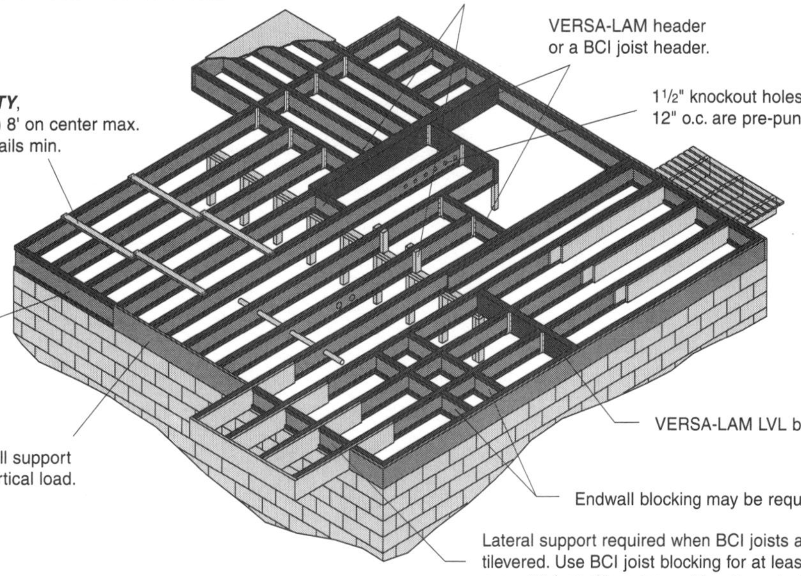

NOTE
The illustration below is showing several suggested applications for the Boise Cascade products.
It is not intended to show an actual house under construction.

NO MIDSPAN BRIDGING IS REQUIRED FOR BCI'S

FOR INSTALLATION STABILITY,
Temporary strut lines (1x4 min.) 8' on center max. Fasten at each joist with 2-8d nails min.

BCI rim joist (where bearing length allows) will support 2000 lbs per lineal foot of vertical load.

VERSA-RIM 98 rim board will support 4000 lbs per lineal foot of vertical load.

BCI joist blocking or 2x4 "squash" block on each side required when supporting a load-bearing wall above.

VERSA-LAM header or a BCI joist header.

1½" knockout holes at approximately 12" o.c. are pre-punched.

VERSA-LAM LVL beam.

Endwall blocking may be required.

Lateral support required when BCI joists are cantilevered. Use BCI joist blocking for at least 4 feet of every 25 feet of bearing wall length and at least 4 feet on each end of cantilevered area.

Residential Floor Span Tables

O.C. spacing	400 SERIES - 1½" FLANGE WIDTH			45 SERIES - 1¾" FLANGE WIDTH				60 SERIES - 2⁵/₁₆" FLANGE WIDTH				
	9½"	11⁷/₈"	14"	9½"	11⁷/₈"	14"	16"	11⁷/₈"	14"	16"	18"	20"
★★★ THREE STAR ★★★												
12"	16' - 3"	19' - 4"	22' - 0"	17' - 0"	20' - 2"	22' - 11"	25' - 5"	22' - 2"	25' - 1"	27' - 9"	30' - 0"	30' - 0"
16"	14' - 10"	17' - 8"	20' - 1"	15' - 6"	18' - 5"	20' - 11"	23' - 2"	20' - 2"	22' - 10"	25' - 4"	27' - 8"	29' - 11"
19.2"	14' - 0"	16' - 8"	18' - 11"	14' - 8"	17' - 5"	19' - 9"	21' - 10"	19' - 0"	21' - 7"	23' - 10"	26' - 1"	28' - 3"
24"	13' - 1"	15' - 3"	16' - 10"	13' - 8"	16' - 2"	18' - 3"	19' - 1"	17' - 5"	18' - 3"	19' - 1"	24' - 3"	26' - 3"
32"	11' - 3"	12' - 11"	13' - 8"	11' - 10"	13' - 1"	13' - 8"	14' - 3"	13' - 1"	13' - 8"	14' - 3"	21' - 7"	22' - 11"
★★★★ FOUR STAR ★★★★												
12"	12' - 9"	15' - 2"	17' - 3"	13' - 3"	15' - 10"	17' - 11"	19' - 11"	17' - 4"	19' - 8"	21' - 9"	23' - 9"	25' - 9"
16"	11' - 7"	13' - 9"	15' - 8"	12' - 1"	14' - 4"	16' - 4"	18' - 1"	15' - 9"	17' - 10"	19' - 9"	21' - 7"	23' - 5"
19.2"	10' - 11"	13' - 0"	14' - 9"	11' - 4"	13' - 6"	15' - 4"	17' - 1"	14' - 9"	16' - 9"	18' - 7"	20' - 4"	22' - 0"
24"	10" - 2"	12' - 1"	13' - 8"	10' - 7"	12' - 7"	14' - 3"	15' - 10"	13' - 8"	15' - 6"	17' - 2"	18' - 10"	20' - 5"
32"	9' - 2"	10' - 11"	12' - 5"	9' - 7"	11' - 5"	12' - 11"	14' - 3"	12' - 5"	13' - 8"	14' - 3"	17' - 1"	18' - 6"
★ CODE APPROVED ★												
12"	18' - 0"	21' - 5"	23' - 11"	18' - 9"	22' - 4"	25' - 4"	28' - 1"	24' - 6"	27' - 9"	30' - 0"	30' - 0"	30' - 0"
16"	16' - 3"	18' - 9"	20' - 8"	17' - 2"	20' - 4"	22' - 5"	24' - 4"	22' - 4"	25' - 4"	28' - 0"	30' - 0"	30' - 0"
19.2"	14' - 10"	17' - 1"	18' - 11"	16' - 1"	18' - 6"	20' - 6"	22' - 2"	21' - 1"	22' - 10"	23' - 10"	28' - 5"	30' - 0"
24"	13' - 3"	15' - 3"	16' - 10"	14' - 5"	16' - 7"	18' - 3"	19' - 1"	17' - 5"	18' - 3"	19' - 1"	25' - 5"	27' - 0"
32"	11' - 3"	12' - 11"	13' - 8"	11' - 10"	13' - 1"	13' - 8"	14' - 3"	13' - 1"	13' - 8"	14' - 3"	21' - 7"	22' - 11"

Span tables assume that sheathing is glued and nailed to joists.
Spans represent the most restrictive of simple or multiple span applications.
Span tables are based on a residential floor load of 40 PSF live load and 10 PSF dead load, and a clear distance between supports.
★★★ *Live Load deflection at L/480.*
★★★★ *Live Load deflection at L/960 to provide a floor that is much stiffer for the more discriminating purchaser.*
★ *Code allowed live load deflection at L/360.*
Span tables are based on shear values for minimum bearing lengths without web stiffeners for BCI depths of 16 inch and less. See BC Calc software for higher shear values using web stiffeners or wider bearing lengths.

Courtesy: Boise Cascade

Figure 4-25A
Floor framing

Floor Details

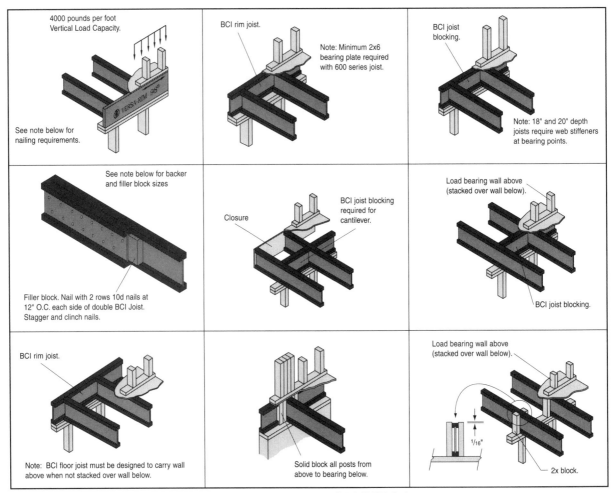

Additional details available with BC FRAMER Software.

Floor Notes

MINIMUM BEARING LENGTH
- 1³/₄" is required at joist ends; 3¹/₂" at intermediate supports.

NAILING OF SHEATHING TO TOP FLANGE
- Space 8d, 10d and 12d box nails and 8d common nails no closer than 2" o.c. per row; space 10d and 12d common nails no closer than 3".
- Maximum spacing of nails is 18" o.c. for BCI 400 Series, BCI 450 Series joists and 24" o.c. for BCI 600 Series joists. If more than one row of nails is used, the rows must be offset at least ¹/₂." 14-gauge staples may be substituted for 8d nails if the staples penetrate the BCI joist at least 1".

NAILING REQUIREMENTS
- Nail BCI Rim, Blocking or Versa-Rim to bearing plate with 8d nails at 6" o.c. When used for shear transfer, nail to the bearing plate with the same nailing as the plywood shear schedule. Nail joists at bearings with two 8d nails, using one on each side, placed 1¹/₂" minimum from the end to avoid splitting.
- Nail BCI 400, BCI 450 rim joist, or Versa-Rim to BCI joist with two 8d box nails, one each at top and bottom flange. For BCI 600 series rim joist, use 16d box nails.

BACKER AND FILLER BLOCKS
- BCI 400 Series: ¹/₂" plywood backer block, 2 pieces ¹/₂" plywood filler block.
- BCI 450 Series: ⁵/₈" plywood backer block, 2x6 filler block.
- BCI 600 Series: 2 pieces ¹/₂" plywood backer block, 2x8 + ¹/₂" plywood filler block.

WEB STIFFENER REQUIREMENTS
- Web stiffeners are always required at all bearing points for 18" and 20" deep BCI's.
- Web stiffeners are always required for BCI's used in hangers that do not extend up to the top flange of the joist.
- Web stiffeners for 9¹/₂" through 16" BCI's are not required unless used in hangers that do not extend up to restrain the top flange of the BCI joist or as shown in roof details.

RIM JOISTS OR BLOCKING
- 2000 PLF vertical load transfer for each BCI blocking panel or rim joist; 18" and 20" joist blocking requires stiffeners.
- Where Versa-Rim 98 is used, BCI joist solid blocking is not required. Versa-Rim 98 will support 4000 PLF vertical load transfer.

Courtesy: Boise Cascade

Figure 4-25B
Floor details

BCI Joists

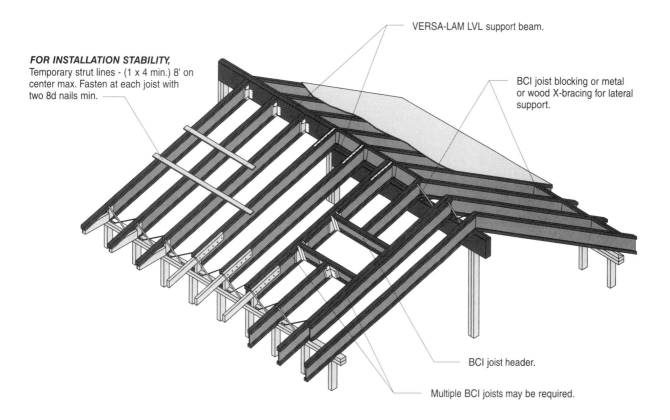

FOR INSTALLATION STABILITY, Temporary strut lines - (1 x 4 min.) 8' on center max. Fasten at each joist with two 8d nails min.

VERSA-LAM LVL support beam.

BCI joist blocking or metal or wood X-bracing for lateral support.

BCI joist header.

Multiple BCI joists may be required.

Roof Notes

NAILING OF SHEATHING TO TOP FLANGE

- Space 8d, 10d and 12d box nails and 8d common nails no closer than 2" o.c. per row; space 10d and 12d common nails no closer than 3".

- Maximum spacing of nails is 18" o.c. for BCI 400 Series, BCI 450 Series joists and 24" o.c. for BCI 600 Series joists. If more than one row of nails is used, the rows must be offset at least 1/2". 14-gauge staples may be substituted for 8d nails if the staples penetrate the BCI joist at least 1".

SAFETY WARNINGS

SERIOUS ACCIDENTS CAN RESULT FROM INSUFFICIENT ATTENTION TO PROPER BRACING DURING CONSTRUCTION. ACCIDENTS CAN BE AVOIDED UNDER NORMAL CONDITIONS BY FOLLOWING THESE GUIDELINES:

DO NOT ALLOW WORKERS ON BCI JOISTS UNTIL ALL BLOCKING, HANGERS, RIM JOISTS AND TEMPORARY BRACING ARE COMPLETED AS SPECIFIED IN ITEMS A THROUGH C.

A) A lateral support, such as an existing deck or a braced end wall must be established at the ends of the bay. Alternatively, a temporary or permanent deck (sheathing) may be nailed to the first 4' of joists at the end of the bay.

B) All hangers, blocking, X-Bracing and rim joists at the end supports of the BCI joists must be completely installed and properly nailed.

C) Temporary strut lines of at least 1x4 must be nailed to the sheathed area or braced end wall as in item A above and to each joist at no more than 8' o.c. Otherwise, buckling sideways or roll over is likely under light construction loads.

D) Sheathing must be completely attached to BCI joist before removing temporary strut lines.

E) The ends of cantilevers must be temporarily secured by strut lines on both the bottom and top flanges.

F) The top flanges must be kept straight within 1/2" of true alignment.

Courtesy: Boise Cascade

Figure 4-25C
Roof framing

Roof Details

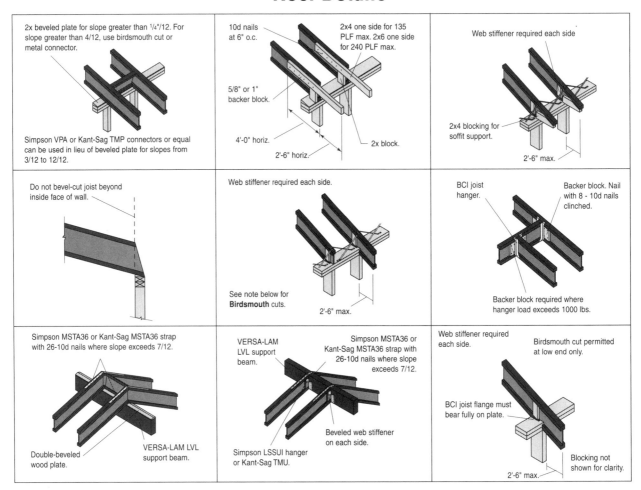

Additional roof details available with BC FRAMER Software.

Roof Notes

MINIMUM BEARING

- Joist ends require a minimum bearing of 1³/₄"; intermediate supports require 3¹/₂".
- 18" and 20" depth joists require web stiffeners at bearing points.

LATERAL SUPPORT

- All roof joist end bearings must be laterally supported using BCI joist blocking or X-bracing.

BIRDSMOUTH CUTS AT BEARING

- The flange of BCI joists may be birdsmouth cut only at the low end of the joist. The birdsmouth cut BCI joist flange must bear fully on the plate, rather than overhanging the inside face of the plate.

MAXIMUM SLOPE

- All roof details are valid to 12/12 slope max., unless otherwise noted.

VENTILATION

- Perforated or drilled 1¹/₂" knockout holes at 12" o.c. may be used for cross-ventilation of joist space. For specific requirements, consult a ventilation expert.

PROTECT BCI JOISTS FROM THE WEATHER

- BCI joists are intended for protected applications and should be kept dry.

Courtesy: Boise Cascade

Figure 4-25D
Roof details

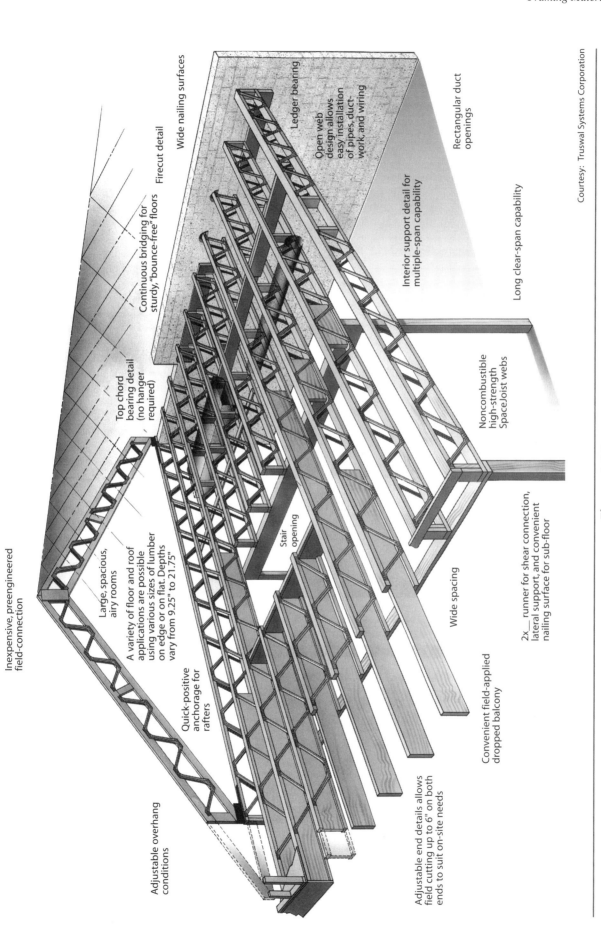

Inexpensive, preengineered field-connection

Adjustable overhang conditions

Large, spacious, airy rooms

A variety of floor and roof applications are possible using various sizes of lumber on edge or on flat. Depths vary from 9.25" to 21.75"

Quick-positive anchorage for rafters

Top chord bearing detail (no hanger required)

Continuous bridging for sturdy, "bounce-free" floors

Firecut detail

Wide nailing surfaces

Ledger bearing

Open web design allows easy installation of pipes, duct-work, and wiring

Rectangular duct openings

Interior support detail for multiple-span capability

Long clear-span capability

Noncombustible high-strength SpaceJoist webs

Stair opening

Wide spacing

2x___ runner for shear connection, lateral support, and convenient nailing surface for sub-floor

Convenient field-applied dropped balcony

Adjustable end details allows field cutting up to 6" on both ends to suit on-site needs

Figure 4-26
SpaceJoist installed in typical application

Courtesy: Truswal Systems Corporation

A The package contains everything you need

B Trusses are precut, notched and predrilled for a 9-bolt connection

C The completed truss ready to lift into place

Courtesy: Trus Joist MacMillan

Figure 4-27
SpaceMaker Truss is ready to assemble

Knockdown Trusses

Earlier in the chapter I talked about Trus Joist MacMillan's TimberStrand LSL Premium Studs. Well, they've expanded this concept into a ready-to-assemble attic framing system called the *SpaceMaker Truss*. The design of this truss makes it easy to create extra living and storage areas or just an open room to finish as a home office or entertainment room.

When you incorporate a dormer and stairs into the design, you have to double up the trusses. Of course, Trus Joist MacMillan offers a truss doubler kit. Single trusses shouldn't exceed 24 inches on center. The minimum is $3/4$-inch tongue-and-groove floor sheathing for trusses at 19.2 inches on center or less and a minimum of $7/8$-inch tongue-and-groove floor sheathing for trusses at 24 inches on center — using adhesive and screws.

SpaceMaker Truss components can't be cut, notched, or drilled except as indicated in the instruction manual. The only areas you can drill are the bottom chord and vertical support. There are two pitch sizes available, 10:12 and 12:12, and four rated spans: 22, 24, 26, and 28 feet. Just think — a ready-to-assemble truss, made from fast-growing trees into a high-strength engineered product (Figure 4-27).

As for manhours, figure that it will take a two-man crew 15 to 20 minutes to bolt trusses together (the first time around). That's 0.25 to 0.333 manhour per truss. Multiply this figure by 2 for a one-man crew. After that, you might average 7 to 8 minutes (0.116 to 0.113 manhour) per truss. Figure on at least $1^1/2$ hours (1.50 manhours) per truss for a crew of five to lift from the ground to the first floor and three members of the crew to secure the truss in place.

That time could be cut down somewhat if you use lifting equipment. You'll have to use lifting equipment when working at two stories and above.

This chapter has briefly described a variety of framing materials that could easily work on your next project. Perhaps you're ahead of the game and already know about or use some of these products. If so, congratulations. Either you're a conscientious builder or you're willing to try new things — or both. But don't stop here. Keep looking for and trying alternative materials and methods. Become an experienced professional.

For those who've not yet stepped off the conventional construction path, it's time to get your feet wet. You have so many more choices today than I had when I started over 20 years ago. The opportunities are waiting for you, but you have to make that first step. Remember, when some crazy company came out with big sheets of plaster sandwiched between paper and called it wallboard, a lot of builders wouldn't go near it. Lath and plaster works just fine for me, they said, shortly before being put out of business by younger guys who could rock a wall in minutes. Not every product will be right for you, but you won't know that until you try it. The key to success is to be willing to try something different. It's that simple. I hope something in this section was appealing enough that you'll run down to your local supplier and check it out!

Chapter 5 has some interesting roofing products that may fit right in with your next project. Some of the products might sound too good to believe, so contact the manufacturer for a sample. Seeing is believing . . .

5

Roofing Materials

■ ■

With the materials we've discussed in the first four chapters, you could build the entire structure — except for finishing the roof and exterior walls (siding). And uniquely-designed new roofing products, paired with siding products from the next chapter, can make a world of difference in your project's final appearance. Of course, the final result depends entirely on what your customers hope to achieve. Do they want to reflect their own personal tastes or just make their home stand out in the neighborhood? Roofing and siding products can make a strong statement, so it's important that your customers choose products that complement each other in design, texture and color. Together, the statement they make should be positive.

In this chapter I'll introduce some roofing products for you to consider. Some of these products may be new to you and others are new to the market. Either way, the important factor to consider before selecting any of these products is their performance record:

■ Are they suitable for a particular climate?

■ Have they been tested in harsh geographical areas?

■ How long have they been on the market?

These are questions you definitely want to research, because roofing and siding materials are exposed to everything nature can throw at them. They've got to be able to withstand the punishment. Check with the manufacturer for a track record of any product you're considering. If possible, get references so you can check out the installed product firsthand, and talk with the customer.

I'm not trying to discourage you from trying these products, but remember that you have to be a savvy professional about *any* of the products listed in this book. Do your homework before selecting and installing any alternative products. Not only do the manufacturers have to guarantee their products, but you have to guarantee them as well. Callbacks are unpleasant and unprofitable, so choose your products wisely.

Shakes and Shingles

Here on the West Coast, cedar shakes and shingles are readily available and add unique finishing touches to homes. With these materials, a roofing contractor can create a roofing masterpiece for any customer. The shakes are attractive, durable, and provide good insulation from heat and cold. Fire hazards and weather conditions have spurred the development of fire retardant polymers pressure-impregnated into the cells of cedar shakes and shingles for Class C and B shake rating. Class A fire rating can be achieved when the roof decking consists of $1/2$-inch solid sheathing and $1/2$-inch water-resistant gypsum, for example. Shakes and shingles can also be pressure-treated with preservatives and carry 30-year warranties from failure due to fungi.

You can expect 25 to 30 years of life in a dryer climate and roughly 10 to 15 years in moist climates. Cedar's inherent oils form a natural barrier against moisture and decay. But when you're installing shingles or shakes over a solid roof deck, consider using one of the special underlayments currently on the market. Their design creates a continuous air space between the roof deck and the shakes or shingles, allowing the wood to breath. The entire underside of a shingle or shake can then dry, reducing the potential for rotting and warping. I'll discuss one particular product later on in this chapter.

While it takes around 150 years for a cedar tree to grow to maturity, the supply of trees ready for harvesting is sustainable at current production levels. Trees harvested on state lands are replaced with new growth, with three trees planted for every tree cut. The cedar shake and shingle industry takes the salvage and low-grade cedar that lumber mills can't use. That's a good example of effective resource utilization. In fact, much of the material used by the shake and shingle industry consists of short cedar logs, blocks, stumps, and boles salvaged from old logged areas. Few, if any, trees are felled specifically for the shake and shingle market.

Every building product used on this earth comes from a natural resource, but only wood products come from a *renewable* resource. Wood is biodegradable and compostable and the choice for a sustainable future. Growing trees absorb and store carbon and emit oxygen, helping to restore and sustain the ozone layer. This is only true of forests in a growing phase, such as forests managed for a sustained yield.

From raw material extraction to finished product, the manufacture of wood shingles and shakes uses only a small fraction of the energy consumed to produce substitute roofing products. While a ton of finished wood requires only 2.9 million Btu (British thermal units) of energy to extract, manufacture, and transport, a ton of aluminum requires 200.5 million Btus; steel, 50.3; brick, 9.1; and concrete block, 8.8. Wood also insulates much better than these other materials. Inch for inch, wood is 16 times more efficient as an insulator than concrete, 415 times as efficient as steel, and 2,000 times as efficient as aluminum.

To learn more, contact:

Cedar Shake & Shingle Bureau
P. O. Box 1178
Sumas, WA 98295-1178
604-462-8961

If you work in areas at high risk for fires or are prohibited from applying cedar roof shakes, you can use alternative products that resemble cedar shakes. There are so many different types of roofing products on the market made of recycled materials that your customers have plenty of choices.

Cedarlite

This product doesn't contain any recycled materials, but it's the closest there is to a nonwood shake replacement. Made of lightweight concrete tile (5 pounds for a $13^{1}/_{4} \times 13$-inch tile, or about 595 pounds for 100 square feet) Cedarlite looks like the real thing (Figure 5-1). Because of its light weight, you can reroof most structures without structural reinforcement. Each tile has unique baffles on the top surface that give proven protection against wind-driven moisture. The color is also consistent all the way through, so when you cut a tile, you won't see raw concrete in the saw cut.

Being made of concrete, *Cedarlite* carries a Class A fire rating, which is especially helpful in areas prone to brush, grass, or forest fires. The company (Monier, Inc.) offers a 60-year transferable nonprorated warranty. And there aren't many roofing products that can last longer than 25 years! The company also offers valley flashings, trim tile, and specialized

Courtesy: Monier, Inc.

Figure 5-1
Cedarlite Shakes look like the real thing

fittings for different climatic conditions, all designed to complement the finished product.

The pitch of the roof determines whether or not the roofer needs to wear a harness or other safety equipment while working on the roof.

Construction Notes — Cedarlite tiles have five different shake detail surfaces. Install them in a random-bond (offset) pattern to enhance the traditional shake appearance.

These are only recommendations and high points to consider when installing this product. Monier doesn't warrant any method of installation of their product. The roofing contractor must follow all applicable building and roofing codes, and use the proper roofing practices:

▌ Install on a minimum 3:12 pitch roof.

▌ The recommended substrate is plywood or equivalent, $^1/_2$ inch thick, with adequate support at laps to prevent deflection. Or use plywood clips.

▌ You must use one of three different starter courses and 30-pound ASTM (American Society for Testing Materials) asphalt felt.

▌ Rolled hem valley or shake valley flashing must be a minimum 28-gauge corrosion-resistant or galvanized metal at least 24 inches wide. They may have either a single or double diverter. In areas of extremely high humidity, use copper or stainless steel flashings and valleys.

▌ In areas of extreme freezing, snow, or cold, apply a single layer of ice and water guard a minimum 24 inches in from the eaves plate line and 18 inches on each side of the centerline of the valley.

Manhours — When estimating your manhours, there are two factors that'll have a big impact on your time. The first is the pitch and design of a roof. The second is your crew size. Will you use a two-man or a three-man crew? Figure a crew of two at 0.822 manhours per square (SQ) to load tile and accessories up on the roof. To install the product on an average roof with few cutups, figure a crew of two at 3.25 manhours per square. If you install 10

squares with a crew of two, how long will it take to install each square? Calculate it this way:

3.25 × 10 SQ = 32.5 MH ÷ 8 hours =
4.0625 days ÷ 2 (crew) =
2.03125 days × 8 hours =
16.25 hours ÷ 10 SQ = 1.625 hours per SQ

Or you could use this alternative calculation:

3.25 × 10 SQ = 32.5 MH ÷ 2 (crew) =
16.25 hours ÷ 10 SQ =1.625 hours per SQ

Or simply:

3.25 MH ÷ 2 (crew) =
1.625 hours per SQ for a crew of two

Eco-shake

Manufactured by Re-New Wood, Incorporated, *eco-shake* was specially designed to resemble and replace wood shakes. Even in upscale neighborhoods, eco-shakes can easily pass for the natural look of wood shakes. Look at the luxury home in Figure 5-2. The manufacturer recommends their use in both the residential and commercial markets.

Courtesy: Re-New Wood, Incorporated

Figure 5-2
Eco-shakes fit in an upscale neighborhood

They're fabricated from 100 percent recycled post-industrial (not post-consumer) waste materials: polyvinyl chloride (PVC), which could include garden hoses and surgical tubing, and sawdust from sawmills and cabinet makers. PVC makes the product very flexible. In fact, it has received certification from Underwriters Laboratories (UL) for impact (hail) resistance.

Eco-shakes are available in 22-inch lengths and in random widths of 5 inch, 7 inch, and 12 inch, which makes them look more like real wood shakes. They're available in four standard colors (teak, umber, driftwood, and charcoal) to resemble weathered wood shakes. The color is consistent all the way through. You can nail, cut, score and split them just like wood. The ridge cap is premolded and flexible enough to adapt to most roof pitches. It's very lightweight, approximately 250 pounds per square, and requires no reinforcement to the roof trusses.

The fire rating for eco-shake is Class A when installed to manufacturer's specifications (over approved decking). It has passed the Wind Driven Rain Test — 110 mph winds with a driving rainfall of approximately 9 inches per hour — required by Dade County, Florida, since Hurricane Andrew. Other tests include over 5,000 hours of accelerated UV weathering and freeze-thaw cycling. The results were acceptable: insignificant changes in color and no cracking. It's highly resistant to moisture, won't deteriorate like wood, and isn't susceptible to insect infestation, fungus, mold, or mildew. Eco-shakes carry a 50-year transferable warranty against defects in material and workmanship degradation due to mold, mildew, fungus, or rot.

Construction Notes — Eco-shakes are as easy to install as wood shakes. They can be scored and split and require no special equipment. Each shake needs three nails, either hand nailed or power driven. The manufacturer is in the process of retesting and revising application instructions for their Class A fire rating with Underwriters Laboratories. The revised installation instructions require an underlayment of a fiberglass mat with a waterproof coating (available where eco-shakes are purchased). You use this underlayment in place of $1/2$-inch water-resistant gypsum board and 30 pound felt for a Class A fire rating. If you have any questions about application, get in touch with the manufacturer.

Manhours — It's a little more difficult to calculate manhours for eco-shake because it installs faster than shakes but slower than composition. To come up with an estimate, I've added the manhours to install both shakes and composition, then divided by 2 to arrive at an average 2.68 manhours per square for 10-inch exposure. When installing hip and ridge caps, figure 20 linear feet (10-inch exposure) can be installed in one manhour. For installing 30-pound felt, figure 0.160 manhours per square. A crew of two or three will speed production.

Steel Shakes

Gerard Roofing Technologies markets Gerard Shakes — a steel roofing product that simulates the look of traditional hand-split wood shakes. It's available in six earth-tone colors, and has a clean, contemporary appearance. A roof done with this material looks pretty impressive, and should add considerable value to your customer's home.

You can usually install it right over an existing roof or spaced sheathing with no added reinforcement, since it's among the lightest roofing materials available (Figure 5-3). It weighs only 1.4 pounds per square foot, or 140 pounds per square, yet it's proven to withstand winds up to 120 mph and hailstones up to 1.75 inches in diameter. You can install steel shakes directly over failing wood shakes in a technique approved by building codes nationwide. Installing directly over an existing roof lets the building withstand adverse weather while the new roof is installed. It can also save tearoff and disposal costs, which can be significant. See Figure 5-4.

The tiles' steel base is composed of 30 percent recycled steel. Crushed and graded stone granules are bonded to the steel, then topped with a clear acrylic overglaze. It's then oven-cured to produce a very durable roofing product. It's also fire- and weather-resistant. Gerard recycles all of its steel offcuts, some 30 tons a year. Each shake tile panel is approximately $44^3/_4$ inches long by $15^1/_2$ inches wide, with four recessed and five raised shake impressions. It comes 23.2 shake panels per square.

Courtesy: Gerard Roofing Technologies

Figure 5-3
Steel shakes install over existing roof

Courtesy: Gerard Roofing Technologies

Figure 5-4
Steel roofing panels over an old roof

Gerard also offers other product lines in a wide range of colors to suit different architectural styles, and all have a limited lifetime weatherproof warranty. Roofing panels are manufactured from materials that can't burn or support combustion. According to the manufacturer, they have a Class A fire resistance rating by Underwriters Laboratories (when installed with specified underlayments as part of a complete system).

Construction Notes — Gerard has a detailed manual about the correct installation of their products, both for reroofing and new construction. There are some concerns and considerations you need to keep in mind when working with steel roofing:

▌ Try to use the manufacturer's accessories: "Z" bar, chimney saddles, valley metal, fascia metal, and so on. Using their items helps create an integrated system.

▌ Always order a touch-up kit. It's great for repairing any surface damage that might occur during installation. It also provides a great final finish to the fasteners, which are applied horizontally through the nose of each panel.

▌ You can make counter battens of wood, horizontal 1 × 4s or vertical 2 × 2s (Figure 5-3), or use 22-gauge "hat" channel steel.

▌ You need benders and attachments to bend panels, for hip and valley installations, for instance. And you need a cutter to cut the panel both lengthwise and sideways for those areas. These are tools you can't work without.

▌ When walking on the material, they recommend that you wear rubber-soled athletic shoes and step directly on the panel where it's supported by a batten.

Manhours — On an average roof with few cutups, figure a crew of two can install battens and 12 square of panels in 8 hours. For installing "W" type metal flashing, figure 0.030 manhours per linear foot, and 0.024 manhours per linear foot for installing hip, ridge, and rake caps.

Cement Fiber Shakes

Louisiana-Pacific has created *Nature Guard*, a roofing product that truly resembles a wooden shake. The detail is unbelievable. They feature a tapered

Courtesy: Louisiana-Pacific Corporation

Figure 5-5
Nature Guard cement fiber roof shakes

profile roughly 22 inches long and in widths of 5, 7, and 12 inches to closely mimic natural wood shakes when installed (Figure 5-5). Nature Guard's available colors (dark and light browns and dark and light grays) further enhance the wooden shake appearance.

Nature Guard shakes are made from recycled materials, including wood fibers and fly-ash, as well as cement. At 580 pounds per square, they're considered a lightweight roofing product (compared to clay tile, for example). That means they're light enough to use for reroofing without restructuring or truss supports. However, be sure to consult a structural engineer if you're uncertain of the load tolerances of any structure. Nature Guard is classified as a Class A fire-rated product and carries a 25-year limited transferable warranty.

Construction Notes — This product installs about like wood shakes. At the eave edge, install a starter course of 30-pound felt interlayment over sheathing, minimum 18 inches wide. Use a maximum 8-inch

exposure for a $1^1/_2$-inch shake overhang. Install a $1/_4$- by $1^1/_2$-inch wood lath strip over the interlayment and flush with the eave edge. Place the starter course face down over the wood lath.

Here are some additional points to keep in mind while handling and working with Nature Guard:

I Shake bundles are slippery, so install all battens and make sure they're secure prior to bringing shakes up on the roof. Even though the shakes are engineered to be extremely strong and walkable, don't walk on the unsupported smooth end of the shakes. No point pushing your luck. And don't ever walk on the shake roof if it's wet! You'll be eating your customer's grass before you know it.

I Cutting shakes produces cement dust that needs to be regularly removed throughout installation. The tiles aren't the same color all the way through, so you have to coat exposed cuts with the manufacturer's color touch-up paint. This is for aesthetic purposes only. The cut surface doesn't need protection.

I Hand nailing isn't recommended, and your local building department may not approve the use of staples. Use a pneumatic nailer instead.

I Installation of shakes on roofs with pitches under 3:12 isn't recommended except for appearance only. Install it over an approved sealed-membrane system.

I Install with a 10-inch exposure and in a random width sequence, as shown in Figure 5-6.

Manhours — Estimate that you can install 20 linear feet of hip and ridge caps (at a 10-inch exposure) in one hour (1.0 manhour). For 30 pound felt, figure 0.160 manhours per square. To install the Nature Guard, figure on 3.52 manhours per square. That works out to $1^3/_4$ hours for a crew of two to install one square, assuming an experienced crew. The first time around, you may want to figure on 2 to 3 hours per square with a crew of two. To speed production, use a crew of three.

Composition Shingles

Composition asphalt shingles, also called 3-tab (shingles), are by far the most popular residential roofing material. I won't go into any details on brand name products, because they're so common. But I do have a word of advice for contractors who frequently bid less-expensive roofing products. That low bid may win you the job, but lack of quality can lead to premature shingle failure. Poor raw materials, faulty design, and inconsistent manufacturing quality can all contribute to early shingle failure. Here are some common problem areas:

1. Shingle matting: Is the mat tough enough to resist tears from wind or routine handling and installation?

2. Tab fly-away: How flexible is the shingle? Can it withstand Mother Nature's strongest wind and driving rain?

3. Nail pull-through: How much force does it take before a nail pulls through? This is directly related to the shingle mat's strength and resistance to blow-off.

Do these problems crop up with the roofing products you install? So, how can you prevent shingle failure? It's simple. Offer your customers only high-quality roofing products. Educate them on the quality differences between roofing products, and encourage them to select a reliable shingle manufacturer who offers a solid warranty. Since such a large part of a roofing job is labor, it doesn't make sense to try to save a few bucks by installing cheap materials.

The most common base for an asphalt shingle is fiberglass. However, a few companies offer organic felt-based shingles, depending on your area of the country. Organic felt-based products are made of recycled paper and reclaimed mineral slag with a recycled content between 20 and 25 percent. Here are some of the factors to consider when you purchase your next asphalt shingle:

▌ Check the construction of the product. Is it laminated or not? Architectural or designer shingles are constructed as laminates. That adds weight, depth and dimensionality to the shingle.

▌ Check the weight. If it climbs into the 357 to 400 plus pounds per square range, then you have a "heavyweight" product. They're stronger, offering more resistance to wind, tearing and blow-off.

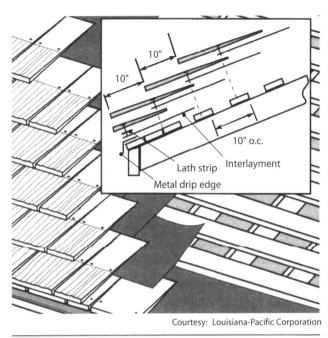

Courtesy: Louisiana-Pacific Corporation

Figure 5-6
Installing Nature Guard shakes

▌ Check the manufacturer's warranty. Is it up to 30 years or is it a lifetime limited transferable warranty? Look for warranties that cover materials and labor in the first three to five years after application. Material defects are likely to show up within this time period.

▌ Check the fire resistance rating carried by the product. Make sure it's appropriate for the application.

For manufacturers of fiberglass-based shingles and organic felt-based shingles, check with your local roofing distributor. Here's one other point I should bring to your attention: When you're considering a product you don't plan to install yourself — if you plan to use a roofing subcontractor, for example — ask if he or she has experience with similar products. Different products often need different installation techniques. It's important to have an experienced subcontractor who knows the product.

Metal Systems

Met-Tile is a metal panel (26-gauge zinc/aluminum alloy-coated steel) that has the look of traditional clay or concrete tile. It'll fool most people. It's 36

Courtesy: Met-Tile, Inc.

Figure 5-7
Met-Tile metal panels span from eaves to ridge

inches wide, and available in lengths from 2 to 20 feet. The long panel length lets you install them from eave to ridge, eliminating seams (Figure 5-7).

At only 125 pounds per square, you can use it on virtually any type of house with no need for further structural reinforcement. The steel panel also contains a percentage of recycled steel and is finished with an environmentally-friendly water-based coating.

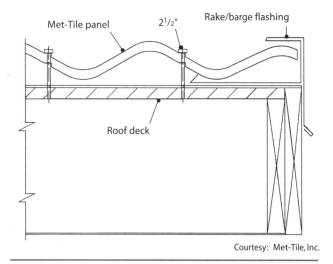

Courtesy: Met-Tile, Inc.

Figure 5-8
Met-Tile standard rake/barge detail

Met-Tile is also a "problem-solver" in areas with special hazards or weather conditions. Consider it in fire-prone states, tropical areas (it has proven resistance to 230+ mph winds), and snow country (metal has an inherent ability to shed snow and ice). It's also great for remodeling projects because you can install it directly over an existing roof. It requires a minimum pitch of 3:12, or a possible slope of 2:12 if the side laps are caulked. It carries a 20-year warranty against chipping, splitting, flaking, cracking, or peeling of the finish under normal conditions. Met-Tile is available in eight designer colors, and special colors are available on request. It's a complete roofing system and comes with screw fasteners, closures, and a full complement of accessories.

Construction Notes — A detailed instruction manual is available, but here are some highlights to keep in mind when working with Met-Tile:

▮ Wear rubber-soled footwear for safety purposes and to avoid damage to the surface of roofing panels. Work gloves are also a good idea when carrying and handling the panels.

▮ The panels are coated with a very thin film of protective wax to minimize nicks and scratches during installation, so use extra caution at all times when handling or walking on panels.

▮ If small scratches do occur, use the manufacturer's touch-up paint as directed.

▮ Panels can be installed over plywood sheathing, open steel purlins, or wood trusses.

▮ Fastener dimples are built into the panels for uniform fastener location. Figure 5-8 shows $1/4$-inch hex head carbon steel fasteners combined with appropriate sealing washers. Use bi-metal stainless steel fasteners in salt-air climates and other corrosive environments.

▮ An electric nibbler is recommended for field cutting of panels. But metal slivers from the cutting tool pose a special hazard. They may become embedded in the panel, causing rust spots or streaking of the roofing surface. To prevent this, use a large magnet to remove metal slivers from all surfaces.

▮ A light annual cleaning is recommended in most situations. Roofs located in coastal areas or in high pollution zones may require more frequent cleaning. In locations where trees block the natural cleansing effect of rainwater (allowing buildup of leaves, dirt, and other matter on the roof), more frequent cleaning is advised.

▌ The factory-applied finish is designed to outlast standard house paint by many years. However, after prolonged weathering, it may eventually require repainting. Refinishing may also be needed if extremely severe weather conditions cause excessive damage to panel surfaces. In these situations, first follow the manufacturer's recommended cleaning procedures. If cleaning doesn't restore the roof's appearance, refinish the surface with a good grade of commercially-available air-dry auto paint using materials and procedures recommended by the manufacturer.

Manhours — For manhours, figure 0.030 manhours per linear foot for "W" type metal valley flashing. For panels that range from 2 to 20 feet, use 0.022 manhours per square foot for first-time installation. Each 3- × 20-foot (60 SF) panel takes a crew of two about 40 minutes to install:

0.022 MH × 60 SF = 1.32 ÷ 2 [crew] =
0.66 × 60 = 39.6 minutes

For installation of sidewall and endwall flashing, figure 0.035 manhours per square foot. A crew of three could speed up production.

Fiber-Cement Tiles

MaxiTile has the feel and look of clay tile, but it's about 65 percent lighter (340 pounds per square). That makes it ideal for reroofing projects. Figure 5-9 shows their *SuperTile*, which has been in production since the 1920s. The tiles are composed of portland cement, silica sand, and treated cellulose fibers (no recycled materials), and they're colored all the way through.

The cellulose fibers from harvested trees are fabricated into strands, laid out in a grid pattern, and laminated to the required thickness. This process gives it strength, so the tile can be thinner and lighter than clay or concrete tiles. To achieve overall structural strength, the tiles are steam cured under pressure in an autoclave (pressurized chamber).

MaxiTile carries a 50-year transferable limited warranty, but it's not warranted against freeze-thaw damage. The company prefers that you call for geographical clarification. You can get a Class A fire rating when you install a 72- or 90-pound granulated felt with the granules up. The maximum roof pitch is 21:12 (I think I'd sub that one out!) with the minimum 3:12. Below that, it requires an approved sealed membrane. MaxiTile is available in the form of a shake, slate, and Spanish tile in a variety of earthtone colors.

Construction Notes — Here are a few pointers to consider when working with MaxiTile:

▌ Over wood shakes, remove any old trim and cut the overhang at the eaves. Apply a minimum 40 pound felt, and then install vertical nailing strips (1 × 4) over rafters and horizontal battens (1 × 4) on top of the vertical battens at 20 inches on center.

▌ Over wood shingles, make the same preparations as for wood shakes except you can choose either nailing strips and battens, or fasteners long enough to penetrate $^3/_4$ inch through the existing sheathing.

▌ Over rock roofs, spud and sweep off all rocks so the 90 pound felt won't be punctured. Install tiles using fasteners long enough to penetrate the substrate at least $^3/_4$ inch.

Courtesy: MaxiTile, Inc.

Figure 5-9
SuperTile

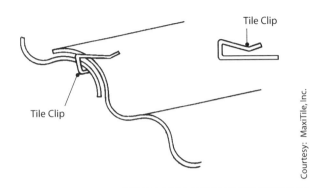

Courtesy: MaxiTile, Inc.

Figure 5-10
Tile Clip resists separation of tiles

▌ In areas where maximum wind speeds are below 60 mph, use MaxiTile's square cap nail; for wind speeds from 60 to 110 mph, use their $1^5/8$-inch screw and washer system. Each tile must also have a dab of adhesive $1/4$ inch thick by 1 inch wide applied to the side of each barrel. Special screws are available for steel framing.

▌ For areas with winds above 110 mph, use a $4^3/8$-inch screw and washer system and fasten externally through the headlaps at the top of each barrel. The headlap is the finish tile cap used at the ridge, hip, and rake. Apply a dab of silicone over the head of each fastener.

▌ Tiles don't require predrilling.

▌ In areas prone to high winds, a simple solution could be to use the $1^1/2$-inch Tile Clip (Figure 5-10). It provides increased resistance to separation or lifting of the tile at side laps during heavy wind.

Manhours — When installing these tiles, figure 4.55 manhours per square and plan on at least a three-man crew. This manhour estimate doesn't include loading materials up onto the roof. For accessory items, consider the following:

1. 30 pound felt: 0.050 manhours per square foot

2. Eave closure or birdstops: 0.30 manhours per linear foot

3. Metal flashing: 0.306 manhours per square

4. Hip, ridge, and rake caps: 0.047 manhours per linear foot

Corrugated Fiber Roofing

When I think of corrugated roofing materials, the fiberglass panels normally used for patio covers come to mind. My experience working with these products taught me that premium grades are the only way to go. I must confess that I never thought of the product as a true roofing material until I saw a sample of *Ondura* from Nuline Industries. This thick corrugated fiber/asphalt roofing panel is available in eight striking colors and has the feel of tile. Post-consumer waste products make up 50 percent by weight of its raw materials. This includes newspapers, corrugated cardboard, and coated stock which is taken out of the waste stream. They're all combined, run through a corrugator, and then impregnated with melamine and asphalt followed by two coats of latex paint.

Ondura seems ideal for reroofing projects because the 4 × 6-foot panels weigh only 18 pounds each, which makes for easy installation over two layers of composition roofing. Look at Figure 5-11. The corrugated design allows the panels to withstand snow loads up to 55 pounds per square foot and wind

Courtesy: Nuline Industries

Figure 5-11
Installing Ondura panels over
existing roof saves tearoff and disposal costs

forces over 100 mph. It would also be great for the agricultural markets, especially barns. One of Nuline's accessory items is a translucent fiberglass panel with the same configuration as the roofing panels — great for building skylights (Figure 5-12). Other accessory items include ridge caps, colored washer nails, pipe flashings and boots, and closure strips, vented or not. Vented strips work great when used in conjunction with a vented ridge vent. Don't rule out application for the residential and commercial markets.

Construction Notes — Some of things you should know when working with Ondura include:

▌ Minimum pitch is 3:12.

▌ It takes 4.5 sheets to equal 100 square feet.

▌ The sheets can be scored with a utility knife and flexed back and forth for a clean break.

▌ When using sheets for reroofing, use nails long enough to penetrate 1 inch into supports of decking beneath.

▌ Nails should be installed as shown in Figure 5-13. It's important not to nail into the valleys of the panels. And don't overdrive or underdrive those nails. Slight hand pressure on the ridge where it's been nailed shouldn't separate the panels. If a nail has to be removed, use a 1$^1/_2$-inch wood dowel or pipe placed in the valley to give the claw hammer some support when pulling the nail.

▌ Figure 5-14 shows a side view of the purlins spaced 24 inches on center. The panel requires a 7-inch overlap that lands on a purlin. For heavier snow loads or potential snow drift, reduce purlin spacing to 18 or 12 inches on center, not 16 inches.

Manhours — A two-man crew can install 156 square feet in about 1$^3/_4$ hours, based on 0.022 manhours per square foot:

0.022 MH × 156 SF = 3.432 ÷ 2 [crew] = 1.716 [0.716 × 60] = 1 hour, 43 minutes

For hip, ridge, and rake caps, figure 0.024 manhours per linear foot and 0.014 manhours per linear foot for closure strips. These manhour figures are

Figure 5-12
Translucent fiberglass panels
are part of the roofing system

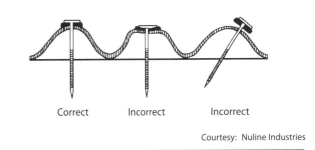

Correct   Incorrect   Incorrect

Figure 5-13
Nail Ondura correctly

based on hand nailing, which is the only method used for this system. Incidentally, 156 square feet equals 6.5 panels.

Miscellaneous Items

I mentioned that roofing and siding materials go hand and hand, but just pairing up two that match isn't enough to ensure a complete working system. Accessory items provided by manufacturers help finish off the project and allow the system to work as a whole. I suggest that you use the manufacturer's accessories rather than products that aren't designed to work with the materials you're installing.

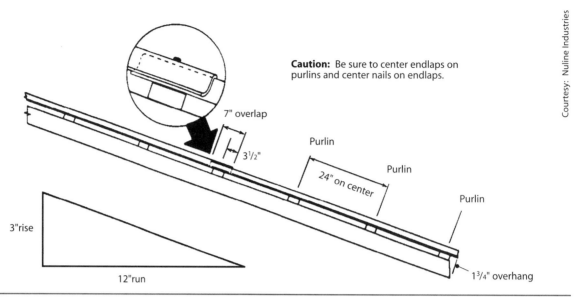

Caution: Be sure to center endlaps on purlins and center nails on endlaps.

7" overlap

3¹/₂"

Purlin

Purlin

24" on center

Purlin

3"rise

12"run

1³/₄" overhang

Courtesy: Nuline Industries

Figure 5-14
Purlin spacing for Ondura

However, there are products on the market that work with siding and roofing materials to help make these products last longer.

What about waterproof underlayments for those areas on the roof that are likely to leak? Do you use them or not? What about ventilation? It's important! Are rain gutters part of the roofing system? If so, do you install metal or PVC? It's not enough to just install roofing and siding products without taking some precautions. You need to consider the big picture — how will these products hold up once you leave the job? This section should give you some insight into products that will help you install a complete working system. Your customer will benefit, and you can confidently guarantee your work.

Artificial Thatch

Have you ever been to California or Florida and seen how some of the great entertainment, recreation, and resort roofs were thatched? Have you ever wondered if they were the real McCoy or not? Tropic Top has a 0.020-inch thin aluminum sheet that looks like the real thing. It's 14 inches wide by 3 feet long with a baked-on beige colored finish. The advantage of this product is that it won't deteriorate or catch fire like natural thatch. The longer Tropic Top is exposed to the environment, the more realistic it looks.

Construction Notes — Tropic Top installs about like other conventional roofing materials. When the sheets arrive, you simply pull them apart and nail them along the solid edge. After installing the first piece, make sure the nailing surface is close to the eave (which provides quite an overhang). Once you've installed the product, return to the overhang. Using leather gloves, place a dowel under the material and pull the dowel toward you while at the same time using your other hand on top of the material to roll the material over the dowel toward the ground so it looks natural (Figure 5-15). That's all there is to it. Now you have the look of the real thing without the maintenance and fire-hazard headaches.

Manhours — With such an unusual roofing material, how do you estimate manhours? Very carefully! First, there's a learning curve when installing this product. Second, there are no set rules for manhours because the material is used for specially-designed applications that are out of the norm of traditional building materials. Generally speaking, it requires a lot of hand work to achieve professional-looking final results. For the purpose of bidding, I suggest you use 0.0616 manhours per square foot. This works out to around 12 minutes to install each half sheet (14 inches × 3 feet) after the sheet has been pulled apart. This product can easily be installed by one person, but consider assembling a crew if you have a large area to cover. It could help speed up production, especially since this is tedious work.

Cedar Breather

How do you get shakes or shingles to dry out naturally once they're installed over solid decking? When sheathing is used as a solid decking material, the life span for wood shakes isn't as long as when they're installed over purlins. The open areas between purlins allow air to circulate to the underside, which helps dry out wet shakes. So how do you accomplish this with solid decking? Benjamin Obdyke Incorporated is one company that has found a solution to the problem. They make *Cedar Breather*, a lightweight (0.967 ounces per square foot) three-dimensional nylon matrix underlayment. It's available in rolls 39 inches wide and 61.5 feet long. That's enough to cover approximately 200 square feet.

The way it works is simple. Once you've installed it between the decking and the shakes, air circulates freely throughout Cedar Breather. This air movement helps to dry the shakes or shingles from the underside. The application is a little different for shakes than it is for shingles:

▮ For shake installation, the manufacturer recommends applying a 30-pound felt starter at the eaves, then installing Cedar Breather over the felt starter and the entire plywood decking. Install the shakes with interleaved 18-inch felt.

▮ Under shingles, first install the felt over the decking surface, then Cedar Breather, and finally the shingles.

Figure 5-16 shows a valley flashing system. On a cedar roof with Cedar Breather, apply the flashing so it's in direct contact with the underside of the shake or shingle and on top of the felt and Cedar Breather.

You can also use Cedar Breather for sidewall application (Figure 5-17). First, install an appropriate vapor barrier on the warm side of the wall insulation to reduce moisture movement from the inside to the outside of the structure. Follow it with Cedar Breather, and then the shingles. You can also use Cedar Breather over rigid foam sheathing once

Figure 5-15
Artificial thatch is easy to install

building paper has been attached. For both roof and wall applications, use 0.360 manhours per square and figure on using a crew of two for installation.

Waterproofing Underlayment

Some roofs aren't protected by a good waterproofing underlayment. The continual thawing and refreezing of melting snow or the backup of frozen slush in gutters can cause water to seep between the roof deck and shingles. Wind-driven rain can have a similar effect. So how do you protect the deck from damaging ice dams and wind-driven rain? Owens-Corning found a way with *WeatherLock*, a rolled underlayment with a self-adhesive backing that makes for easy installation. The product is comprised of four main sections:

1. Surface: textured polyethylene sheet

2. Interior: fiberglass mat reinforcement

3. Adhesive: mixture of Styrene-Butadiene-Styrene (SBS) and asphalt

4. Removable backing: poly-film backer sheet

WeatherLock is both a moisture and vapor barrier and is available in lengths of 33.3 feet or 67.7 feet by 3 feet. Because of its composition, the underlayment automatically seals around nail holes for maximum protection. The product is normally used in areas where water has a tendency to collect or flow.

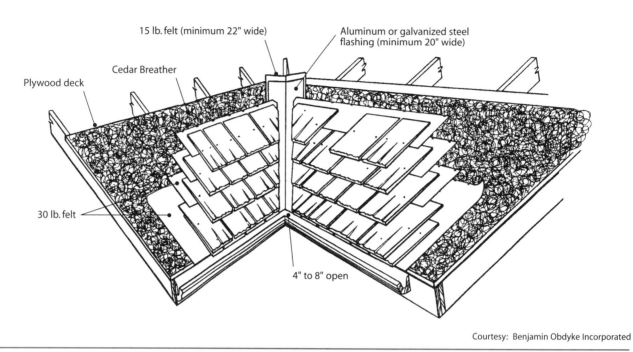

Courtesy: Benjamin Obdyke Incorporated

Figure 5-16
Installing Cedar Breather under shakes

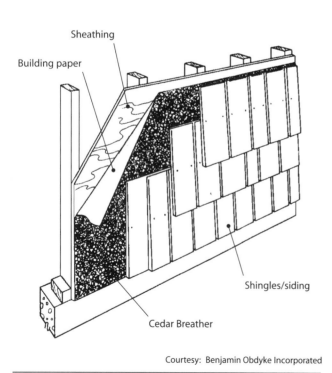

Courtesy: Benjamin Obdyke Incorporated

Figure 5-17
Cedar Breather under sidewall shingles

Figure 5-18 shows those areas, which include eaves, valleys, vents, chimneys, and skylights. WeatherLock is backed by a five-year limited warranty.

When installing on a new roof deck or a reroofing project, the manufacturer recommends that you get down to the existing decking. That means you have to remove any old roofing materials. For horizontal installation, start from the eaves ($^1/_4$ inch overhang) to a point 24 inches inside the exterior wall. This puts the underlayment above the maximum ice dam buildup line in most areas. Consult your local building department for specific requirements. Ends should overlap by 6 inches and courses overlap 3 inches to the marked ply line. You can install valleys in one piece, 18 inches on either side of the valley. The minimum pitch is 1:12 to install the underlayment. The ideal temperature for installation is between 40 and 100 degrees F.

They also offer WeatherLock GS (granulate surface) for those who prefer a granulate surface. Some roofing contractors feel more comfortable walking on the underlayment when the surface is granulated. The granulated surface also provides a little more resistance to ultraviolet radiation. WeatherLock shouldn't be exposed for more than a couple of weeks without covering it with shingles. The granulated underlayment costs about 10 percent less than poly-faced WeatherLock. For installation of both products, use 1.25 manhours per square. Because there's a release backer that has to be removed to expose the adhesive, consider making this a two-man operation. Additionally, a crew of two could possibly install a square in 0.625 manhours, or 37 minutes.

Ventilation

Have you ever been up in the attic during summertime? A home that isn't properly ventilated is very uncomfortable to live in. Venting hot air from the attic makes the entire home feel cooler. Trapped hot air in an unventilated or improperly ventilated attic will defeat the best efforts of air conditioning, insulation, and shade trees to help keep a home cooler in summer. As heated air rises from the living area and the sun heats the roof, hot air accumulates in the attic. If the attic isn't properly ventilated, the warm air becomes trapped in the home. This effect both reduces the comfort level and invites mold and mildew. During winter months, the accumulation of moisture in the rafters and insulation could cause rot, lower insulation efficiencies, and bring water back in the home from ice dams.

There are many ways to solve this situation. One way is to use a power ventilation system. On existing homes this may be difficult, especially if power isn't available in the attic. I should know — I learned this just last year trying to install one in our home. The system was a snap to install, but the electrical, on the other hand, was something else. I put this system in because it was a quick solution. But in new construction, a passive ventilation system is probably the most effective system for removing hot air. A passive system includes soffit vents, ridge vents, and attic baffles.

It's important to combine all three of these ventilation components. Homes can benefit by increasing the number of intake and exhaust vents. Homes with finished attics or cathedral ceilings can also benefit from proper ventilation. Finished attics require ceiling and ventilation channels to route air to the upper attic from the lower attic. In cathedral ceilings, ventilation channels run continuously from the eaves to

Courtesy: Owens-Corning Fiberglas Corporation

Figure 5-18
WeatherLock waterproofing underlayments

the ridge to keep air flowing across the underside of the roof deck. So where do you start?

1. Begin by installing continuous soffit ventilation strips under the eaves to allow fresh, cooler air to enter the attic. When choosing a vent, consider plastic or aluminum to prevent rusting.

2. From inside the attic, install attic baffles between the rafters to maintain an unobstructed air channel over the insulation. Without these ventilation channels, you won't get a good flow of air up and out the ridge vent. The channels also prevent windblown insulation from plugging or blocking the air flow from the soffits.

3. A continuous ridge vent allows hot air to exhaust from the attic. If a ridge vent is impractical to install, then place traditional roof vents as close to the peak as possible. Check with your local building department for the minimum requirements for your area.

4. For a solid roof deck, consider using an underlayment with built-in channels to help move air.

Attic ventilation is vital, no matter what season. Hopefully, the following products will help with your next project.

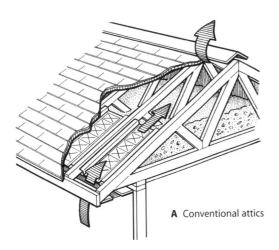

A Conventional attics

B Cathedral ceilings

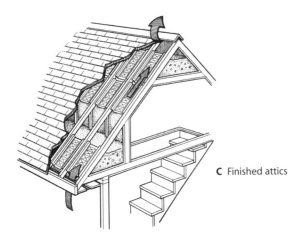

C Finished attics

Courtesy: ado Products

Figure 5-19
proVent attic ventilation chute

Attic Baffle

You need the correct number of unobstructed vents to make sure the ventilation system works properly. One area that needs a lot of attention is the spot in the attic located right above the exterior walls where the ceiling and rafters come together. This area frequently gets filled in with insulation, especially if it's blown in, which plugs up the air flow from the soffit vents. One way to eliminate this barrier is to install baffles between the rafters. There's a rigid plastic attic ventilation chute, *proVent*, that's made from 100 percent post-industrial recycled plastic (factory scraps). It fits between rafters 16 and 24 inches on center and is easy to install in both new and retrofit applications.

Construction Notes — You slide the plastic baffle over the top plate of the exterior wall, then staple it to the underside of the roof sheathing. Now you're probably wondering how to get into such a tight area. Good question! For new construction, install the baffles before the ceiling. In an existing home, the proVent is 4 feet long, so you can usually staple enough of the baffle to secure it. But since you can't staple the entire length, the end could flap during a windstorm. So staple as far as you can, and roll up insulation on the backside of the baffle to hold it in place in this tight space. It doesn't interfere with the home's R-value and helps eliminate ice buildup in this area.

Figure 5-19 shows three places you can install proVent. In *conventional attics* (houses with continuous soffit intake and ridge vents), ventilate every rafter space with a baffle. For *cathedral ceilings*, run a continuous baffle from intake to exhaust, leaving a 1-inch space between bends for removal of trapped moisture. You can easily cut it to fit any joist space in *finished attics*.

Manhours — On a new home, figure on 0.220 manhours per baffle, either 16 or 24 inches on center. One person can handle this task, but two could speed things up a bit. On existing homes, you may not have total access to install baffles. Personally, I'd hire a subcontractor who specializes in this type of work. Depending on the access and what's up in the attic (like blown-in insulation), I estimate at least 0.50 manhours per unit. Also keep in mind that you may

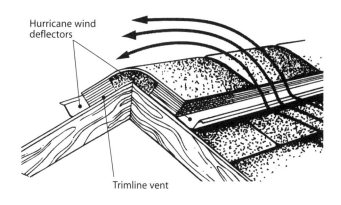

Courtesy: Trimline Roof Ventilation System Inc.

Figure 5-20
Trimline vent with hurricane wind deflectors

be responsible for the cost of replacing the insulation. Be sure to work this out with the customer before signing a contract.

Ridge Vents

You still need one more component to make the roofing system complete: a ridge vent. The one I have in mind is the *Trimline* by Trimline Roof Ventilation System Inc. They offer four models: Standard, Hurricane, Low Profile, and Rigid Roll.

▌ The standard vent is made of formed corrugated plastic 4 or 8 feet long and 9 or 11 inches wide. On each side of the vent are six stacks of corrugated strips 3 inches wide. Including the top, this has a total height of 1 inch. The center is left hollow so hot moist air can enter the cavity and filter through the cells of the corrugated strips. It comes with its own fasteners.

▌ The hurricane uses the same vent except it has aluminum wind deflectors on both sides of the vent. Weep holes along the bottom of the deflector allow water and melted snow to drain safely away from the vent. This unit is recommended for use in areas where rainstorms, hurricanes, or blizzards are common. It has been tested to withstand 100 mph winds accompanied by 8.8 inches of simulated rainfall. It also passed a 70 mph blizzard test, keeping out ice and snow (see Figure 5-20).

▌ The low profile is the same as the standard, except it's four layers and the total height is $5/8$ inch. An optional wind deflector is also available.

▌ The new 20-foot Rigid Roll is the only roll-type ridge vent recommended for use with a coil nailer. It's the same height as the low profile and is available in 8 and 11 inch widths. An optional wind deflector is also available.

Construction Notes — The Trimline ridge has a net free ventilation area of 18 square inches per linear foot. To operate properly, you must have at least 25 to 50 percent additional soffit ventilation above the amount in the ridge. For example, with 1,000 square inches of ventilation in the ridge, it's recommended that you install at least 1,250 to 1,500 square inches of ventilation in the soffit for proper air flow.

The amount of ventilation is controlled by the length of the slot cut along this ridge. Cut a $3 1/2$-inch slot at the ridge or approximately $1 3/4$ inches on each side of the center line of the ridge. Make it as long as the amount of linear ventilation that's required for the attic (Figure 5-21). For example, if you need 1,000 square inches of ridge ventilation and there are 18 square inches of ventilation per linear foot of the ridge vent, cut 56 linear feet of ventilation slot. If the

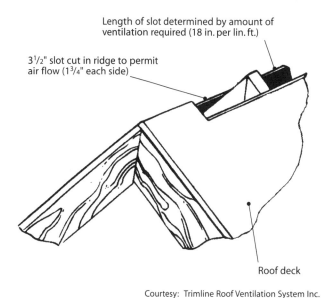

Length of slot determined by amount of ventilation required (18 in. per lin. ft.)

$3 1/2$" slot cut in ridge to permit air flow ($1 3/4$" each side)

Roof deck

Courtesy: Trimline Roof Ventilation System Inc.

Figure 5-21
Ridge slot

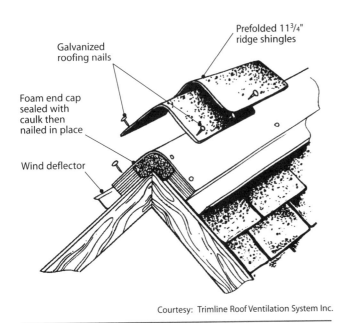

Galvanized roofing nails

Prefolded 11³/₄" ridge shingles

Foam end cap sealed with caulk then nailed in place

Wind deflector

Courtesy: Trimline Roof Ventilation System Inc.

Figure 5-22
Installing the Trimline ridge vent

ISO-VENT

How do you vent a solid roof deck? Generally, you don't! I've had my share of problems with steel roof decks, cathedral ceilings, and post and beam construction. But I think I've found a product to relieve some of the headaches. *ISO-VENT* by Johns Manville is a polyisocyanurate foam insulation panel that's truly unique in design. First, it has a moisture-resistant glass-reinforced facer bonded to both sides. Second, it's designed with a flat bottom and formed channels on the top surface. It's these channels that make the panel special. Once the panel is covered with plywood or OSB and fastened into the substrate, the channels on the top provide a ventilated air space. With the proper soffit and ridge vent design, you now have a complete ventilation system. The purpose of the panel is to add R-value and create ventilation to provide airflow and reduce the heat buildup over solid decking.

A solid deck doesn't mean you can eliminate the vapor barrier. Look at Figure 5-23. Notice that ISO-VENT has been applied over a barrier to help protect

total roofline is 75 feet, that would leave 19 feet of ridge line that isn't slotted. But don't just blindly follow the formula; consider the overall appearance. If you only need to cut 56 feet of the 75-foot total ridge line, consider installing the vent the total length of the ridge to make a more attractive roofline. Figure 5-22 shows how to install the ridge vent. Both the vent and the ridge shingle require roofing nails 2¹/₂ inches long.

Manhours — For installation, figure 0.086 manhours per linear foot to install their vent. When adding the wind deflector, add 0.022 manhours per 8-foot length per side. These figures are based on new construction and hand nailing, but don't include installing the ridge caps. Their Rigid Roll installed with a nail gun could speed production by 30 percent. On remodeling projects you need to factor in labor for cutting in the ridge slot, but estimating the manhours may be difficult. The object is to get to the decking surface to cut in a slot, but the layers of existing roofing materials can make a big difference on your labor. Another factor to consider is just how old the roofing material is. You may end up opening a real can of worms. Under the circumstances, you might be better off to walk away from the project or consider installing roof vents or gable vents instead.

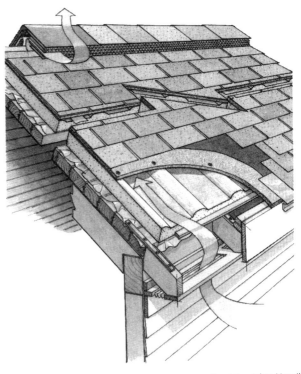

Courtesy: Johns Manville

Figure 5-23
ISO-VENT insulation panel

roofing components when high interior humidity is a factor. Check with an engineer or ventilation consultant to determine the need for and placement of a vapor retarder/barrier.

You can use ISO-VENT in new construction and reroofing projects on residential, commercial, and industrial buildings. The next time you have to build a compact roof system and you want a vented insulated substrate, consider this dual-function panel.

The panel is $48 \times 87^3/_4$ inches and is available in the thicknesses and thermal values shown in this table:

Thickness (inches)	Average R- Value
2.5	14.0
3.0	18.6
3.25	20.0
3.5	22.3
4.0	25.7

Manhours — To install double 2×4 nailers along eaves and rake, use 0.046 manhours per linear foot. To install their thickest 4-inch panel over solid decking and vapor barrier, use 1.29 manhours per square. This includes the installation of screws and plates. Once this is all in place, you still need to apply top sheathing, so figure 0.009 manhours per square foot based on $^1/_2$-inch CDX plywood. Plan for a crew of three for this type of installation.

Remember, roofing and siding products need to complement each other in design, texture, and color. Some roofing companies also manufacture and supply siding materials and have gone to great lengths to ensure that their products complement each other. Whatever products you and your customer choose, keep in mind they need to make a statement. So take your contractor's cap off and replace it with your designer cap. Continue on to the next chapter, on siding, and learn how to create that masterpiece for your customer!

Siding Materials

Are you still wearing your designer cap? Some of the products I'll introduce in this chapter should give you good ideas for designing projects that are sure to please your customers. Remember, roofing and siding products have to work together, to complement each other and the surrounding environment. These new products are great ways to get creative. You can mix products for a new and unique look, or consider dressing up those exterior walls in the style of yesterday.

The exterior of any building is its most visible and vulnerable part. Your goal is to make it attractive, while also considering maintenance requirements. There's a section on siding maintenance at the end of this chapter.

Wood has been the traditional siding for years, but there's a new breed of products on the market that may be better suited to your geographical area. Climate is a big consideration in choosing the types of materials you install. Moisture, heat, and cold can take their toll on most siding materials. Using solid-sawn wood with natural weather-resistance (like cedar or redwood) puts some strain on harvested forests. Personally, I like the look of machine-grooved shingles, fancy butt red cedar shingles, and redwood lap siding. These natural wood products can make an attractive finishing touch to any home, and I've installed my fair share over the years. But there are ways you can conserve on the use of these products.

One way is to use the natural materials only on the face of the home, and side the rest with a man-made product that closely resembles the natural material, such as one that contains wood fibers, virgin or recycled materials. Some of the products we'll discuss are made up of these materials, and I'll point those products out. To assist you, I've divided this chapter into five categories that will help you zero in on a particular product. Those categories include:

▌ Wood products

▌ Fiber-cement type products

▌ Vinyl-type products

▌ Masonry products

▌ Exterior insulation and finish systems (EIFS)

With that in mind, let's begin with the wood products.

Wood Products

Let's face it, customers still like wood, and so do many builders. For those of you who like wood side-wall shingles, for instance, you can get a panel with natural cedar laminated to 8-inch-wide plywood, one to three courses high. Not only do you save on labor, but you also cut down on waste. Oriented strand-board panels and lap siding use wood fibers from smaller, less expensive trees. And then there's hard-board siding with grain textures.

Courtesy: Shakertown 1992, Inc.

Figure 6-1
Fancy Cuts cedar siding panels

My guess is that most of you have installed one (or all) of these products at one time or another. You know that the success of these products depends on the installer's expertise. It's essential that you follow the manufacturer's installation manuals. In fact, you should have manuals in your library for all the products you install, and contact the manufacturers periodically for any updates. You don't want to miss any technical bulletins they've issued.

Sidewall Shingles

There are few things more enjoyable to a builder than to stand back and admire a project finished in cedar shingles. I really like the look of real wood, especially grooved sidewall shingles. But there are a couple of disadvantages. First, estimating accurately was always a problem. Second, product quality and consistency vary. When I opened a box of shingles, I never knew what I was going to get. It seems like I always had to straighten or square up a lot of the shingles, wasting a lot of time and material.

If you like the feel and smell of cedar but don't like wasting labor and materials, check out Shakertown. They make cedar siding panels in a

collection of shingle designs laminated to $5/16$-inch exterior plywood. The plywood is 8 feet long with a height from one to three courses high and exposures of $4^{1}/_{2}$, 7, and 14 inches. Figure 6-1 shows a round design from the Fancy Cuts series, two courses high, with a 5-inch exposure. They're also sold individually (96 pieces to the carton). One carton will cover 25 square feet when shingles are installed at a $7^{1}/_{2}$-inch exposure. The Fancy Cuts series also includes diagonal, half cove, arrow, fishscale, octagon, diamond and hexagon patterns. You can mix and match to create a conversation piece.

Because of plywood's square butts and the design of the overall panels, they automatically self-align and lock together once nailed. Panels can accommodate stud spacing either 16 or 12 inches on center. You won't lose anything to trim. You pay for coverage instead of board feet, which eliminates up to 30 percent overlap waste. The red cedar is 100 percent clear, No. 1 Grade heartwood, kiln-dried. And the manufacturer does the prefabrication, so there's no hand-sorting or wasted product due to soiled, split, or knotty shingles. Just imagine — you can save money on both labor and materials!

Construction Notes — Prebuilt corners make installing the system a snap. Actually, you install the corners first and then butt the panels up against the caulked corner edges. Color-coordinated nails are included. Place them 1 inch down from the top edge on every stud, including both ends. Also install nails along the lower edge (the self-aligning groove). Stagger panel end joints so no two vertical panel joints fall on succeeding courses. The manufacturer recommends a $1/16$-inch gap between panel ends to allow for expansion due to humidity. Since this isn't a waterproof siding system, apply a weather-resistant barrier before installing the panels. You can achieve one-hour fire resistance if you apply the siding material over $1/2$ or $5/8$-inch gypsum sheathing board, depending on stud spacing. For areas like a gable end, installing a panel creates waste, so you can get loose shingles for just this kind of situation.

Manhours — Installing panelized shingles saves about 30 percent in labor time over installing loose shingles. For estimating purposes, use 0.003 manhours per square foot for installing building paper and 0.028 manhours per square foot for the 7- and 14-inch exposure panels. For the $4^{1}/_{2}$-inch exposure, you may want to increase your time estimate by 20 percent. One person can install the panels, but a crew of two may be more efficient.

OSB Siding

OSB siding was first developed in the mid-1980s by applying an overlay to oriented strandboard panels and embossing them with a cedar texture. Now Louisiana-Pacific has developed a much more sophisticated process. After extensive research and development, L-P came out with their SmartSystem line of *treated engineered wood siding* (that's what they call it instead of OSB siding) and exterior products. The wood substrate is engineered for strength and durability, from the special conditioning and handling of the logs to the precise temperature drying of the waferized wood strands to equalize moisture. It's a combination of fiber from fast-growing trees and specialized ingredients to provide resistance to rot and insects. This combination, plus technical expertise, actually improves on traditional building products in a number of ways, including making them more affordable and available.

SmartLap

Available in 6- and 8-inch widths ($^7/_{16}$ inch thick) and in 16-foot lengths, SmartLap features an embossed texture with the appearance and workability of natural cedar (Figure 6-2). The $^5/_{16}$-inch-thick SmartLap II gives the same quality look with a narrower profile and a lower price than cedar. It's engineered to resist cupping, splitting and warping. A special coating provides weather resistance, and a zinc borate preservative protects against termites and fungal decay. Finally, it's preprimed and ready for paint. It carries a limited 25-year transferable warranty, with a five-year 100 percent repair/replacement feature.

Construction Notes — Here are a few things to consider when installing a SmartLap product:

▌ Always begin with a starter strip and be sure the siding-to-grade clearance is a minimum of 6 inches. The siding should also extend at least 1 inch below the plate (Figure 6-3). Because of moisture exposure, don't let the siding or starter strip come into contact with concrete, stucco or any other masonry products.

Courtesy: Louisiana-Pacific Corporation

Figure 6-2
SmartLap engineered wood lap siding

▌ Locate joints over studs with one nail per stud whether or not you're fastening over nailable sheathing. Stagger joints at least one stud apart and leave a $^3/_{16}$-inch space between the jointing pieces. Fill the space with a nonhardening paintable sealant with a manufacturer's stated service life of at least 25 years.

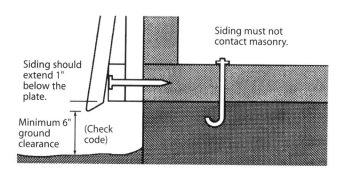

Courtesy: Louisiana-Pacific Corporation

Figure 6-3
SmartLap requires minimum 6-inch ground clearance

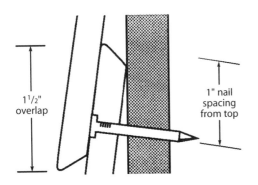

Figure 6-4
SmartLap overlaps 1 1/2 inches

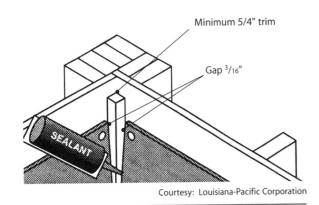

Figure 6-5
Allow 3/16-inch gap at inside corners

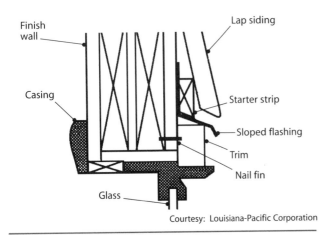

Figure 6-6
Sloped noncorrosive flashing at doors and windows

◼ The maximum spacing for siding is 24 inches on center. When fastening to nailable sheathing, the nailing pattern can be 8 inches on center or one nail per stud. When nailing directly to studs, be sure to first apply a vapor barrier to the studs. Nail penetration needs to be at least 1 1/2 inches deep.

◼ Face nailing voids the warranty. Install it by blind nailing using stain- and corrosion-resistant nails. Where exposed nails are necessary to correct building irregularities and at cutouts and under the eaves, use galvanized nails (stain- and corrosion-resistant) and seal completely with sealant. Always begin nailing at one corner and proceed to the other end; don't nail toward the center from both ends.

◼ Blind nailing means to nail 3/4 inch down from the top edge of the siding. The next course overlaps the previous lap by 1 inch for 6- and 8-inch siding (Figure 6-4). That hides the nails and holds the siding tight. For 6-inch siding, nail 1/2-inch from the top edge.

◼ On inside and outside corners, always hold the panels back 3/16 inch from trim and seal. The corner trim shown in Figure 6-5 is a minimum of 5/4-inch square. Consider priming all four surfaces before installing.

◼ When working around roof rakes, provide a 1 1/2-inch clearance from the finished roof surface and the underside of the siding following up the rake. The flashing on the house in this type of application should be 4 inches minimum. Paint the cut edge.

◼ Protect siding from direct roof water runoff and ground splashback with adequate flashing, gutters, and eaves.

◼ Around doors and windows, apply sloped noncorrosive flashing (drip cap) behind siding and over horizontal trim. The slope will help to drain water away from the siding (Figure 6-6). Allow a 3/16-inch gap when siding ends at the trim around doors, windows and other openings. Be sure to properly seal these areas.

Manhours — When working with this product, it's very important to follow the manufacturer's specifications. Don't lop off any steps to save time. All cut edges have to be sealed and caulked, which will take additional time. Don't forget to factor these tasks into your manhours. Figure at least 0.030 manhours per linear foot for applying sealant (caulk) and 0.035 manhours per linear foot for priming cut ends. A crew of one can handle both tasks. For applying the siding itself, either 6 and 8 inches, use 0.032 manhours per square foot. Use a crew of two to install the siding.

SmartPanel

If you want to speed things up a bit, panel siding is the way to go. Louisiana-Pacific has a treated engineered wood product that's just the ticket. Like SmartLap, it features an embossed grain that looks like cedar. Available in two thicknesses, $^7/_{16}$ and $^{19}/_{32}$ inches, the 4-foot-wide panels come in four choices of length: 8, 9, and 10 feet, and 100 inch. SmartPanel II is $^3/_8$ inch thick, offering an even more affordable option. The SmartPanel product comes with shiplap or square edges and a 4 or 8-inch channel grooved pattern. Or you can select the ungrooved panel. All the features and benefits of SmartLap apply to SmartPanel as well (Figure 6-7).

Construction Notes — The installation of SmartPanel siding is similar to SmartLap siding with a few key exceptions:

- Installing horizontally (grooved only) and using staples can void the warranty.

- The panel is surface nailed, so nails need a minimum $^1/_4$-inch head. Choose the correct size nail (stain- and corrosion-resistant) for the panel thickness: 6d for $^3/_8$ and $^7/_{16}$ inch and 8d for $^{19}/_{32}$ inch.

- Panels can be installed directly to studs or over optional sheathings. Install nails 6 inches on center around the perimeter and 12 inches on center on intermediate supports.

- For sound building procedures, install a continuous vapor barrier on the warm side of the wall cavity.

- On the first panel, and after succeeding panels are aligned, lightly tack each corner. Then install the first row of nails at the edge. Once this first row is in, remove tacking nails and continue to nail the rows (intermediate studs), working your way out to the opposite edge. Installation is complete once the panel has been fastened to the top and bottom plates.

Courtesy: Louisiana-Pacific Corporation

Figure 6-7
SmartPanel 8-inch channel grooved siding

- Position succeeding panels against the preceding panel and, again, tack the corners. Figure 6-8 shows an alignment bead built into the shiplap joint which automatically leaves a $^1/_8$-inch gap. Maintain that gap when butting panels up against doors, windows, sills, and trim. Nail through the shiplap joint but don't nail in the channel groove. Place the nails $^3/_8$-inch from the panel edges. Finish by sealing with caulk.

- Don't overdrive the nails. Drive them so the head is just in contact with the siding surface. Seal any nails driven below the overlaid surface. Remember, all cut edges need to be primed or sealed.

- The horizontal butt joint between two panels needs special treatment. Protect this joint from moisture by applying flashing to the backside of the upper panel that extends over the top edge of the lower panel. This flashing is commonly referred to as "Z" flashing. The flashing needs a bent drip edge (out from the siding) like that shown in Figure 6-6. Be sure not to crush or damage the "Z" flashing. An alternative is to overlap the bottom panel a minimum $1^1/_2$ inches with the top panel, making sure the bottom edge is painted.

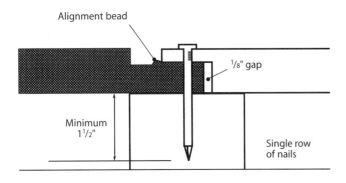

Alignment bead

⅛" gap

Minimum 1½"

Single row of nails

Courtesy: Louisiana-Pacific Corporation

Figure 6-8
Proper nailing procedures for SmartPanel

▌ Some builders like the look of a "belly band," or band board trim. But don't install this trim directly over the joint. Instead, after applying flashing, install the band out away from the wall's surface using plastic spacers (with a hole) behind the band and up against the siding. With two nails above each other, locate the spacers so each nail will pass through the center. Spacers should be a minimum ³⁄₈ inch deep. Again, don't overdrive the nails.

Manhours — To estimate panel installation, consider a crew of two to three. Estimate 0.025 manhours per square foot for 4 × 8-foot sheets. For 9 and 10-foot sheets, increase the manhours by 10 to 15 percent. To install sheet metal flashing, use 0.040 manhours per linear foot, and for belly bands from 4 to 10 inches wide, use 0.030 manhours per linear foot.

Hardboard

When you think of hardboard, what's the first thing that comes to mind? Furniture, perhaps? If it's not furniture, it's probably interior paneling. But now you can use hardboard for other products, like exterior siding.

Hardboard is made of fibers from wood chips and board trimmings. In the past, these were lumber industry wastes. The wood chips may come from fast-growing trees, other sawmills, or from logs not suitable for dimension lumber. The fibers are combined with binders and bonded under heat and pressure to form grainless panels with either smooth or textured surfaces. But before I get into what they have to offer, I just want to warn you about hardboard's worst enemy: moisture. Here are some guidelines for handling any hardwood product.

When the product is dropped off at the job site, stack it on 2 × 4 sleepers to keep it off the ground and away from moisture which would cause direct damage. Always cover the stack, but make sure the cover isn't in direct contact with the material. Place 1 × 2 strips (called *stickers*) on the stack to help keep the cover off the surface of the panels. Don't "tent" the stack because that tends to trap moisture. Make sure there's air circulation. Here are some other suggestions:

▌ Never stack it on green concrete.

▌ Allow at least 6 inches of space between the siding and the ground.

▌ Keep shrubs and plants away from exterior surfaces by at least 1 foot.

▌ Never use it on any structure where the siding will be in direct contact with excessive moisture.

▌ Apply a vapor barrier under the siding.

▌ Avoid attaching it to green or crooked framing members.

▌ Cover crawl spaces with 6 mil polyethylene film to control moisture and overlap all joints by 12 inches. Weigh it down so it doesn't blow away.

▌ Be sure attics and crawl spaces are cross-ventilated to the outside with a minimum of 1 square foot of net free ventilation for every 150 square feet of floor area. Attics can go to 1 square foot of net free ventilation for every 300 square feet of attic area if you provide both high and low ventilation. You should check ventilation requirements with your building department.

▌ Lastly, follow the manufacturer's specifications.

If you take the proper care, hardboard siding can be durable as well as attractive. You can find it in a wide variety of appealing siding sizes, styles, textures and finishes. I'll introduce a couple of manufacturers that have a few patterns you may find interesting.

Masonite

The Masonite Corporation offers so many hardboard siding profiles that I can't describe them all. Their wide variety of styles, sizes, and textures come ready to paint or fully finished in a range of colors. The manufacturing process (intense heat and pressure) makes Masonite somewhat harder and more dense than real wood. That's why their motto is "Wood made better." Because they use fibers, the finished product is free of knots and other natural flaws. This engineered product is designed to resist splitting, cracking, checking, and delaminating. That's why they can offer a series of limited warranties.

From a distance, you can't tell if the siding is the real McCoy or not. Even at close range the deep-embossed grain pattern looks and feels like the real thing. One product I especially like is Colorlok, a lap siding with a weathered pine texture and unique built-in self-aligning feature. A spline made of plastic running the full 16-foot length of the backside rests on the top edge of the previously-installed

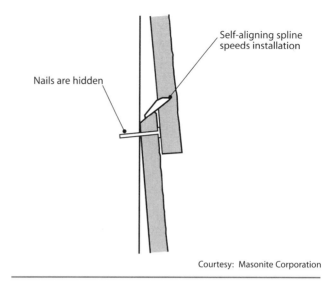

Courtesy: Masonite Corporation

Figure 6-9
Blind nailing Colorlok lap siding

course (Figure 6-9). This could really speed up production. Another timesaver, one that could eliminate a subcontractor, is the fact that the siding is prepainted. And, the manufacturer carries a full line of color-matching aluminum accessories to help detail the project (Figure 6-10).

① Joint molding (6" and 8")
② Individual outside corner (8")
③ Starter strip (10")
④ Continuous inside corner (10")
⑤ J-trim (12')
⑥ Caulk
⑦ Colored nails (1¼")
⑧ Touch-up paint
⑨ Lok-On corner post system (10')

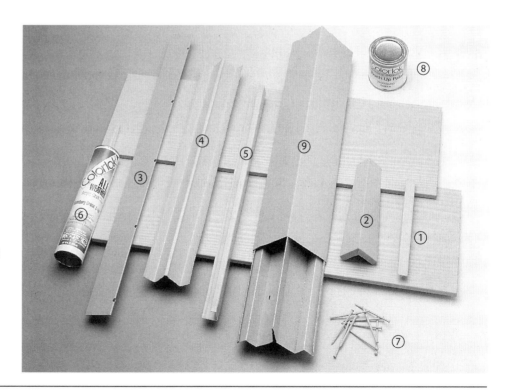

Courtesy: Masonite Corporation

Figure 6-10
Accessories for Colorlok lap siding

The $7/16$-inch thick siding is available in 6- and 8-inch widths and exposures of $4^7/8$ and $6^7/8$ inches. The 16-foot lengths allow it to cover more surface area. Colorlok carries a 15-year limited warranty on the factory finish and a 25-year limited warranty on the substrate. Eleven colors, from Adobe White to Vicksburg Green, are available. You can mix colors and sizes to create many unique designs. Doesn't that get your creative juices flowing?

Construction Notes — Installing this system is very similar to OSB siding except that there's no need to prime or seal cuts (a real timesaver!). Here are some other things to remember:

▌ Level and install the specially-formed starter strip along the base of the sheathing or the mudsill. Allow at least 6 inches between this bottom edge and the finish landscape grade or horizontal surfaces such as steps, patio slabs, porches and decks. Apply the starter strip using 6d box head nails spaced 12 inches on center. Drive the nails into the mudsill (Figure 6-11 A).

▌ Position the first course of siding so the back of the siding locks into the starter strip. Secure the siding by holding down firmly and nailing it $5/8$ inch down from the top edge at each stud location, nailing from one end to the other. Don't nail toward the center from both ends or force or spring siding into place. Nail spacing can't exceed 16 inches on center. Whether you hand or power nail, be sure not to overdrive the nails (Figure 6-11 B).

▌ Install horizontal shim strips for continuous support behind the siding above or below openings (Figure 6-11 C). Face nailing of siding is required immediately above and below windows and above doors. Space nails 12 inches on center.

▌ Butt joints must fall over a framing member. Fasten siding on both sides of the joint (Figure 6-11 D) and stagger joint location from one course to the next. Plan ahead so your joints don't fall in a "stairstep" pattern. That's not very appealing.

▌ Once the inside corner has been applied, hold back the siding on both sides of the centerpoint by $3/16$ inch and then caulk (Figure 6-11 E).

▌ Install individual outside corners as each course meets at the corner. As an alternative, you can use the Lok-On cornerpost system (Figure 6-11 F). Attach the base to all outside corners before installing siding. Nail into the slots provided in the base. Apply the siding, butting ends to the channel in the base and leaving a $3/16$-inch gap. After the siding is installed, hook one side of the Lok-On cap over the base edge. Snap the other side into position and lock into place using finger pressure, about every 18 inches.

Manhours — Here are some recommendations that should give you a good idea about how the entire system comes together. Before installing this product, read the manufacturer's application instruction #203 (and check with Masonite Corporation to make sure it's current). After the initial learning curve and using a crew of two or three, you could possibly cut 10 to 25 percent off the following manhours for this system, which are based on the first installation:

▌ To apply the starter strip, use 0.030 manhours per linear foot.

▌ For J-trim, outside corner, and inside corner, use 0.033 manhours per linear foot.

▌ To install siding, joint molding, and caulking, figure 0.032 manhours per square foot.

ABTco Stucco

ABTco (ABT Building Products) offers different profiles and designs in both hardboard laps and panels, but their stucco panel gives the true detail of a hand-troweled finish (Figure 6-12). The deep-textured stucco trowel markings are embossed directly into each 4 × 8-foot and 4 × 9-foot shiplap edge panel. The $7/16$-inch thick panels are prefinished in a white thermoset acrylic or primed and ready for the final color coat. ABTco also supplies touch-up paint, caulk and nails to color-match the panels.

The siding is made from Pro-1 (Roring River Process), which is exclusive to ABTco. This process combines four extra steps into the production:

1. A natural fiber filament layer is fused to the surface of all their siding products. This layer is impregnated with resins and linseed oil, providing protection against the elements.

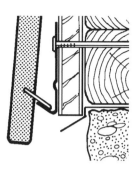

A Bottom course detail

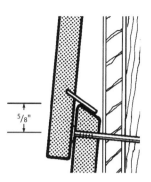

B Lap detail — nail placement

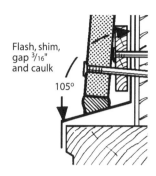

Flash, shim, gap 3/16" and caulk

105°

C Door and window treatment

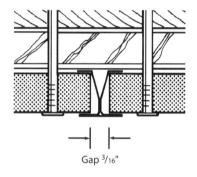

Gap 3/16"

D Butt joint detail

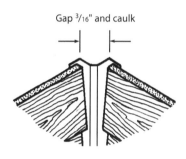

Gap 3/16" and caulk

E Colorlok inside corner

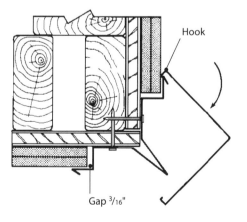

Hook

Gap 3/16"

Snap

F Colorlok outside cornerpost

Figure 6-11
Details for installing Colorlok

A Close-up shows texture

B ProTrim Trimboard adds contrast

Courtesy: ABTco, Inc.

Figure 6-12
Pro-1 stucco panel

2. Next, they press in the profile. Their molding process, with matched top and bottom plates, uses high heat and pressure to naturally emboss texture and profile into each product.

3. Then the product is oven baked to remove all natural and process moisture and to cure the wood one more time. This process improves overall strength and enhances the siding's physical properties.

4. Finally, using a controlled procedure, they bring each product back to a constant moisture content. This final moisture balancing step insures consistency as well as enhanced dimensional stability.

Four separate layers of thermoset primer are also applied to the edges of each siding product. The exposed face provides a great painting surface because of the natural fiber filament layer and the heat-cured primer. The product carries a 25-year substrate limited warranty, and the finish carries a 5-year limited warranty.

Construction Notes — ABTco products are generally installed just like other hardboard products, with a few exceptions:

▌ Panel siding may be applied to studs spaced 16 or 24 inches on center where local building codes permit. They can be installed directly over studs once building paper (non-vapor barrier type) has been applied.

▌ In all insulated or heated buildings, install a continuous vapor barrier of 6 mm polyethylene or less (like polyethylene film or foil-backed gypsum board) to the interior side of studs to prevent damaging condensation from accumulating within the walls.

▌ To install over masonry walls, first frame out the walls with furring strips spaced 16 inches on center and thick enough to accept full-length nails ($1^1/_2$-inch penetration into framing member). Then apply a continuous vapor barrier between the framing and masonry walls. Siding must not come in contact with the masonry wall.

▌ For slab-on-grade foundations, a vapor barrier is required.

▌ This system is surface nailed, so use their color-matched nails and be careful not to overdrive.

▌ Double-nailing at vertical joints is required 6 inches on center (Figure 6-13). On intermediate studs, space the nails 12 inches on center.

Manhours — It's important to follow the guidelines laid out in the manufacturer's instructions so you don't void the warranty. As for manhours, figure that you'll need a crew of two or three to handle these panels efficiently. To install panels of either size, use 0.032 manhours per square foot, plus 0.026 manhours per linear foot for exterior trim.

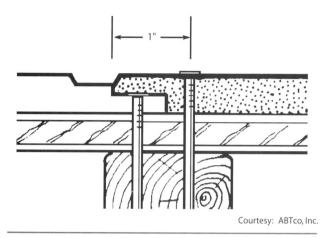

Figure 6-13
Pro-1 panel joint

Fiber-Cement Composites

If you're looking for a siding product that doesn't rot, is immune to termites, and withstands high impact (to mention a few of the hazards out there), then fiber-cement composite siding may be the answer. Its main ingredients are cement, silica sand, and cellulose fibers from small diameter fast-growing trees or reclaimed wood processing waste. To achieve their overall structural strength, they're steam-cured under pressure in an autoclave (pressurized chamber). They're noncombustible and asbestos-free, which makes them especially valuable in fire-prone areas. They're available with different profiles and textures, including stucco and wood grain that resembles the real thing. For those customers who enjoy the simple things in life, there's the smooth face.

The following manufacturers have a variety of products to choose from. I can't cover all the available products, so contact the company directly and ask for product literature to learn more what each company offers. The addresses are in the Appendix.

MaxiPlank

This exterior horizontal lap siding board is available in a smooth or textured surface with a length of 12 feet. Widths are 6, $7^1/2$, $8^1/4$, $9^1/2$, and 12 inches, with a thickness of $5/16$ inch. They're a natural light gray and you can paint them on-site with water-based acrylic exterior grade paints.

According to the manufacturer, you can apply MaxiPlank to wood or metal studs spanning up to 24 inches on center without a vapor barrier, where building codes permit. To give you a rough idea of its weight, a $7^1/2$-inch $\times$ 12-foot board weighs around 17 pounds. To complement the siding, the company offers 10-foot lengths of 26-gauge metal trim as well as internal and external corners, door and window trim, and a concealed jointer. However, they're not primed or painted. The product is covered by a 50-year limited transferable, prorated product warranty.

While you can cut MaxiPlank with a power saw (using a mask and goggles for protection), you can also make cuts using their scoring tool, a specially-designed cutter, or electric hand or pneumatic shears that eliminate dust. For notches, cut in the vertical lead using a hand saw or power tool, and then score your horizontal line.

For round and square openings, use the Bosch Rotary Cutting Tool, as shown in Figure 6-14. Besides cutting ceramic tile, it works nicely on MaxiPlank providing you use the $1/8$-inch (not the $1/4$-inch) carbide tip. It makes cutting circles a snap. Weighing in at about 2.4 pounds, this versatile tool will cut most of the products listed in this book. Just make sure to use the appropriate bit. It's available in a kit that includes a complete set of bits and the circle cutter. When cutting a surface that has a texture, cut from the back side.

Figure 6-14
Bosch rotary cutter

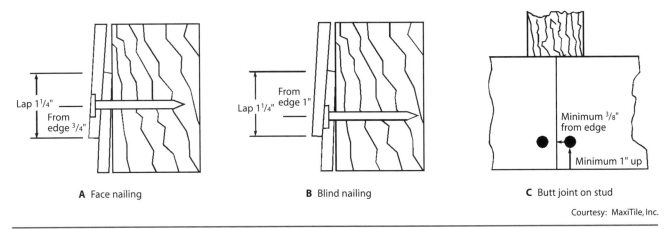

A Face nailing   **B** Blind nailing   **C** Butt joint on stud

Courtesy: MaxiTile, Inc.

Figure 6-15
Fastening MaxiPlank

Construction Notes — The installation of MaxiPlank follows the usual rules for lap siding, with a couple of exceptions:

▌ For surface nailing in wood studs, use 6d galvanized common nails (*never* staples) through both laps (Figure 6-15 A). Hold the materials firmly against studs when fastening. The nail must penetrate at least $1^1/_4$ inches, and be no closer than $3/_4$ inch from the bottom edge.

▌ For blind nailing in wood studs, the manufacturer recommends an #11 gauge, $1^3/_4$-inch galvanized roofing nail through the top of the undercourse lap. Laps should overlap the lower course by $1^1/_4$ inches. To maintain this measurement, make a couple of templates or use the Maxi gauge clips or off-stud jointers. The nail should be placed 1 inch down from the top edge (Figure 6-15 B).

▌ For metal studs, use an 8 × $1^5/_8$-inch self-drilling and self-embedding screw. For blind fastening, use 8 × $1^1/_4$-inch screws. You must have at least three threads into the stud.

▌ For butt joints, fasten a minimum 1 inch up or down from the top or bottom edge and $3/_8$ inch in from the end (Figure 6-15 C). If you're fastening closer to the top or bottom edge or end, predrill oversize holes. Stagger the joints a minimum of 2 feet. And make sure to drive the nails *straight*.

Manhours — Before installing this product, read the manufacturer's application instructions. I recommend cutting a few pieces to get a feel for it. (This also applies for fastening.) There's definitely a learning curve with this system. Figure a crew of three or four.

▌ To apply the starter strip, estimate 0.030 manhours per linear foot.

▌ For internal and external corners and door and window trim, use 0.033 manhours per linear foot.

▌ To install siding, jointer, and caulking, figure 0.040 manhours per square foot. It's possible after the third time around you may be able to bring this figure down to 0.032 manhours per square foot.

Hardisoffit

James Hardie Building Products has fiber-cement composition products that include laps and panels, but here I'll focus on their soffit material. As you know, the soffit is one area on a home that takes a lot of abuse. One way to control damage is to install a Hardisoffit cementitious soffit that carries a 50-year limited product warranty. It's dimensionally stable and won't crack, rot, or delaminate under normal conditions. These resilient fiber-cement panels resist damage caused by extended exposure to soffit enemies: moisture, humidity, sunlight, snow, and salt air.

Since Hardisoffit is noncombustible, it's a good choice for areas that are prone to fires. The manufacturer points out that all their products are free of asbestos, fiberglass, and formaldehyde.

A wide variety of sizes is available to help cut down on labor costs, including widths of 12, 16, 24, 36, and 48 inches, all 8 feet long. It weighs less than 1.9 pounds per square foot and is $1/4$ inch thick. Because the product is so thin, it's important to protect edges and corners from chipping after it's delivered to the job site. Stack it on edge or lay it flat on a smooth level surface and under cover to keep it dry. If it should get wet, let it dry thoroughly before installing. This product comes primed and the manufacturer recommends finishing it with a high-quality, exterior grade, 100 percent acrylic latex paint. Hardisoffit panel is available in smooth or woodgrain textures.

Hardisoffit's thinness allows you to cut it to size with the score-and-snap method. If you choose to use power tools, wear a mask to avoid inhaling dust. You'll need to wear safety glasses even if you don't use power tools. The company offers a tungsten-tipped knife specially designed for the score-and-snap method.

Construction Notes — Even though you can nail it, screws may be a good fastener to consider. When placing a screw, hold back $3/8$-inch from the edge. Use a self-tapping tip screw when applying it to metal. However, if you find yourself too close to an edge or corner, consider a pilot hole before installing any fastener. The product is designed to span 24 inches on center. Joints are normally held apart $1/4$ to $3/8$ inch and then caulked, but you can use a vinyl "H" molding for a cleaner look (Figure 6-16 A). There are a number of accessory items currently available, but sales reps from James Hardie can provide names of companies that supply both metal and vinyl venting and connector molding designed especially for Hardisoffit.

Also available are vinyl soffit vents that make the project go smoothly (Figure 6-16 B). Figure 6-16 C shows how the soffit vent and "H" molding complement each other.

Manhours — With this type of installation, you need at least a crew of two, but three may be more efficient depending on the size soffit panel being

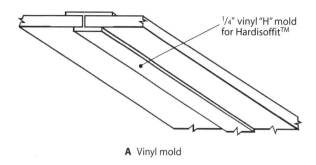

A Vinyl mold

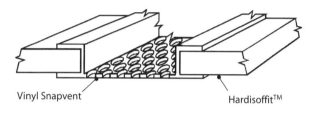

B Snapvent

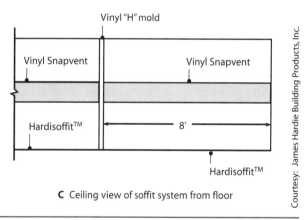

C Ceiling view of soffit system from floor

Courtesy: James Hardie Building Products, Inc.

Figure 6-16
Installing accessories for Hardisoffit

installed. For labor estimates, which includes soffit material, moldings, and vents, use these for installing the most common sizes:

- 12 inches × 8 feet: 0.057 manhours
  per linear foot

- 16 inches × 8 feet: 0065 manhours
  per linear foot

- 24 inches × 8 feet: 0.070 manhours
  per linear foot

Vinyl-Type Products

Products made from vinyl can offer another alternative to wood. Today, companies can create siding products that give the feel and look of cedar shingles, hand-split shakes, brick, and hand-cut stone. They also reproduce vintage products like the Victorian fishscale and sunburst designs that can dress up a gable. But most importantly, they offer solutions to areas that take a lot of abuse from the weather: eaves and the undersides of the overhang. Fascia material and soffits with full ventilation can sure help these troublesome areas.

While some products are made from vinyl — PVC (polyvinyl chloride), polypropylene, acrylic, urethane, and thermoplastic resins — they're not high on the recycle list. And they don't use recycled materials except for a few siding companies that recycle about 10 to 20 percent of their own post-industrial scraps back into the manufacturing process. When dealing with such products, it's important to check out the warranty. Look for the following:

▌ How long is the warranty? Is it prorated, and does it cover 100 percent of the labor and materials?

▌ Is the product warranty automatically transferable to the next homeowner? Are you required to register it before you can transfer it?

▌ Are service fees charged for claims and/or complaints?

▌ Is there coverage from salt air and spray, and wind-driven sand?

▌ Is the siding product warranted against blistering, chipping, corroding, cracking, flaking, peeling, pitting, splitting, and warping?

▌ What about fading and hail coverage?

▌ Does the product require maintenance, like painting?

▌ Is the product warranted against contraction and expansion?

You need to know these things because it's your name that's attached to the job. If there are any problems, they'll fall on your shoulders before the manufacturer. So it's important to protect your interests when it comes to installing any product. Do your homework. Know the manufacturer, understand the product, realize what's in the warranty, and communicate with your customer before signing a contract and purchasing the products.

Here are some unique products that can help give a creative twist to any project. They can be mixed with wood products. It's up to you and your customer to choose the mix.

United States Vinyl Shakes

United States Vinyl Shakes, LTD. (USVS) designs, develops, manufacturers, and sells vinyl shake siding with a proven track record. They don't manufacture the vinyl itself; instead they use the Geon Company (a former division of B.F. Goodrich) exclusively for their product. Geon Vinyl has over 40 years of successful long-term weathering performance in the siding industry. Their special additives to the PVC resin make the vinyl unique and they offer a full 50-year warranty, nonprorated, on labor and material. As far as I know, USVS is the only vinyl shake on the market today that offers three profiles (Figure 6-17) in an authentic shake look:

▌ Patrician Shake: 10-inch exposure (resembles $1/2$ to $3/4$-inch hand-split shakes)

▌ Normandy Perfections: 10-inch exposure (resembles grooved shingles)

▌ New Englander: double 5-inch exposure (again, resembling grooved shingles)

Available in beige, blue, clay, driftwood, ivory, and white, the shakes have the same color all the way through the product so they don't ever need painting. And all three patterns interlock with each other or with vinyl clapboard siding to create special designs.

Construction Notes — When installing the siding, USVS recommends these steps:

▌ Use a rigid insulated foam board as an underlayment sheathing to create a level surface and to act as an insulation barrier.

▌ Always start the first course using a vinyl or aluminum starter strip.

Courtesy: United States Vinyl Shakes, Ltd.

Figure 6-17
USVS vinyl shake siding

▌ To make the system complete or as a unit, use "J" channel around the doors and windows. Use vinyl corner posts as well. USVS doesn't sell accessory items, but the accessories are industry standard and are available in matching colors.

▌ Install panels from left to right. A vertical positioning lip allows for hidden seams and a four-sided lock.

▌ After you've applied the first row, cut one-third off the first panel to start the second row. After completing this row, cut two-thirds off the first panel for the third row. Start the fourth row with a panel that has been cut in half. In the fifth row you should be back to a starting full panel and then you repeat the pattern again. This method provides added high-wind resistance to the side locks and a more authentic variation to the pattern.

▌ Panels are $27^3/4$ inches long and $10^3/4$ inches high. While the panel includes 13 nail holes, it only needs two nails per full panel. Use the third nail hole in from each end. The key to successful installation is not to nail it too tight. Leave the head of the nail approximately $^1/8$ inch out from the locking lip. If you nail too tightly, you'll have difficulty locking in the next row of panels.

▌ The top row can be face-nailed (using rust-resistant nails only), but holes should be slightly bigger than the shank of the nail.

Manhours — This product can be installed by one person. In order to speed production, a crew of two or three can give it a real working rhythm. For installing insulated foam board (USVS recommends up to $^3/4$-inch board), use 0.010 manhours per square foot. For installing any of the three siding profiles USVS offers, use 3.35 manhours per square foot.

Nailite

Nailite International makes injection-molded siding from thermoplastic resins for the residential and commercial markets. You can use them both indoors and out. After they're finished with acrylic paint, they look so much like real wood, brick, and stone that you might have a hard time spotting the difference. Figure 6-18 shows how well they've reproduced the feel, texture, and details. Their authentic replicas, available in eight different colors, include hand-split shake, perfection-plus cedar, hand-laid brick, and hand-cut stone (Figure 6-19).

Here are approximate panel weights: shakes, 3 pounds; brick and stone, 4.5 pounds; and cedar, 2.5 pounds. The shake, brick and stone require about 20

Courtesy: Nailite International

Figure 6-18
Nailite molded resin siding

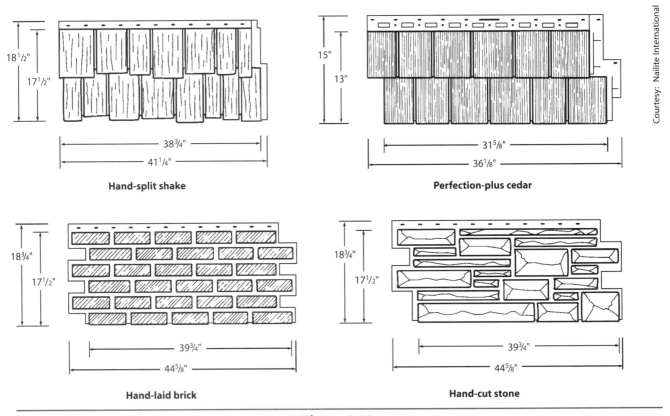

Hand-split shake

Perfection-plus cedar

Hand-laid brick

Hand-cut stone

Courtesy: Nailite International

Figure 6-19
Nailite replicas of natural materials

panels per square, and 36 panels per square for the cedar. They're all about 0.090 inches thick.

Nailite also created all the accessories to complete the job, including a wide variety of outside corners in pattern or texture designs, brick and stone ledge trim, mortar fill (for brick and stone), J-channels, aluminum universal starter strips, and touch-up paint.

The panels are designed to be installed in both the residential and commercial market with an elevation up to 40 feet. The application can be for the entire building or to add architectural accents to the following:

▌ foundations

▌ mansard roofs

▌ entryways

▌ gables

▌ flower boxes

▌ chimneys

▌ skirting

Nailite panels are manufactured from high-quality resins with special additives to help protect against the damaging effects of the sun's rays. Several coats of high-quality paint are applied to the surface of each panel during the manufacturing process to further protect them from the elements and to give the panels their unique look and texture. But expect natural weathering to occur. While that weathering doesn't harm the quality of the panel itself, periodic touch-ups or refinishing will help restore the panels to their original look and finish.

Nailite International has made this product since the late 1970s. Their company policy has always been that the responsibility for installing their siding panels in accordance with local building codes lies solely with the contractor or installer. When you read their Installation Guide (and I recommend it strongly), remember that the recommendations are only suggested application instructions. The manual simply tries to summarize good siding practices as well as some of the industry standards for installing panelized siding systems. They were developed over a period of time from actual trade practice and the requirements of various building code agencies — but always check your specific building code.

Construction Notes — With that in mind, these are some helpful hints:

▌ Install the panels from left to right.

▌ If you'll install them on more than one wall, finish one wall first before starting the next.

▌ Use a minimum of five nails evenly spaced per panel. The cedar panel requires one additional nail along the side nailing fin. See Figure 6-19.

▌ Do the face nailing in inconspicuous areas, such as a mortar line or between shakes or shingles. The head of the nail can be covered with mortar fill or painted.

▌ It's important to engage all clips and fingers on the back sides of panels (Figure 6-20).

▌ In temperatures of 45 degrees F or lower, the product will contract. If you're installing it in cold weather, make sure to allow an additional $^1/_8$ inch for material expansion.

▌ At temperatures below 40 degrees F, store panels indoors before application to keep them pliable for easier installation.

▌ Use a house wrap that breathes. A foil-faced material isn't acceptable with this product.

▌ When working with outside corners, be sure to install the first corner 1 inch below the starter strip. Square off the left side of the first panel so it will fit into the corner (Figure 6-21). Install each corner as you apply each row, and adjust as necessary where two corners overlap to maintain the horizontal alignment. You have an adjustment flexibility of $^3/_4$ inch. Once you've cut the panel, hook it over the starter strip and slide it into the corner. When approaching the end of a course, you may find it easier to lock two panels together, square off the right-hand side of the panel, and then install the two panels as one into the corner.

Manhours — You'll need at least three installations to get comfortable with installing the product. Even three projects may not be enough. Figure on a crew of two to speed production. To estimate installation

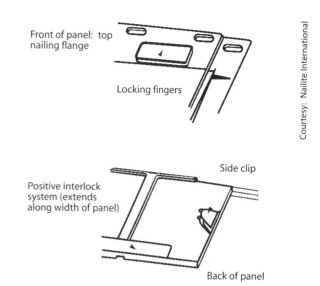

Figure 6-20
Nailite positive interlock system

including corners, use 4.25 manhours per square. This time will probably come down when you've had enough experience to get used to the installation procedures.

Wolverine

For over 40 years this company has offered a complete line of vinyl siding products that let you improve any project, whether historical or modern, large or small, new or remodeled. It comes in a wide

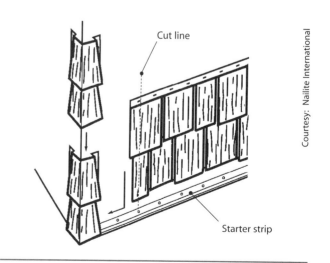

Figure 6-21
Corner installation for Nailite siding

Courtesy: Wolverine Vinyl Siding

Figure 6-22
Wolverine soffit system

range of styles, finishes, and colors and includes accessories for soffits, fascia, gables, inside and outside corners, and trim for doors and windows. They also offer a lifetime limited transferable warranty. We all know Wolverine's siding product, but they also have soffit material that isn't so well known, but which really works well.

Every home needs to breathe, whether it's to keep the attic cool in the summer and dry in the winter, to extend the life of the roof, or to help your customers save on their utility bills. A fully-ventilated soffit material can help accomplish this (Figure 6-22). With 16 colors to choose from, you can easily blend Wolverine's soffit into the customer's preferred color scheme. And it works well with other siding products, like wood and brick.

Don't limit yourself to just the soffit area. You can also use Wolverine as an exterior wainscoting for the inside of a porch or to detail a porch ceiling. That leaves your customer with no painting to worry about.

Construction Notes — Installing the material is pretty simple. The main thing to check out before starting is construction of the rafters and/or truss overhang. Figure 6-23 illustrates some common construction methods. Here are some things to consider:

▊ The recommended nailing for soffit material is 16 inches on center. On spans of 24 inches or more, add nailing strips (Figure 6-23 D).

▊ At corners, be sure to cut and install the channel to allow $1/4$ inch for expansion at each of the adjoining walls.

▊ To turn the corner, measure from the channel at the wall corner to the channel at the corner of the fascia board. Subtract $1/4$ inch for expansion. Cut and install the soffit double channel lineal (Figure 6-24). You might need a backing for the lineal. Miter the corner soffit material and install.

▊ You may need to flex the panel a bit to install the soffit material into the second channel.

▊ If you are using a fascia cap, don't worry about flexing the panel because the cap will be installed last (Figure 6-25).

Manhours — While you can install soffit material from a ladder, it's more effective to use scaffolding. Assess this before bidding the job so you can include the scaffolding in the bid if you plan to use it. Consider this a project for at least a crew of two. To install a 6-inch fascia, J-channel, and perforated or solid soffit panels in the following widths, use these suggested manhours:

▊ 12-inch: 0.047 manhours per linear foot

▊ 18-inch: 0.055 manhours per linear foot

▊ 24-inch: 0.060 manhours per linear foot

Style-Mark

For over 20 years, Style-Mark Inc. has produced high-density urethane decorative trim. Their catalog is full of authentic and attractive decorative millwork, door and window trims, louvers and moldings. They're easy to handle and lightweight, which helps save on labor costs. Urethane resists insects, decay, and weather. Style-Mark trims are durable, easier to

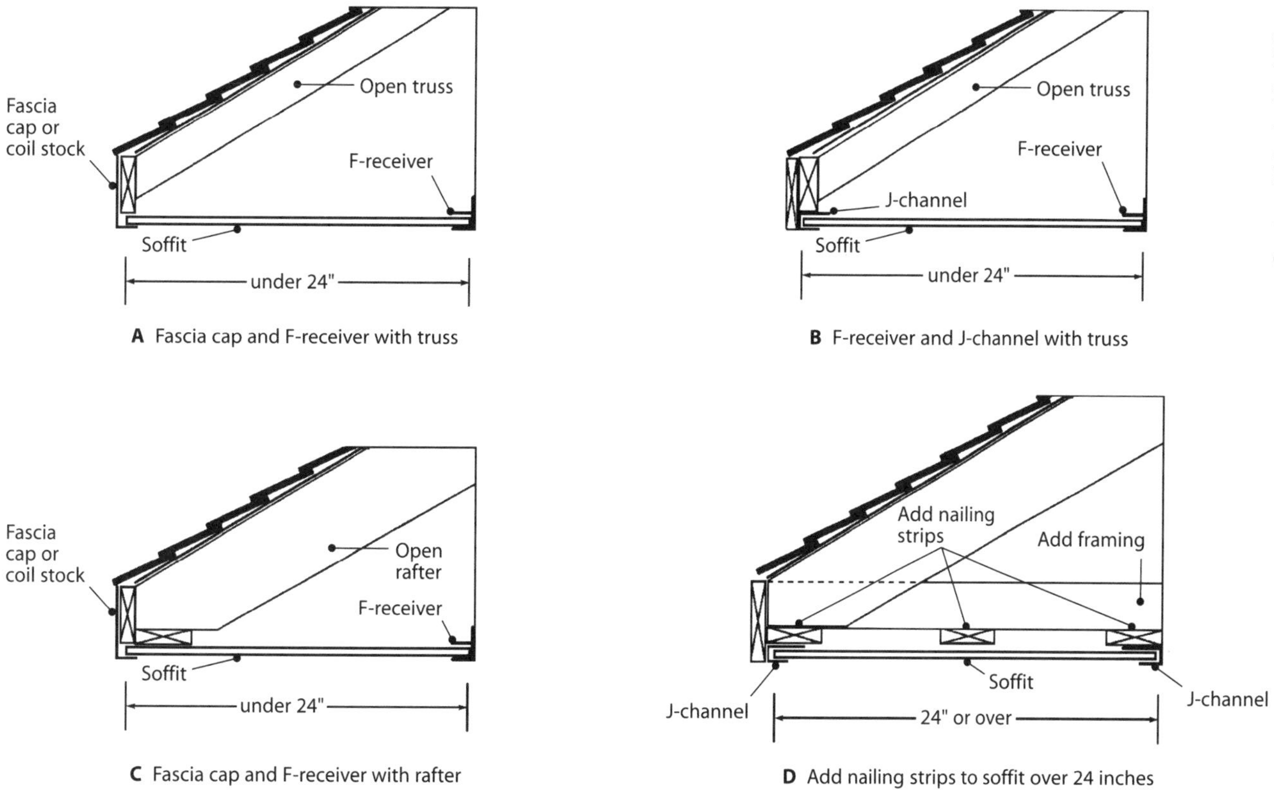

Courtesy: Wolverine Vinyl Siding

A Fascia cap and F-receiver with truss

B F-receiver and J-channel with truss

C Fascia cap and F-receiver with rafter

D Add nailing strips to soffit over 24 inches

Figure 6-23
Wolverine F and J Channels

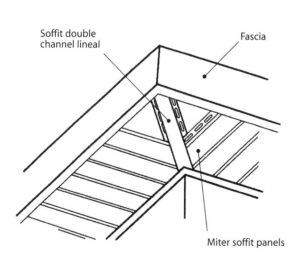

Courtesy: Wolverine Vinyl Siding

Figure 6-24
Soffit double channel lineal

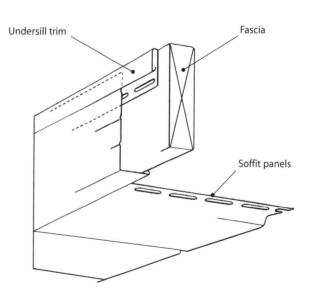

Courtesy: Wolverine Vinyl Siding

Figure 6-25
Vinyl fascia cap

maintain than wood, and will add value to a home. I've used the product and it's quite impressive once installed.

Because urethane is impervious to moisture, painted surfaces aren't subject to the chipping and peeling so common with wood. When the product leaves the factory, it's finished with two protective coatings and doesn't require any additional finish. If you do decide to paint it, use a good quality latex or oil-based paint. You don't need a primer coat. Another option is to stain the product. Once the stain has completely dried and cured, apply a high-quality exterior-grade clear top coat containing ultraviolet (UV) stabilizers.

One product I want to mention is the fishscale panel or *pediment*. It's one way to add traditional charm to a gable end or to dress up the front entrance. The panel is 17 inches high by 103 inches wide by $^3/_4$ inch deep. It's made in an interlocking system that features a half-lap joint. The panels interlock and overlap, end to end and top to bottom. You can easily apply this system in less time than it takes to apply individual cedar wood shakes, plus you don't need to worry about the product rotting or decaying.

Construction Notes — Here's a basic outline for installing the panel:

▌ First install the $^3/_4$- or $^7/_8$-inch J-channel along the top and/or sides of the wall area to accept the panel.

▌ Panels install from right to left. After measuring, cut them with a circular saw, as shown in Figure 6-26. (But personally, I'd secure that panel with clamps before making the cut.)

▌ Trial-fit the panel into the $^3/_4$-inch channel (Figure 6-27).

▌ Apply adhesive to the wall surface and to the lock panel and caulk the seam after joining the panels. Or you can caulk first and allow the caulk to ooze out when the two panels come together, then remove the excess caulk. In Figure 6-28, notice the urethane-based construction adhesive on the lock panel. The channel has been held down from the soffit and a filler board added, creating space for trim.

▌ Panels can be surface nailed or stapled. Be sure to fill with painted caulk. And finally, paint if desired. See Figure 6-29 for the finished product.

Courtesy: Style-Mark Inc.

Figure 6-26
Cutting Style-Mark decorative trim

Courtesy: Style-Mark Inc.

Figure 6-27
Fitting the panel into the $^3/_4$" channel

Courtesy: Style-Mark Inc.

Figure 6-28
Fitting the trim into the channel

Courtesy: Style-Mark Inc.

Figure 6-29
The finished job shows the trim on the soffit material

Manhours — This product can be installed by one person, but for projects high off the ground, a second person will speed production. For installing the J-channel, use 0.033 manhours per linear foot, and for installing the panel use 3.34 manhours per square. Depending on the number of cuts the project requires, you may want to add 25 percent to the 3.34 figure.

Masonry Products

Masonry products bring to a home a touch of nostalgia and class that's unmatched by any other product, with the possible exception of EIF systems (which I'll discuss next). Personally, I just like the look of brick and stone. When you mix it with redwood or cedar, it makes quite an impact on the overall design. You can often find used bricks salvaged from demolished commercial buildings. These bricks have character that's hard to reproduce. Not only do they look great in siding, but consider using them in interior walls, fireplaces, and outside barbecues.

Some companies produce traditional brick veneer that's so realistic you can't tell it's not full-sized brick. Installing a product like this reduces material and labor costs. One company has created a panelized brick system. The panel increases insulation values, holds the brick in place while the adhesive sets, and keeps the veneer brick straight for a great-looking grout line.

Even imitation stones are so close to a natural stone that you may find it impossible to tell them apart! I've used them in many projects, not just because of their realistic appearance but because they're lightweight and easy to handle. If you haven't had an opportunity to use masonry products, either real or synthetic, I encourage you to consider them. Visit your nearest salvage yard and let your imagination run wild!

Cultured Stone is the brand name for a man-made "stone" manufactured by Cultured Stone Corporation, a company that has offered a wide variety of stone and brick products for more than 30 years. The product was designed for use in nonstructural applications, such as a veneer facing on masonry, metal, or frame construction. This includes:

■ New construction, both exterior and interior walls

■ Remodeling and redecorating of existing walls

■ Fireplaces (around openings, hearth, exterior chimney or chase finishes)

■ Landscaping

They create flexible molds from natural stones of varying sizes and textures. Then they pour a mixture of portland cement, lightweight aggregates, and iron oxide pigments (for color) into the molds, then vibrate them. While still in the molds, the stone

backs are raked to provide a textured surface to assist in bonding to mortar. This process creates random (and realistic) sizes, shapes, and textures.

All Cultured Stone is covered by a 30-year limited warranty and, just like natural stone, it's noncombustible. Since it's lightweight and doesn't require additional foundations or supports, it's also less expensive to install than natural stone. Tight quality control ensures that product color and texture are maintained from lot to lot and year to year.

The product line of wall stones consists of more than 70 different colors and styles in 18 texture and shape families, all with matching 90-degree corners. In addition, they supply stepping stones, pavers, capstones, quoins, watertables, sills, and hearthstones. Also look for a line of complementary door and window molding (Figure 6-30). You can cut or shape the stone to conform to any arch radius. They have so many different products that you'll never run out of design ideas.

Construction Notes — Installing Cultured Stone is fairly simple, but as always, I recommend that you read the manufacturer's installation manual. They're packed full of ideas and techniques concerning their product. Here are the highlights:

■ A weather-resistant barrier must be used on all exterior and interior mortar applications except for those over masonry, concrete, or stucco.

■ During installation, place small stones next to large, place heavy-textured pieces next to smooth, and put thick stones next to thin. Also, mix stones from different boxes to create a desirable balance of individual stones on the finished project.

■ During hot or dry weather, moisten the back of each piece with a fine spray of water or a wet brush to prevent excessive moisture absorption from the mortar.

■ Over concrete, masonry or a scratch coat substrate, dampen the substrate surface area before applying mortar.

■ Protect the work site from temperatures below freezing, as mortar won't set up properly under such conditions.

Courtesy: Cultured Stone Corporation

Figure 6-30
Cultured Stone window molding

Courtesy: Cultured Stone Corporation

Figure 6-31
Installing Drystack Ledgestone on mortar scratch coat

Courtesy: Cultured Stone Corporation

Figure 6-32
Applying precast stone at corners

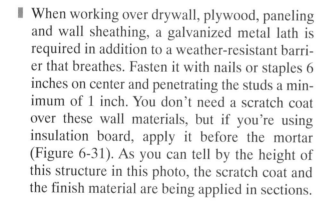

▌ When working over drywall, plywood, paneling and wall sheathing, a galvanized metal lath is required in addition to a weather-resistant barrier that breathes. Fasten it with nails or staples 6 inches on center and penetrating the studs a minimum of 1 inch. You don't need a scratch coat over these wall materials, but if you're using insulation board, apply it before the mortar (Figure 6-31). As you can tell by the height of this structure in this photo, the scratch coat and the finish material are being applied in sections.

▌ Don't spread mortar over more than a workable area (5 to 10 square feet) to prevent the mortar from "setting up" before the stone is applied.

▌ Normally you would work from the bottom up. But I found it easier and cleaner to work from the top down. That way you don't drip mortar on previously-applied stone.

▌ To ensure complete coverage between the mortar bed and the back surface of the stone, apply mortar to the entire back of the stone.

▌ When applying corners, alternate long and short legs in the opposite directions (Figure 6-32). When required, apply the corner pieces first.

▌ If additional mortar is required, use a grout bag to fill in joints.

▌ Once mortar joints have become firm or thumbprint dry, point them up with a wood stick or metal jointing tool. Rake out excess mortar, compact, and seal edges around stones.

Manhours — Because so many products are available, these labor estimates are based on three basic categories. The figures in parentheses represent an estimate of the amount of material that can be installed in an 8-hour day by a crew of two. The type of stone you choose and the complexity of the project can easily affect your manhours. Be sure to keep that in mind before using the figures below. These figures include mortar based on $1/2$-inch joints but don't include installing scaffolding.

▌ Prefitted types (160 square feet): 0.100 manhours per square foot

▌ Brick veneer (120 square feet): 0.133 manhours per square foot

▌ Random cast type (120 square feet): 0.133 manhours per square foot

Exterior Insulation and Finish Systems

I was first introduced to EIFS (exterior insulation and finish systems) a few years back when it was used on a friend's home. It's one of the most unique exterior systems on the market today. I was amazed how quickly the entire system went together, and how the system let the contractor create interesting architectural details (Figure 6-33). It can turn homes into stylish, elegant structures overnight.

Initially, EIFS were introduced as commercial cladding products, but now they're moving into the residential market, including geographic areas traditionally dominated by stucco, brick, and other traditional sidings. Multilayered EIF systems resemble stucco but provide levels of energy efficiency, moisture protection, and design freedom unmatched by other cladding products.

Some homeowners have experienced moisture problems with EIFS. Possibly due to incorrect installation, moisture became trapped between the insulation board and the substrate, resulting in damage from dry rot and mildew. Consequently, some states have placed restrictions on its use, or banned it entirely, and many stucco manufacturers will now only install the drainable version of the system (page 134).

The entire system is field-applied. Architectural shapes (quoins, keystones, arches) are prefabricated using insulation board and attached to the walls by skilled applicators. Then the exterior is completed by applying the base coat, reinforcing mesh, and finish coat over the entire insulation board surface.

EIF systems usually consist of five components (Figure 6-34):

1. An adhesive and/or mechanical attachment to attach the insulation board to the substrate or existing wall surface (usually plywood)

2. An insulation board to reduce heat flow through the wall

3. A base coat on the face of the insulation board that functions as the primary weather barrier

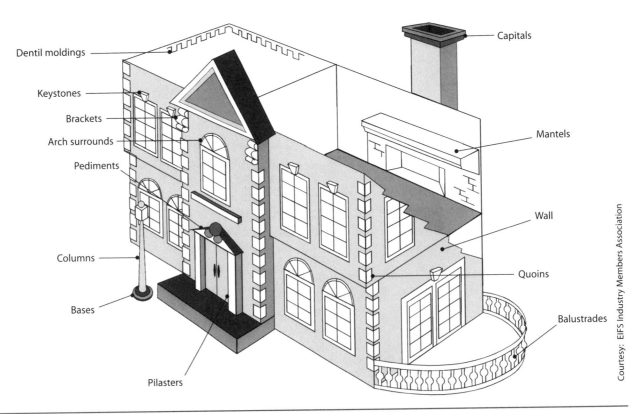

Labels: Dentil moldings, Keystones, Brackets, Arch surrounds, Pediments, Columns, Bases, Pilasters, Capitals, Mantels, Wall, Quoins, Balustrades

Courtesy: EIFS Industry Members Association

Figure 6-33
Architectural shapes created with EIFS

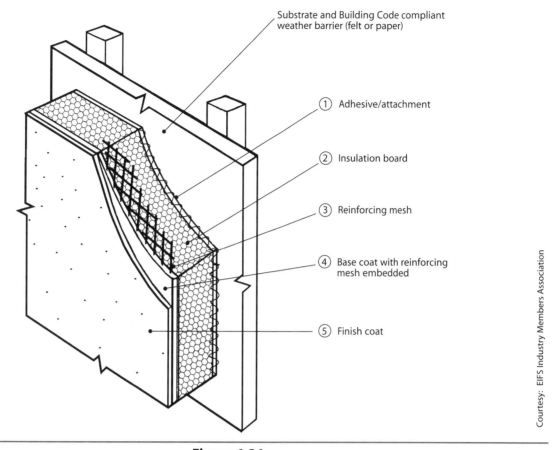

Substrate and Building Code compliant
weather barrier (felt or paper)

① Adhesive/attachment

② Insulation board

③ Reinforcing mesh

④ Base coat with reinforcing
mesh embedded

⑤ Finish coat

Courtesy: EIFS Industry Members Association

Figure 6-34
Components of an EIF system

4. A glass fiber-reinforced mesh fully embedded in the base coat for added strength and impact resistance

5. A finish coat to protect the entire system. It can be applied in a wide variety of colors and textures to create virtually any "look" imaginable, including stucco or natural stone. You can style an EIFS finish to create the special looks that characterize different regions of the U.S.

To help you better understand these systems, here's some information provided by the EIFS Industry Members Association (EIMA). It's a non-profit trade group representing professionals associated with the EIFS industry and manufacturers who offer such systems. They say that major advantages include design versatility (new shapes, colors, and textures in a durable, crack-resistant surface that requires only routine upkeep) and energy efficiency.

EIFS provide an unbroken layer of insulation which significantly reduces air infiltration. Typically, EIFS add between R-4 and R-5.6 (per inch of insulation board) to a home. Combined with standard wall cavity insulation, this extra layer can significantly improve energy savings.

EIFS are designed to perform in all regions, during all seasons. The overall system effectively insulates homes in both hot and cold climates. The insulation board reduces heat flow into and out of a home, to retain heat in cold temperatures and to keep heat out in hot temperatures.

There are two basic EIFS classes: PB and PM. The PB (polymer based) system, which has a base coat of varying thickness with a minimum base coat application of approximately $1/16$-inch (1.6 mm), is used for most homes and commercial buildings. It accounts for more than 95 percent of U.S. applications. In high-traffic areas, like the first floors of malls, schools, theme parks or some commercial buildings, architects often select the more impact-resistant PM (polymer modified) system. It features a base coat of a uniform thickness that ranges from $^1/_4$ inch (6 mm) to $^3/_8$ inch (9 mm).

EIFS are sometimes referred to generically as "synthetic stucco" because they're similar in appearance to stucco and both are installed by the plastering trade. But the similarity ends there. EIFS virtually eliminates the problems associated with stucco. The material is flexible yet durable, with less tendency to crack, fade, or show other signs of wear. They're installed in much less time (and with less disruption) than other exterior materials, so they save time and money and require only routine maintenance.

The expanded polystyrene (EPS) insulation board commonly used in EIFS is environmentally safe. It's been produced and used as thermal insulation for over three decades with no harmful health effects. Fire tests and experience in actual fires have shown that EIFS cause no significant increase in hazard.

Construction Notes — In most cases, you'll have a subcontractor installing an EIF system, but you need to understand the system and hire an experienced sub. After all, you're the one who's responsible to your customer. For the best possible results, insist that your applicators follow the three-part inspection checklist recently developed by EIMA. These guidelines are a key component of EIMA's education program for applicators and other construction professionals.

Here are some of the association's recommendations:

1. *Inspection* — Before starting any EIFS job, applicators should inspect the exterior wall surface for cracks or other damage, algae, chalk, dirt, dust, efflorescence (growth of salt crystals), oil stains, fungus, grease, mildew or other foreign substances. A moisture meter will determine if the surface is dry enough to accommodate EIFS. Carefully record any moisture damage, the location and width of all cracks, and the condition of all sheathing.

2. *Substrate preparation* — Remove any contaminants from the wall surface, replace weather-damaged sheathing and repair damaged or cracked surfaces before applying the system. To guard against moisture penetration, install expansion joints where dissimilar materials or construction meet, along floor lines in multi-level wood frame construction, where joints already exist in the substrate, and other areas where significant movement is likely.

3. *Backwrapping* — After applying insulation boards, adhere reinforcing mesh to the board surface. This mesh must wrap around any exposed edge of the insulation board. (This is called *backwrapping*.) It ensures that exposed edges of insulation boards will be adequately protected. Apply a minimum $2^1/_2$ inches of mesh to the base of the wall, using an appropriate adhesive. Allow excess mesh to hang down. Then apply the insulation board to the wall surface between the top and bottom mesh strips. Wrap any excess mesh and embed it on the face of the insulation board. At least $2^1/_2$ inches of excess mesh should wrap around the top and bottom edges of the board. Take care to prevent the adhesive from collecting on the mesh that wraps the insulation.

Backwrapping must be used wherever a continuous stretch of insulation board is interrupted by a door or window frame. Insulation for most EIFS jobs is provided by $^3/_4$- to 4-inch-thick EPS boards with a maximum size of 2×4 feet. Before adhering them to the wall, check the boards to ensure that they meet manufacturer's specifications. Typically, use a hot, sharp knife guided by a square to cut EPS board. Hold the knife at a low angle to ensure a clean slice. If available, you can use a circular saw, router, table saw, band saw or hot wire machine (see "Tools" in Chapter 2) to cut the board.

4. *Adhesive application* — Apply the adhesive to insulation board with a trowel, using a notched trowel or ribbon and dab methods. Hold the notched trowel at a minimum 30-degree angle to produce the correct ribbon size. Firmly press the trowel when forming ribbons to prevent excess adhesive from collecting between the ribbons. Keep the trowel clean to prevent adhesive buildup in the notches.

5. *EPS board application* — When applying the insulated board, it's important to begin from a level base line. Scrape excess adhesive from the edges of the boards before applying them to the wall to prevent the adhesive from creating unwanted "thermal bridges" between the boards. That can lead to future problems. Board ends should butt tightly together to prevent thermal breaks in the system. Fill any spaces wider than $1/16$ inch between the boards with slivers of EPS insulation. Also, it takes the right amount of pressure to ensure that the adhesive will grab. If you're in doubt, press firmly or tamp the boards to ensure a good bond. Request that applicators use a large wood block or board to ensure uniform pressure on the entire EPS board. Insulation boards should always be placed so that vertical and horizontal joints are staggered and bridge sheathing substrate joints.

6. *Rasping* — To ensure flat uniform board surfaces, each EPS board should be leveled and shaped using a rasping board. Rasp the entire surface of the boards, not just the edges, to avoid a wavy look to the wall. Once the boards are in place, protect them from sun and water damage.

7. *Base coat application* — Embed the mesh in the base coat until no mesh color is visible. Trowel from the center to the edges of the mesh to prevent possible wrinkles in the base coat. Any excess base coat should be troweled off the surface. Embed mesh in the base coat vertically or horizontally in 40-inch strips, with strip edges overlapping or butting together, depending on the manufacturer's recommendations. Install *butterflies* (small diagonal strips of mesh) at the sills and headers prior to application of field mesh for additional protection around doors and windows.

8. *Primer application* — Primer can provide a color base, improve resistance to weather, enhance the appearance of the finish coat, and prevent efflorescence in cement-based exterior walls. Apply the primer with either a brush or paint roller.

9. *Finish application* — As the general contractor, insist that your subs schedule finish applications so there are enough applicators on hand to complete an entire section of wall at one time. It's recommended that two people (three are better) be assigned to this part of the job, one to apply the finish and the other to float the finish to the desired texture. The finish should be mixed with a clean, rust-free mixer, with small amounts of clean water added as needed. It's imperative that the newly-applied base coat and primer be dry before the finish coat is applied. The finish should be applied in one continuous operation at a time when the surface is out of direct sunlight.

10. *Floating textures* — There are many EIFS finishes on the market, but the two most common types are pebbled and random textures. Pebbled texture is applied to the thickness of the pebbles in the finish and distributed evenly with a plastic trowel. Random texture is applied to the thickness of the largest stone in the finish, then scraped down until it's no thicker than the largest stone and floated to produce a random texture effect. But the finish shouldn't be allowed to set too long or "burn" marks will appear in the texture.

11. *Sealants* — Whenever the insulation system or EPS boards meet another material, the joint must be sealed. Typical areas of concern include door and window frames, decks, roofs, pipes, wires, meter boxes, and exterior faucets. Sealants installed between two surfaces stretch back and forth as those surfaces move. When applying sealants, it's important that they bond to only two surfaces, for instance the back-wrapped EPS board edge and a window frame. Sealants must *never* be bonded to a third surface (like the substrate) or the sealant can't stretch and may fail. To ensure the sealant joint is wide enough, push a "backer rod" (closed-cell polyethylene type) into the joint. This produces the desired joint size while providing a third surface to which caulk won't adhere. For adjoining surfaces that aren't sufficiently deep to accept a backer rod, you can substitute "bond breaker" tapes. Always leave a space between the EPS board and the adjoining material for this option.

EIMA offers some additional pointers to ensure the best possible results on an EIFS job. First, protect EIFS from dust, dirt, precipitation, freezing, and continuous high humidity during the installation. Also immediately cover the tops of walls with the final trim (or temporarily protect them) to prevent water infiltration behind the system. Install cap flashing as soon as possible after the finish coat has been applied. Finally, protect open joints from water intrusion during construction with backer rod or a temporary covering until they're permanently sealed.

Manhours — EIFS must be installed by manufacturer-certified applicators. If you're not certified and want to install this system, hire a certified subcontractor. However, to estimate costs, figure that a crew of three can field apply 400 square feet of exterior wall insulation and finish systems in an 8-hour day. These manhour estimates are for structures not over three stories high and don't include surface preparation, wall cost, or scaffolding:

- Adhesive mixture, 2 coats: 0.013 manhours per square foot

- Glass fiber mesh: 0.006 manhours per square foot

- Insulation board, 1 inch thick: 0.013 manhours per square foot

- Texture finish coat: 0.24 manhours per square foot

- Total with 1- to 4-inch insulation: 0.056 manhours per square foot

Drainable EIFS

Twelve manufacturers currently offer two residential EIF systems that allow moisture drainage (the manufacturers are listed in the Appendix, along with EIMA). This is in response to new building code requirements in North Carolina, as well as growing customer demand for a drainable system that will prevent the damage caused when moisture becomes trapped between the surface of the exterior plywood and the backside of the insulation board. This type of damage is most prevalent in areas that experience high humidity, such as the South.

On the surface, it's hard to tell the drainable EIF system from the original. The difference is in the design of the insulation board. In one system that provides drainage, there are grooves in the back of the insulation board (called *drainage board*). Before the board is attached to the substrate, a vinyl drainage track is fastened at the starting point of the substrate around the perimeter of the home. Then building paper is attached to overlap the drainage track. When completed, the drainage board slides into the track and is held in place by a corrosion-resistant screw with a specially-designed plastic washer.

The second drainable system, which accounts for 80 percent of the market, uses a drainage mat that is attached to the substrate over the weather barrier with $1/2$-inch staples. The insulation board doesn't need grooves because the drainage mat is thermally preformed, in a zigzag configuration (made of polymide). A corrugated V pattern, oriented vertically, allows the moisture to drain down and out. The mat is approximately $1/8$ inch thick and flexible.

Figure 6-35 shows many of the products used in the drainable EIFS. The drip edge Starter Trac is sloped and has weep holes to carry incidental moisture away from the wall cavity. An angled window drip edge used over windows and doors diverts moisture away from openings. You can also use it at the bottom of the assembly, but it must be backwrapped. The self-furred ULTRA-LATH can carry unwanted moisture down to and out through the Starter Trac. (ULTRA-LATH is installed in accordance with ASTM C1063.)

If you prefer to use a drainable system, you'll still have to flash, seal, and caulk around doors and windows, deck connections, and other moisture entry points. Detailing the installation properly will improve performance and success of the overall system. While drainable systems vary in design from manufacturer to manufacturer, they can all help reduce the risk of moisture damage to the wood sheathing behind the EIFS.

Depending on the design, installing an EIFS with drainage could increase your manhours by 15 to 20 percent, so don't forget to factor this in when using the system. Builders who want additional

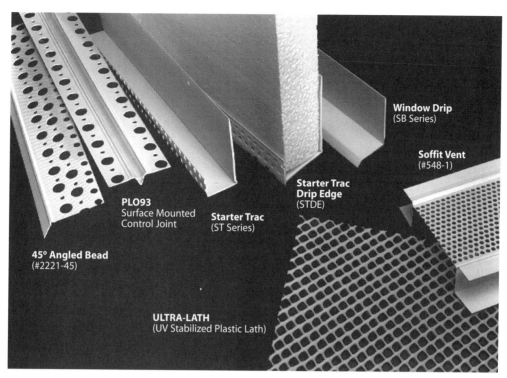

Figure 6-35
EIFS water management control components

information that will help their applicators achieve a trouble-free EIFS installation can call EIMA at 1-800-294-3462.

The information I've covered in this chapter should give you and your customer plenty of ideas. I've only touched on a few of the products on the market —- there's a wide variety out there to select from. Of course, before you can apply any siding, you need to install the doors and windows. In the next chapter I'll cover insulation for the interior and exterior walls. In Chapter 8 we'll cover doors, windows, and both interior and exterior moldings and trim. But first, let's look at how to maintain that new siding you've installed.

Siding Maintenance

The exterior of any building is its most visible and vulnerable part. Your goal is to make it attractive, but you also have to consider the maintenance requirements for your customer.

Mildew

Mildew is a big maintenance problem, and it's a special area of customer concern. It should also be a concern to you — especially if you're asked to paint a home that you suspect has mildew, and mildew can grow on any exterior surface. Here's some information about mildew from the American Hardboard Association that should be helpful to you and your customer.

Sometimes paint will have a dirty, streaked appearance due to mold growth or mildew. Mildew is a fungus growth on the painted surface, the result of airborne spores attaching to the surface. Paint and other organic surfaces may deteriorate and become permanently stained if mildew grows on them. The surface must be treated before repainting because painting over mildew won't control its growth. It'll just grow through the new paint.

Identification of mildew is the first step in its elimination. Because most mildew is black, it's frequently confused with dirt. One way to confirm the presence of mildew in the field is to apply a drop of 5 percent sodium hypochlorite solution (common household beach) to the stain. Mildew will usually bleach out in one or two minutes. Dirt won't.

You can remove mildew with one of the many commercial mildew washes specially formulated for this particular task. Your local paint dealer can usually recommend a good one. It's important that you follow the label instructions carefully and heed all precautionary warnings. An alternative method for removing mildew is to scrub the affected surface thoroughly with the following solution:

▌ $2/3$ cup trisodium phosphate

▌ $1/3$ cup detergent containing no ammonia (Tide brand or equivalent)

▌ 1 quart 5 percent sodium hypochlorite (Clorox brand or equivalent)

▌ 3 quarts warm water or enough to make one gallon

Wear rubber gloves and goggles when applying this solution. Scrub with a fairly soft brush, then rinse thoroughly with fresh water. Avoid splashing the solution on shrubbery or grass.

If conditions are right, new mildew growth can appear in a few days or weeks, so don't wait too long to paint the surface. Soon after the surface has been cleaned and dried, apply a paint recommended by the manufacturer as mildew-resistant. Supplemental mildewcide can be purchased at most paint stores to mix into the paint for additional control.

Other Stains

If a waxy or oily stain or discoloration persists after you've cleaned to remove surface dirt or mildew, you can usually remove it with hot soapy water. A hot detergent solution, applied by brush or with a steam cleaner, can remove the stain. (Steam cleaners operating at approximately 300 degrees F and a tip pressure of 200 to 400 psi are often successful. Consult the manufacturer's instructions and recommendations for a wax stripper or degreaser detergent.) Rinse with clear water until all traces of detergent are removed. It may take further detergent steam cleaning if the clear water beads up on the siding surface. A steam cleaner, rather than a cold water pressure spray device, is more effective at removing both dirt and waxy or oily accumulations with less possibility of damaging the surface.

Now on to that information on insulation I promised.

7

Insulation Materials and Radiant Heat

▪▪▪

Normally I don't get too excited about insulation. I don't suppose you do either. But it's a key factor in making a structure into a complete and comfortable energy-efficient home. Properly installed, insulation with the R-value required by the building code helps keep a home warm in the winter and cool in the summer. Of course, that helps keep those utility bills down. Perhaps that's why some customers *do* get excited about insulation.

What about housewrap? Does it increase the R-value? Is it better than the 15-pound felt that we're all used to? Building paper helped make a structure weathered-in (waterproof from wind-driven rain) before the siding was installed. It also increased the effectiveness of the insulation and reduced drafts from the outside. Air infiltration and moisture buildup are major problems common to many homes, especially older ones.

Today, modern housewraps act as a weather barrier, yet they're permeable to release moisture vapor trapped within wall cavities. These materials actually allow a home to breathe, letting moisture vapor escape before it leads to condensation and eventually mildew and rot. Housewrap still stops penetration by wind-driven rain, as well as air or wind penetration through seams, cracks, or directly through a wall cavity. Temperature changes within the walls mean more energy to heat homes because the walls are lower in temperature. And in warm climates (or during summer), warm air moving inside a wall cavity means that more energy is required to cool the house.

To create a high-performance insulating system, first apply caulk or foam sealant to fill all seams, cracks, openings around the top of the foundation, and around any water or electrical lines that pass through exterior walls, floor, and ceiling areas. Then install the insulation. This creates a tighter building envelope and prevents air and wind penetration. It may cut down air infiltration by 50 percent compared to homes where these sealants haven't been applied.

Insulation

You'll usually select an insulation product based on resistance to heat flow (R-value), resistance to air infiltration, cost, and availability. Other considerations include moisture and fire resistance and toxicity. Today, some manufacturers encapsulate insulation batts to control fiberglass irritation and improve indoor air quality by preventing small glass fibers from escaping during installation. One manufacturer has developed longer curled fibers that are easier to handle and virtually itch-free. I sure remember when the insulation was installed the same day the wallboard went up. Oh man, the itching!

Recycling also plays an increasingly important role in the production of insulation products. Some brands of fiberglass insulation contain recycled glass, and other insulation products contain newspapers.

And don't overlook mineral wool insulation made from furnace slag (waste from melting ores) from the steel industry or made directly from basalt (volcanic) rock.

Most likely, the type of project you're working on will dictate whether you use batts or loose-fill insulation. On existing homes, a blown-in product may be the only option. Some remodeling projects may require loose-fill, and for new construction, batts may be an easier solution — but don't rule out a combination of both. You'll also have to decide whether to hire a subcontractor who specializes in insulation. I used to handle the smaller jobs and hire a sub for larger projects. Now that I look back on it, I wonder why I didn't use a sub on all my jobs.

Customers today are better educated about building products and concerned about the high costs of unnecessary energy loss. That leaves you no choice but to become an expert on new products that can help customers feel comfortable in their homes. If not, you'll be left behind while potential customers seek out the more savvy professionals. On that note, let's learn about some insulating products that can improve the thermal efficiency of a structure.

Encapsulated

Johns Manville (formerly Schuller International) manufactures *ComfortTherm* encapsulated fiberglass insulation. It contains 25 to 40 percent recycled glass, mainly post-consumer glass from recycled household food and beverage containers. Their fiberglass process has been adapted to use mixed-color glass which generally can't be recycled into bottles. Each day approximately 30,000 tons of glass containers are thrown away, two-thirds of which end up in landfills. In just one year that adds up to over 7 million tons of nonbiodegradable material. However, each truckload of insulation produced by Johns Manville saves $1^{1}/_{2}$ tons of landfill. That's 150 million recycled glass bottles each year.

ComfortTherm fiberglass insulation was first introduced to the market in 1994. It's user-friendly because it's poly-wrapped to create less irritation and dust during handling. The white facing film serves as the vapor barrier required by most local building codes for exterior walls. And the facing is more tear-resistant than traditional kraft facing and twice as resistant to moisture penetration. Stapling tabs make installation easy and secure. Any moisture trapped in the wall cavity escapes through perforations in the gray polyethylene backing film.

The product is available in these R-values:

- R-11 for sound control use in interior walls and basements — 2 × 4 studs
- R-13 and R-15 high-performance —2 × 4 studs
- R-19 and R-21 high-performance —2 × 6 studs
- R-25 — 2 × 8 studs

Construction Notes — The precut feature makes installation fast and easy, something I believe most builders (myself included) enjoy. Even though the product is encapsulated, wear a long-sleeved shirt, long pants, gloves, eye protection, and most important of all, a good quality dust respirator. In standard-size cavities, fit the batt into the stud space with the printed surface facing you. Before stapling, ensure that the batt hits the top and bottom of the cavity. Staple on the left side (top to bottom), then pull the facing tight and staple the right side (top to bottom).

There are two stapling methods: face and inset. The manufacturer prefers face stapling because it provides a more continuous vapor barrier across the wall without compressing the insulation. On the other hand, inset stapling leaves the framing members clearly visible, providing an adhesive surface for wallboard installation. In either case, place the staples 6 to 8 inches apart. In steel-stud construction, the insulation is generally friction-fit into the cavities. If necessary, use spot adhesive or tape the flanges to hold the insulation in place until wallboard has been hung. When insulating a narrow cavity, cut the insulation 1 inch wider than the space to be filled.

You may want to consider using the tool shown in Figure 7-1. The Ziggy Fiberglass Insulation Cutter actually compresses fiberglass insulation thickness up to 12 inches as you cut through it. That lets the knife slice through both the insulation and the foil, kraft or poly-wrapper. The cutter keeps your hand and wrist from contacting the insulation, and its clean cut produces fewer airborne particles.

For electrical service, simply cut a horizontal slit on the back to allow the wire to pass through the insulation without overly compressing it. For

Figure 7-1
The Ziggy Fiberglass Insulation Cutter

electrical switches and outlet boxes, I recommend that you pull the insulation behind the boxes and then cut partially into the insulation along the box sides so it expands around the box. Leave the pipes exposed to the interior side of the home; if there's enough space, pull the insulation to the back side of the pipe. Treat floors and ceiling joists the same as you would for traditional wall installation. Here's the key thing to remember: Minimize compression of the insulation.

Manhours — A crew of one can handle the installation, but consider using a crew of two to speed the process. Larger framing member spans make the work go faster than smaller ones, so I've given different manhour estimates for 16 and 24 inches on center. You can expect rolled insulation to take 40 to 50 percent longer to install than precut batts due to the cutting required.

To install batts, including wall and ceiling applications, here are my manhour recommendations:

1. Framing members 16 inches on center

   ▮ 2 × 4 — 0.007 manhours per square foot

   ▮ 2 × 6 — 0.007 manhours per square foot

   ▮ 2 × 8 — 0.008 manhours per square foot

2. Framing members 24 inches on center

   ▮ 2 × 4 — 0.004 manhours per square foot

   ▮ 2 × 6 — 0.005 manhours per square foot

   ▮ 2 × 8 — 0.005 manhours per square foot

For crawl space installation for 2 × 4s to 2 × 8s, figure 0.020 manhours per square foot. For wire rods (one per square foot to support the batts), add 0.001 manhours per square foot.

Itch-free

It always seems to work out that insulation and wallboard are hung the same day. And with all the fiberglass particles and wallboard dust floating around, itching is a nuisance that's hard to avoid. Whether you were the one installing the insulation or not, the particles linger. Before the day is out, you feel like an old dog with fleas. Perhaps that's why Owens Corning invented MIRAFLEX — the newest form of glass fiber insulation introduced in nearly 60 years. Unlike traditional straight glass fibers, theirs is composed of two fibers, fused together into a single filament and randomly twisted. That makes MIRAFLEX fibers flexible, soft-to-the-touch, and virtually itch-free (Figure 7-2). In fact, they feel like cotton.

It's now sold for home attic installation under the brand name PINK*PLUS*. It comes in ultra-compact rolls that are half the size of conventional rolls of insulation but cover the same square footage (Figure 7-3). This makes it easier to handle in crawl spaces and attic areas. Unpacked, the insulation reverts to its original form. This product was specially designed for projects that don't require a vapor barrier. Perforations in the poly-wrap release any moisture captured within the insulation.

Conventional PINK*PLUS* insulation uses recycled contents. Owens Corning has recycled more than 5.2 billion pounds of glass in the last ten years. That's quite an accomplishment!

PINK*PLUS* R-25 is an $8\frac{1}{2}$-inch thick insulation featuring MIRAFLEX fiber made from virgin material, sold for home attic installation. You can get it in 25-foot rolls and in cavity widths of 16 and 24 inches.

Construction Notes — Because of MIRAFLEX's unique characteristics, it expands and conforms to attic irregularities when it's unrolled. When

A A single filament

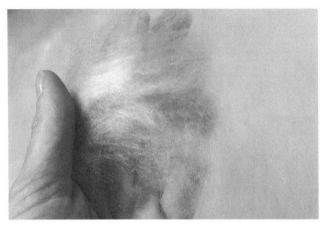

B It's soft and itch-free

Courtesy: Owens Corning

Figure 7-2
MIRAFLEX glass fiber insulation

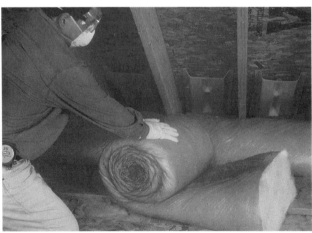

Courtesy: Owens Corning

Figure 7-3
Poly-wrap makes installing MIRAFLEX easier

installing the product, follow all the safety measures I've described for other glass fiber insulation.

Manhours — To calculate manhours for 2 × 10 framing members 16 inches on center, use 0.008 manhours per square foot. For 2 × 10 framing members 24 inches on center, use 0.005 manhours per square foot. For installation in floor joists in a crawl space area, use 0.020 manhours per square foot. Add 0.001 manhours per square foot for the wire rods (one per square foot to support the batts).

Rockwool

While mineral wool is more common in Canada and Europe than in the U.S., it has many qualities to recommend it. Mineral wool is a general category that contains insulation products like *slagwool*, made from waste by-products from steel production, and rockwool, made by melting predominantly basalt rock and some recycled blast furnace slags.

Rock fibers were commercialized after lava fibers, called *angel hair*, were noticed in Hawaii. European technology has improved insulation qualities and manufacturing economics to the point that rockwool products offer distinct benefits and economies compared to glass fiber insulation:

▌ superior fire rating

▌ high service temperature

▌ excellent sound absorbency

Fibrex Inc. manufactures a range of rockwool and slagwool products. I want to bring a couple of them to your attention:

▌ Sound Attenuation Fire Batt Insulation (SAFB) is available in 16-inch and 24-inch by 48-inch batts. For steel stud construction, these full widths fill the entire stud cavity, yielding excellent acoustical, thermal, and fire resistance properties. They're made from basaltic rock that has a melting point in excess of 2000° F. The batts are available in several thicknesses (1$1/2$, 2, 3, and 4 inches). The R-value is 4.0 per inch of insulation thickness. Preformed bonded fibers make the batts self-supporting and semi-rigid. To give you an idea of its weight, a 3-inch batt is 4 pounds. And yes, you can cut it with a utility knife.

▌ FBX 1240 Industrial Board offers the same characteristics as SAFB. Industrial board is available in 24-inch by 48-inch pieces, thicknesses of $1^1/_2$, 2, 3, and $3^1/_2$ inches, and weighs 6 pounds for a 3-inch board.

The major difference between the two insulation products is that 1240 board is designed for its insulation properties and durability. The SAFB, on the other hand, is designed to give superior sound absorption at lower weight. Both products install about like glass fiber insulation. However, their biggest advantage (besides great sound reduction) is that they're friction fitted. Just think — no staples to fuss over. They can even support themselves in a floor joist installation without the use of wire rods.

Manhours — A one-man crew can handle these products. They are itchy, so be sure to wear a long-sleeved shirt, long pants, and a dust mask. For installation in wood studs (regardless of thickness), use 0.003 manhours per square foot. When installing in steel studs, use 0.004 manhours per square foot.

Insulation Board

Tenneco Building Products (formerly Amoco Foam Products Company) manufactures AMOFOAM. It's a closed-cell, extruded polystyrene insulation board with high-density skin surfaces. Unlike some insulation materials, it doesn't contain formaldehyde or CFCs.

AMOFOAM is used in a wide variety of new and retrofit construction applications, including:

▌ slab-on-grade

▌ perimeter foundations

▌ crawl spaces

▌ cavity walls

▌ masonry walls

▌ exterior sheathing

▌ EIFS

▌ siding underlayment

▌ commercial roofing

These products provide long-lasting thermal protection with excellent moisture and soil chemical resistance, making them ideal for below-grade applications. They also feature good dimensional stability, which eliminates the concern for warping, shrinking and swelling.

As with all commercially-available foam plastic insulation, AMOFOAM is combustible and shouldn't be exposed to flame or other ignition sources. It contains a flame retardant additive intended to inhibit a small source fire, but may constitute a fire hazard if improperly used. Therefore, make sure you follow your local building code when using foam plastic insulation products. Before installing them, obtain and read the manufacturer's installation instructions.

The insulation board is also available with a film laminate on both sides with overlapping "shiplap" edges (AMOFOAM-SLX) or with square edges (AMOFOAM-CMX). These products are designed for exterior wall sheathing in new construction where they're installed directly over the framing members. The film laminate lets them resist handling and job site damage (Figure 7-4). The overlapping

Courtesy: Tenneco Building Products

Figure 7-4

Cutting AMOFOAM-SLX rigid insulation with a utility knife

edges of AMOFOAM-SLX provide added protection against air leakage at board edges, enhancing the overall thermal performance of the wall system. The products are available in board sizes of 4 × 8 feet and 4 × 9 feet with R-values of R-3.0 ($^1/_2$ inch), R-3.8 ($^3/_4$ inch), and R-5.0 (1 inch).

Construction Notes — To learn about proper installation of AMOFOAM Insulation Board for various applications, ask Tenneco Building Products for their *Installation Brochure*. It suggests that you use galvanized nails or staples to attach the insulation board in most applications. Fasteners must be long enough to penetrate the framing members or substrate a minimum of $^1/_2$ inch. Typically, you'll space the fasteners 6 inches on center along the perimeter of the wall and 12 inches on center at intermediate locations on framing members. If you use staples, the crown of the staple should be parallel with the framing member. Don't overdrive nail heads and staples as this will damage the insulation.

Finally, install the siding material in accordance with the manufacturer's installation instructions. In all exterior wall applications, install a vapor retarder material like polyethylene sheet or kraft-faced batt insulation on the interior side of the framing members to prevent moisture migration into the wall cavity.

Manhours — This product can be easily installed by one person, but a crew of two speeds production. For all three thicknesses of the 4 × 8-foot sheet, figure 0.010 manhours per square foot. Add 0.001 manhours per square foot for 4 × 9-foot sheets.

Blue Board

One way to dump the "itchy feeling blues" when finishing basements is to install STYROFOAM brand insulation from The Dow Chemical Company. The product, introduced in the 1940s, was the original extruded polystyrene foam. Since that time, it has found its way into more than 1 million homes nationwide. Its blue color makes it clearly identifiable, and it's CFC-free.

The insulation is composed of tiny cells with no gaps between them, so it resists moisture penetration. This is essential to retaining long-term R-value and it's the reason why this product carries a 15-year R-value warranty. I've been told this insulation will retain its thermal efficiency even when stored outdoors or buried in wet earth.

The specific product I want to introduce is WALL-MATE. This cleverly-designed product is an insulation system for basement walls. Not only can it help save time, it also eliminates the itch. It features a 2-inch thick 2 × 8-foot insulation panel that carries an R-value of 10.0. Also available is a 1$^1/_2$-inch thick panel with an R-value of 7.5. The panel edges are slotted at 2-foot intervals for securing to basement walls with 1 × 3 furring strips and masonry screws. If 1 × 3s aren't available in your area, you may need to rip 1 × 4s. The system is more efficient, takes less time to install, and consumes less living space than traditional 2 × 4 framing, fibrous insulation, and wallboard.

Construction Notes — The product is easy to cut using a score-and-snap method. If you do the work yourself, apply a couple of spots of adhesive to the backside to hold the panel in place while you prepare the next panel. After the panels are in place, locate all outlet boxes and electrical runs. Use a hot knife (see "Tools" in Chapter 2) to make all the cuts. (Be sure to have plenty of ventilation when using the hot knife.) Make each cut deep enough so when the wire is in the run, there'll be a 1$^1/_4$-inch clearance from the face of the wire to the face of the insulation. When making horizontal runs, leave adequate clear space behind the furring strips after they're installed. You're really preparing for the electrician, so check with your electrician first to make sure box placement meets code requirements.

Now you're ready to install furring strips. Before you do, there are a couple of things to check for. Make sure that corners have slots to accept the furring strips. Fasten the outside corner furring strips — the same thickness as the insulation board — to the concrete wall. Then attach a second one to the first with wood screws. Both insulation sides butt against the furring strips.

Around doors and windows, slots are needed for furring strips. This is important because the furring strips provide a nailer you use to attach the casing. To install the furring, you'll need a one-piece drill/drive unit which lets you drill through the furring strips and into the concrete. The same tool will also allow you to secure a self-tapping screw specially designed for this type of application. Install five screws, one every 24 inches. When everything is in place (Figure 7-5), install foam insulation around exterior door and window frames (and any other objects installed through the concrete wall) to prevent any outside air from entering. Once the electrical is in place, you're ready to hang wallboard.

Manhours — After studying this system and thinking back over all the basements I've framed and insulated, I believe it is possible to cut your labor time in half. Of course, this is based on a smooth foundation. With a rock foundation, you'll still have to rely on conventional wood framing. The first time around will be a learning experience with this system, and it's important to keep this in mind when bidding. These manhours take that into account:

▌ To install 2 inches of insulation using spot adhesive and to cut in electrical chases, boxes, and furring strip slots, figure 0.033 manhours per square foot.

▌ To install furring strips, figure at least 20 minutes (.333 manhour) per 8 feet for five holes and screws.

Blow-In-Blanket

Ark Seal International has created a Blow-In-Blanket System that's worth checking into. It mixes white blown fiberglass insulation with a thin coating of a safe latex adhesive. They blow the mixture into the cavity behind a fiber netting attached to the face of framing members (Figure 7-6). This process helps to achieve higher R-values, provides a uniform density, eliminates settling and shifting, and fills gaps, voids, and seams. Tests have shown that a 4 percent void area in R-19 insulation causes a 50 percent increase in heat loss. This system eliminates voids and there's no settling. You can achieve the following R-values:

▌ $3^1/_2$ inches — R-15

▌ $5^1/_2$ inches — R-23

▌ $7^1/_4$ inches — R-30

▌ $9^1/_4$ inches — R-38

▌ $11^1/_4$ inches — R-47

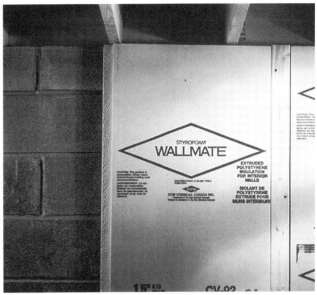

Courtesy: The Dow Chemical Company

Figure 7-5
Installing WALLMATE over a concrete block wall

Courtesy: Ark Seal International, Inc.

Figure 7-6
Blow-In-Blanket fits tightly around electrical boxes

This is something to think about when installing conventional batting in nonstandard cavities or in irregular-shaped framed areas. If you've ever cut insulation batts to fit those areas, you know that no matter how careful you are, you still have gaps. This system's seamless blanket of insulation automatically custom-fits itself to any shape and size cavity. It guarantees a uniform R-value throughout the entire cavity and controls air infiltration. Other difficult areas are around heating, plumbing, and electrical. In addition to tightly filling around such objects, Blow-In-Blanket also helps to muffle sounds.

This system can only be installed by qualified certified contractors using Ark-Seal equipment. The manufacturer tells me that it takes an experienced installer approximately 30 seconds to fill a standard cavity (2 × 6, 8 feet high, 16 inches on center). If you discover there are no subs in your area that handle this product, you might want to contact the company and check into dealership opportunities. It could create a second income for your company.

Cellulose

GreenStone Industries, Inc., a Louisiana-Pacific company, manufactures Cocoon brand cellulose insulation made from 100 percent recycled waste paper treated with a fire retardant. It's manufactured without any asbestos, fiberglass, or formaldehyde. It doesn't cause itching and requires no special handling or labeling to minimize health risks. Using a dust mask is recommended but not required during installation (Figure 7-7).

Cocoon insulation can be installed in wall cavities and attics of both residential and commercial structures, including flat and cathedral ceilings in most climates. If you want to increase the R-value in an attic, for instance, an insulation subcontractor can add Cocoon insulation over existing fiberglass insulation.

The insulation can be blown in dry on horizontal surfaces like attics and crawl spaces. Or it can be professionally installed using a spray method in wall cavities in new construction. Blown or sprayed into attics and wall cavities, it creates a continuous blanket of protection without the gaps, voids and compression typical of batt insulation. This helps to achieve a tighter seal around irregular objects such as wiring, plumbing, and framing materials in attics and walls.

As with any insulation material, don't install insulation over heat producing devices like recessed light fixtures, furnace flues, heating vents or chimneys which penetrate into or through the attic. Use a barrier with at least 3 inches of clearance to allow for proper air flow. You should always check your local building code requirements. For recessed lights, construct a can, open at both ends and high enough to stick above the fixture. You can make this can from flashing material, using pop rivets instead of duct tape on the seam.

Proper ventilation in unheated attic spaces is important for the proper maintenance of any insulation. Soffit or eave vents in conjunction with roof or ridge vents provide an ideal system. The standard guideline is 1 square foot of unobstructed ventilation for every 150 square feet of ceiling area without a vapor retarder (or with electric heat), and 1 square foot for every 300 square feet of ceiling with a vapor retarder (or with gas heat). To avoid blocking soffit vents, the manufacturer recommends using vent chutes or baffles when installing insulation. GreenStone distributes both barriers and vent chutes to professional installers.

You can obtain R-values ranging from R-11 to R-50. For example, to get an R-19 installation for 2 × 6 framing 16 inches on center, it would take 40.1

Courtesy: GreenStone Industries, Inc.; a Louisiana-Pacific Company

Figure 7-7
Cocoon cellulose insulation
made from recycled waste paper

bags to cover 1,000 square feet, using their 25 square foot bags. If you use their 30 pound bags, it would take 20 bags for the same amount of coverage. This is based on horizontal applications and on settled density.

To achieve specific R-values for wall cavities using 30-pound bags (39.6 square feet for 2 × 4 and 25.2 square feet for 2 × 6), allow the following:

▌ R-13.5 — 2 × 4s 16 inches on center by 92.5 inches

14.5" × 92.5" = 1341" ÷ 144" = 9.3 SF

39.6 SF ÷ 9.3 SF = 4.25 cavities filled

▌ R-13.5 — 2 × 4s 24 inches on center by 92 .5 inches

22.5" × 92.5" = 2081" ÷ 144" = 14.5 SF

39.6 SF ÷ 14.5 SF = 2.7 cavities filled

▌ R-21 — 2 × 6s 16 inches on center by 92.5 inches

25.2 SF ÷ 9.3 SF = 2.7 cavities filled

▌ R-21 — 2 × 6s 24 inches on center by 92.5 inches

25.2 SF ÷ 14.5 SF = 1.7 cavities filled

If you're a professional installer, GreenStone can provide equipment and training for both attic and wall applications. Use these manhour estimates for installing Cocoon insulation over ceiling joists (they don't include equipment cost):

▌ R-13 and R-19 — 0.010 manhours per square foot

▌ R-30 — 0.014 manhours per square foot

Housewrap

How do you stop outside air from entering a wall cavity? Probably the most efficient way is to create an envelope around the home. The best way to do that is to install housewrap, because even the most thoroughly-insulated home can suffer thermal inefficiencies. Tenneco Building Products makes AMOWRAP which, when properly installed, seals up sheathing seams and cracks around doors and windows to stop drafts. This product not only blocks

drafts but it allows moisture vapor to pass through, helping to eliminate potentially harmful condensation buildup in wall cavities.

Other features include a built-in UV-stabilizer that allows the wrap to be exposed for up to 12 months in direct sunlight without breaking down or adversely affecting its energy-saving performance. (But it's still not intended for use as a roofing paper or vapor retarder.) Its special cross-woven plastic-coating design gives it superior toughness, even after being punctured by staples or nails, and it holds up under high winds during construction. The signature green color cuts down on sun glare, making it easier on your eyes. Because it's translucent, it's easier to find sheathing and framing members. You can also use it behind brick, stone, masonry, and concrete veneers. And here's something to consider to get more bang for your buck: The company offers a custom print option on the housewrap. That's an excellent way to advertise your company and get your name out in front of potential customers.

Construction Notes — AMOWRAP is available in three sizes: 9 × 100 feet, 4.5 × 195 feet, and 9 × 195 feet. When installing it, follow the manufacturer's recommendations:

▌ Install directly to studs or over structural or nonstructural sheathing.

▌ Begin with the print side out. Align the bottom edge of the roll with the base of the wall, and overlap a corner by 2 to 3 feet, folding several inches of the material under itself. Fasten securely using staples or large head nails.

▌ Wrap the entire building, including door and window openings. When covering nonstructural sheathing, such as foam insulation boards, nail through sheathing and into studs.

▌ When installing a 9-foot roll, place a minimum of three fasteners — at the top, middle and bottom of every stud. To install their 4.5-foot roll, place a minimum of two fasteners, one at the top and one at the bottom on every other stud. Use more fasteners around door and window openings.

▌ At door and window openings, cut an "X" through the housewrap from corner to corner. Pull the four flaps to the interior and fasten to a framing member.

Courtesy: Tenneco Building Products

Figure 7-8
AMOWRAP housewrap needs to overlap 6 inches

▌ When the end of a roll is reached, fold the edge under itself and fasten to the nearest framing member.

▌ If doors and windows are already in place, install as close as possible and seal edges using "sheathing tape" or caulk.

▌ To maximize reduction of air infiltration, overlap at least 6 inches on vertical and horizontal seams (Figure 7-8).

Manhours — If you plan to install this product all by yourself, here's my advice: Don't. You'll need that second pair of hands, so figure on a crew of two or more to speed production. Use 0.006 manhours per square foot using 6-inch overlaps.

Radiant Heat

Besides the insulation, another area that you need to consider for maximum comfort is the floor. Adding insulated board under a concrete slab or polyethylene film on the ground in a crawl space helps to keep the floors warm. But that may not be enough. When it's not, one remedy you can consider is installing radiant heat. That can add comfort to any floor, whether it's covered with resilient flooring, ceramic tile, carpet, hardwood, or even concrete. You can install it on remodeling projects as well as during new construction. Some companies offer an entire package that includes electric cables and a liquid concrete mix.

Radiant heat offers another plus: no air movement, so no drafts. And since you're in the business of selling, consider this another building system that you should bring to the attention of your customers. Don't you think they'd enjoy the comfort of a warm floor?

As an alternative, radiant heat is a real gem because it doesn't involve combustion nor does it blow dust around, so it doesn't aggravate individuals with allergies or asthma. This type of heat keeps temperatures steady and eliminates the cold and hot spots associated with conventional heat. Because the heat is distributed throughout the entire floor, your customers will feel warm at lower temperatures than they're accustomed to with conventional heat.

Step Warmfloor

Electro Plastics has a system, Step Warmfloor, that's unique in design. It's a thin ($^3/_{64}$ inch) conductive plastic mat (a polymer heating element) available in 12-inch by 100-foot rolls. It's preglued (two-sided pressure-sensitive adhesive) and electrically insulated. Installation is easy. Just cut it to the appropriate length with scissors, remove the release liner, and stick the element directly on any stable, clean and dry substrate, wire to a low-voltage power supply and install the finish floor. There's no ductwork to fuss over. If you look carefully at the base of the right-hand wall in Figure 7-9, you can see the copper wires exposed and ready for connection to the black and white wires.

Here are some of the advantages of the system:

▌ You can place it closer to the surface than other floor systems currently on the market.

▌ It reacts immediately to external temperature changes.

▌ All the electrical power is directly transferred to heat — it's 100 percent efficient!

It's self-restricted, meaning it only draws the energy required to maintain the ideal temperature. When the ambient temperature rises, electrical resistance increases and the electric consumption decreases. For this reason, the elements can never overheat.

The heating elements are installed at foot level to provide ideal heat distribution.

The even distribution of heat on the floor reduces heat loss by infiltration.

A continuous low-temperature heating system is more efficient than an on/off heating system.

It requires no maintenance.

Construction Notes — I couldn't possibly explain all the applications for all the different types of floor coverings. Instead, I encourage you to get a copy of the company's handbook so you can see your options and plan appropriately with your customer. There are a few steps to consider before installation:

1. Determine the number of watts per square foot needed for a room. You can figure 3 to 5 watts per square foot.

Figure 7-9
Step Warmfloor ready for
connection of the electrical wires

2. How do you plan to install the wires — in the floor joist, wall studs, or on the surface in the baseboard area?

3. Do you plan to attach the wires to the elements or have an electrician do it?

When installing the product, consider the following:

1. Avoid direct contact between the elements and any conductive material. Keep a distance of 1 inch between elements and at least 2 inches from plumbing fixtures.

2. Elements must be placed in open spaces and not underneath fixed fittings such as bookshelves or cupboards.

3. The maximum run of any element is 23 feet.

4. Connections from wires to elements should be protected with heat shrink tube.

5. Elements must lay perfectly flat. Stretch into place by pulling away the underliner. If necessary, use a hot-glue gun to help fix them firmly down.

Manhours — A 100-foot roll will cover a 200 square foot room. I'd estimate the approximate time for installation in that 200 square foot room at 3 hours — but only after the third time around. For the first installation, increase that by at least 50 percent. On top of this, add 0.50 manhours for an extra person to help with pulling the liner of the elements. You may have to allow at least an hour for prework before installing the element.

INFLOOR

I was first introduced to INFLOOR Heating Systems in the early '90s. It was formerly sold under the company name of Gyp-Crete, now called Maxxon Corporation. This company has provided a variety of underlayment products for over 20 years and has 17 years in the radiant heating business. Maxxon offers both hot water and electric heating, the Warm Floor Kit for area warming, as well as systems for melting snow and ice.

Their INFLOOR Warm Floor Kit (WFK), my main focus, provides a way to warm tile or marble floors with radiant heat. The WFK is intended to be installed in a thin mortar bed or a thin layer of

Therma-Floor, roughly ³/₄ inch thick. Other INFLOOR systems can be used to warm floors covered with wood or resilient flooring, provided the underlayment over the electric cables or hot-water tubing is 1¹/₄ inch thick. The thicker underlayment creates a thermal mass when the finish floor covering is a type that doesn't hold the heat.

The WFK uses low-wattage heating cables attached to the concrete or wood subfloor and covered with Therma-Floor. That's a pourable gypsum floor underlayment. As an alternative, a mortar bed can cover the cables. The cables warm the underlayment and floor covering. A wall-mounted thermostat connected to a heat sensor inside the underlayment regulates floor temperature.

This system can be installed in both new construction or remodeling projects. It's not designed to be the primary source of heat in the rooms in which it's installed but rather a supplement — to take the chill off. Your customer will really appreciate stepping onto a warm tile floor after that morning shower!

The INFLOOR system doesn't waste energy trying to warm large volumes of air, like a forced air system. There are no drafts or hot-air surges so the heat is uniform, with very little temperature difference between the floor and the ceiling. Utility bills for a home covered by radiant heat average 15 to 30 percent less than an identical home using a forced-air furnace. Another advantage is that it uses no registers or cold-air returns to circulate dust or allergens. With health-conscious customers, or the allergy-prone, this feature could be a real selling point.

If your customer is looking for a primary radiant heating system or if you plan to use floor covering products other than what we've discussed — or for residential and commercial heating or snowmelt systems — contact Maxxon Corporation at the address given in the Appendix.

Construction Notes — This information is based on the INFLOOR WFK 120-volt electric system, primarily designed for use in floors. Don't use the heating cable for any other type of application without first contacting the manufacturer for advice. Maxxon offers WFKs in three sizes, each designed to cover a specific range of floor area. Don't alter the length of any heating cable to make it fit a floor area larger or smaller than the recommended range for that cable. Also, the heating cable must not touch, cross, or overlap itself at any point. This could cause the cable to overheat, possibly requiring cable replacement.

To create a warmer floor, especially over a crawl space, insulate the floor joists and install a vapor barrier. This will reduce energy consumption and improve overall system performance. You should contact the manufacturer for complete instructions on how to install it, but keep the following in mind:

1. Select an appropriate location for the 4-inch square electrical outlet (metal) box. Install the box 60 inches off the floor on an interior wall.

2. Below the outlet box, drill ³/₄-inch holes halfway into the side of the bottom plate. Then drill one hole in the top of the plate to accommodate the bending radius of the cable and two holes for the thermostat (Figure 7-10). On the thermostat hole, be sure to chisel off the points in the drilled hole.

3. Remove knockouts in the outlet box and install bushings. Before hanging wallboard, insert pull-cords through the knockouts and the holes previously drilled in the plate. Leave the cords long enough to tie to the cables and thermostat tube. This cord lets you pull the components up into the outlet box after the wall has been finished (Figure 7-11). Don't forget to tie off the cord in the outlet. You don't want the cord to fall through the hole after the wallboard is hung!

4. Mark off the perimeter of the area to be heated (maintaining a 6-inch border between walls, cabinets and fixtures and the heated area). Unwind the cable to expose the "cold" lead so you can tie it to the cord and pull it up into the outlet box (Figure 7-12). Pull the wire so the splice can be located within 12 inches of the guard plate. The factory splice between the "cold" lead and the heating cable must be located on the subfloor.

5. Before running the first length of cable along the border to the opposite side of the room, install a plastic clip on the cable as it exits the hole in the plate (Figure 7-13). When you're installing it directly on concrete, you can glue or nail the clips to the concrete.

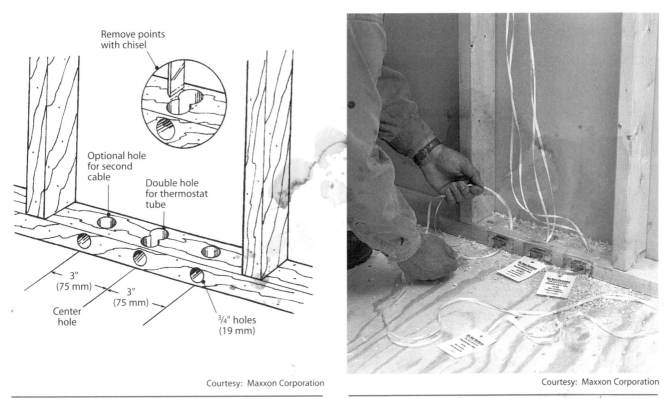

Remove points with chisel

Optional hole for second cable

Double hole for thermostat tube

3" (75 mm)

3" (75 mm)

Center hole

¾" holes (19 mm)

Courtesy: Maxxon Corporation

Figure 7-10
Drill ¾-inch holes for cables and thermostat tube

Courtesy: Maxxon Corporation

Figure 7-11
Insert pull-cords for the cable and tube

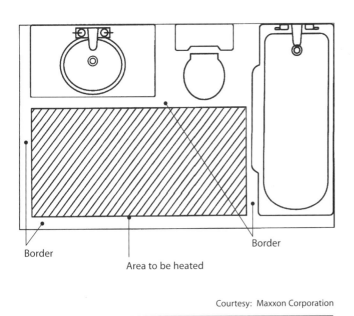

Border

Border

Area to be heated

Courtesy: Maxxon Corporation

Figure 7-12
Don't install heating cables under fixtures

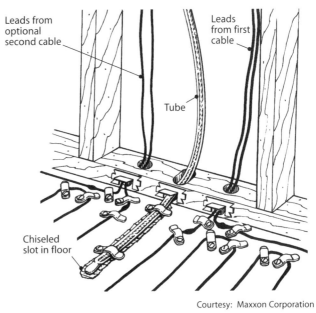

Leads from optional second cable

Leads from first cable

Tube

Chiseled slot in floor

Courtesy: Maxxon Corporation

Figure 7-13
Plastic clips hold the cables

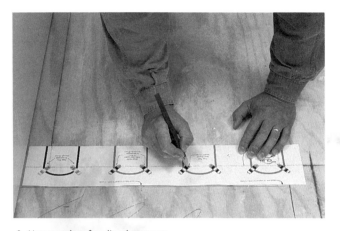

A Use template for clip placement

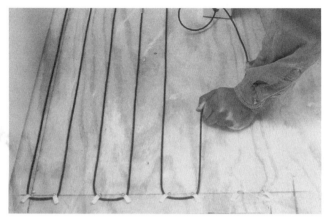

B Weave cable under clips

Courtesy: Maxxon Corporation

Figure 7-14
Installing the cables

6. Now, on the opposite side of the room, fill in the area with cable, working your way back to the original side of the room. Maintain a uniform 4- or 6-inch spacing. Either mark the floor or use a template as shown in Figure 7-14 A. Once you've installed the plastic clips, weave the heating cable under the clips (Figure 7-14 B).

7. When you reach your starting point (in front of the outlet box), tie the end with the second "cold" lead and pull the wire up into the outlet box, again keeping the factory spline on the floor.

8. Install the thermostat tube after the wire (or wires). It's important to place the tube between two heating cables (Figure 7-13).

9. Before pouring any underlayment, check the continuity of the cable to make sure it wasn't damaged during installation. The resistance of the cable should be less than 150 ohms. *Do not energize* the heating cable to see that it works.

10. Cover the cables with ¾ inch of thermal mass, mortar, or Therma-Floor.

11. Now install your tile and connect the wires to the thermostat.

Manhours — Therma-Floor is sold by authorized dealers and installed by factory-trained installers. But a dealer may sell you the product to handle small

jobs, like a bathroom, for instance. This entire system can usually be installed by one person, with the exception of mixing and pouring Therma-Floor. Of course, it depends on room size. You may need an electrician to feed the power and do the final hook-up. The following manhour estimates are based on the WFK 120-volt electric system and experienced installers (having installed the product at least three times). They include the WFK (but not the thermal mass and finish floor covering):

- 12 square feet — 1.5 manhours

- 18 square feet — 2.0 manhours

- 32 square feet — 2.5 manhours

- 48 square feet — 3.0 manhours

- 62 square feet — 3.5 manhours

- 94 square feet — 4.0 manhours

Now that you've become an insulation specialist, let's switch gears and move on to Doors, Windows, and Trim. While we all know the importance of insulation in saving energy, remember that doors and windows are important, too. In the next chapter we'll introduce some interesting products that can give your next project that special spark it needs. Some of the new products not only add to overall comfort but will also add character — something just about any home could use!

Doors, Windows, and Trim

██

It's my experience that customers, for some reason, obsess about three things: having the classiest front door in the neighborhood (usually red), the biggest window (to watch the neighbors admiring their door?), and the fanciest moldings inside and out. They feel these extra touches give their castles that special spark that sets them apart. If this sounds like the type of customers you cater to, you'll be interested in some of the products I'll be describing in this chapter.

Some of these products may not be new to you. Perhaps you've heard of some of them but never tried them. You may just be waiting for an opportunity to work a few of them into your next project. Whatever the case, look at these products with an open mind and decide how using them could benefit your project and your customer.

Doors

Does a customer really want a red door or are they really looking for an energy-efficient door that's maintenance-free? Most customers want all of the above. And some also want a custom look. Unfortunately, solid doors are expensive, and in my experience, they're prone to warping even with sealants and proper curing time. But today's door manufacturers build better frames and doors using finger-jointed material, premium veneers, foam and composite wood fiber cores, and metal or fiberglass skins. They add stiles and rails for a wood grain raised-panel look. After finishing, it's difficult to distinguish it from a real wood door.

These new thermally-efficient doors are a little on the expensive side compared to steel doors, but they're less than wood doors with stiles and rails. But consider their long lifetime and the savings in energy and maintenance they represent. Speaking of energy savings, keep in mind that most heat lost through exterior doors is from air leakage. This is a special problem in doors that aren't manufactured with their own frames (prehung). To maximize energy efficiency, look for frames with compression-type or magnetic weatherstrips, and install a good threshold with a vinyl weather seal. If you find an existing frame that lacks weatherstripping, recommend that your customer add it; it will pay for itself the first year!

The doors I'll discuss have two things in common: They're energy efficient, and they're not made of real wood. They just look like they are. While many customers (and builders/remodelers) appreciate real wood, they'll also appreciate that these products are free of the maintenance associated with real wood doors. Now don't get me wrong. I'll still use wood doors — but I use them *indoors* where they won't be subject to Mother Nature's abuse. Educate yourself on doors and you could work yourself into a full-time sideline to your business!

Jeld-Wen

One of the largest producers of doors and windows for the residential market is JELD-WEN, Inc. They offer a wide variety of door products, from the entrance all the way through the house to the back door. Their line includes door designs in traditional wood materials, molded wood fiber, or metal. Two product lines you may find interesting are the Elite molded and the Challenge steel door series.

Wood Fiber Doors

The Elite product line includes interior doors as well as the Elite Alterna exterior door line. The Elite interior line, in general, offers the following options for rails and stiles in the construction of the frame:

▌ Kiln-dried, finger-jointed Ponderosa pine

▌ Medium density fiberboard (MDF)

▌ A combination of wood stiles and MDF rails

The skins are a molded $^1/_8$-inch wood fiber available with a smooth surface, or in a lightly-embossed Ponderosa pine or oak wood grain. Every door is primed and ready for paint.

The Elite Alterna exterior door, on the other hand, uses a laminated veneer lumber (LVL) inner frame to provide overall strength and to help it resist warping

and buckling. They've also added a solid wood edge strip for improved appearance. The core is a high-density polystyrene that exceeds current energy code requirements. The skin is a molded composite wood fiber with deep wood grain that looks and feels like real wood (Figure 8-1).

Every door comes primed, ready for stain or paint. For best results, use oil-based stains with heavy body and high pigment (like gelled stains). Water-based and low pigment stains, and one-step stains with sealer, don't perform as well. Because the door skin accepts stain differently from wood, you may have to use one stain for the door and a different stain around the frame and moldings. To get a good color match between the two, practice on a small section of the door skin and trim until you get the results you want.

You can buy them prehung or as a slab (door only), but either way you have to insert your own glass. Both the interior and exterior doors carry a five-year limited warranty.

The following Construction Notes apply to the Elite Alterna line.

Construction Notes — To protect your warranty, it's important to follow the manufacturer's specifications for handling, installing and finishing the door. It's also a good idea to get a copy of their finishing instructions to add to your library. Here's a basic outline:

1. The door stiles (i.e., the vertical edges of the door) are made of LVL with a $^3/_8$-inch solid wood edge strip. Keeping that in mind, don't plane the edges for sizing or you may expose the LVL core. Specify the finish size when placing your order.

2. The door must be end- and edge-sealed with a premium quality oil-based sealer before painting.

3. Each door requires three hinges.

4. When installing hardware, apply a bead of sealant between the handset and the exterior surface of the door.

5. Elite Alterna exterior doors are recommended for installation under an overhang that provides adequate protection from the weather.

Courtesy: JELD-WEN, Inc.

Figure 8-1
The Elite Alterna exterior door

Steel Doors

Another interesting product from JELD-WEN is the Challenge steel door. It's a 24- or 25-gauge galvanized steel prehung door. The steel panels are coated with an epoxy primer on the inside to prevent corrosion from the inside out. The wood stiles have two functions: First, they create a complete thermal break. Second, you can plane the stile for easy adjustment after installation. Other features include an ultra-dense polystyrene core for long-lasting insulation, a special door sweep, and two coats of neutral, low-sheen, baked-on enamel primer. That gives a professional-looking finish. You can paint or stain the door, or install it as is.

You can also get a 24-gauge door with a 20-gauge steel edge that has a 90-minute fire-rating. It carries a five-year limited warranty. The collection has different styles and designs available to meet any design requirements, as well as a large selection of distinctive glass inserts.

Their 24-gauge Acclaim steel door is covered with a Plastisol coating embossed with an actual wood grain pattern, giving an authentic stile and rail effect. Staining brings out the rich-looking wood grain. Even though your customers may want the exterior painted, they still have the option to have the feel and the natural look of real wood on the interior.

Construction Notes — To get that wood grain finish, follow the manufacturer's recommendations:

1. For best results, clean the door with paint thinner, not lacquer thinner. Make sure you remove all residue before applying the finish. To enhance the wood grain, they recommend linseed-based stain or artist oils. But you can use a heavy-bodied stain. Apply stain with cheesecloth dampened with paint thinner and begin at the edges, wiping lengthwise. Next apply the stain to the surface of the door using a circular motion, concentrating on the embossed panel first, then on the flat sections.

2. When the door is finished, use the cheesecloth to gently wipe the surface of the door in the direction of the wood grain. This action blends the finish and highlights the wood grain appearance. The depth of color will vary depending on the degree of pressure applied with the wiping action.

3. After the stain has dried completely, spray on a clear, high-quality polyurethane varnish coat to protect the finish. Overlap each sprayed pass, until the door is evenly covered. You can brush on additional coats. Two light coats will give a semi-gloss appearance, while a third coat will give a high-gloss appearance. Allow it to dry for 24 hours; humid areas may require a longer drying time.

Manhours — These manhours apply to both series of doors. One experienced carpenter can handle this job alone. And, you can eliminate many of these functions if you buy the door as a prehung unit:

Hardware:

- Deadbolt: 0.800 manhours per unit

- Handlesets: 1.00 manhour per unit

- Oak threshold with vinyl weather seal ($5/8$" or $3/4$" × 37"): 0.381 manhours per unit

- Hinges ($3^{1}/_{2}$" × $3^{1}/_{2}$"): 0.18 manhours per hinge, which includes both door and jamb

- Weatherstripping: 1.53 manhours per 3'0" door

Finish:

- Staining: 0.006 manhours per square foot per coat

- Finish: 0.007 manhours per square foot per coat

- Light sanding: 0.011 manhours per square foot per 3'0" door

- Painting (handwork): 0.020 manhours per square foot for three coats for both door and frame

Sidelights:

- Installation ($1^{3}/_{4}$" × 1'2" × 6'8"): 1.36 manhours per unit

- Lights: 0.131 manhours per square foot for both sidelights and door

Door:

- Installation: 1.32 manhours up to 3'0" to fit; not including frame, trim, or finish

- Frame: 0.850 manhours per unit up to 3'0" × 6'8" and up to jamb width of a 2 × 6 framing member

■ Trim: 0.250 manhours per side, unfinished up to a 3'0" door

■ Prehung steel, solid core, or molded door: 0.750 manhours up to 3'6"

Therma-Tru

Beautiful exterior doors with all the benefits of wood (and none of the disadvantages) came to the market more than 30 years ago. They're made of energy-efficient materials that their manufacturers claim actually outperform wood — materials that won't warp or split even under the most severe conditions. This step revolutionized the door industry. Today, the traditional concept of high-performance building materials has been refined to include products made of advanced composite compounds. Therma-Tru has developed extruded, compression-molded, and injection-molded materials and technologies. Here are some of their accomplishments:

■ The first residential door with a polyurethane foam core that insulates five times better than wood

■ The first insulated steel door to use wood stiles and rails, creating a full thermal break to prevent the transmission of heat and cold through the steel surface

■ The only manufacturer to use exposed magnetic weatherstripping for the most positive magnetic seal available

■ A self-sealing sill and door bottom system that requires no mechanical adjustments

■ The first adjustable security strike plate, increasing the system's ability to withstand impact equal to 450 pounds

Two of their door designs are unique because of the compounds they use. They include the Fiber-Classic (fiberglass-reinforced) and Classic-Craft (thermoset composite) doors, both of which carry a limited lifetime warranty.

Fiberglass-Reinforced

The Fiber-Classic is manufactured with a panel face of a fiberglass-reinforced thermoset composite compound, compression-molded with a wood grain like natural red oak. Wood stiles and top rail provide a full thermal break and allow the door to be field-trimmed. The frame is primed finger-jointed select pine with compression weatherstripping made of a specially-formulated rubber compound that remains flexible even at low temperatures. Corner seal pads at the base of the jamb legs help protect against air and water infiltration. A moisture-resistant bottom end rail has been added to prevent rot or warping.

There are a variety of sills available: thermally broken fixed (available in either the self-adjusting or the oak adjustable) and public access. The "monumental" grade oak adjustable sill made with a decay-resistant composite substrate is available in bronze, brass, or mill finishes. You can get the oak cap in three grades: monumental, premium, and construction. Other sills are available for inswing or outswing units in aluminum with mill or anodized finish.

The door itself can be stained to resemble wood, or painted. The solid polyurethane core (CFC-free) used between the faces provides an R-value of 13.6. In addition to the 20-minute fire-rated door, they also offer doorlights, sidelights, and doors sized at 7'0"and 8'0".

Therma-Tru prefers to sell a whole door system, but they will sell blanks to replace a damaged door that's been fitted in their own frame.

Thermoset Composite

The Classic-Craft panel faces are manufactured from a proprietary thermoset composite that's compression-molded to look like wood grain. The solid polyurethane foam core has five times greater insulation value than a solid wood door. They've duplicated a hand-crafted red oak door (Figure 8-2) and the door edges are clear northern red oak. The lock side of the door has a 4-inch-wide LVL stile that runs the entire length of the door. Classic-Craft doors are available in 3'0" × 6'8", with sidelights of 12 and 14 inches. You can get removable grilles for the glass, and an oak adjustable sill for the door.

One finishing technique I like is to paint the jamb in a color two or three shades lighter than the casing. The contrasting colors really set off the beauty of the wood grain.

Figure 8-2

The Classic-Craft door system

Construction Notes — Both the Fiber-Classic and Classic-Craft doors can be stained or painted, and the company has made it easy by offering a finishing kit with all the necessary components. The kit includes:

▮ Premixed stain ($^{1}/_{2}$ pint)

▮ Mineral spirits (4 ounces)

▮ Clear satin top coat (1 quart)

▮ A 3-inch China bristle brush

▮ Two white cotton rags

▮ One pair of protective gloves

▮ Two stirring sticks

▮ Complete finishing instructions

▮ Two door skin samples

The kit contains enough to finish both sides of a complete door unit with two sidelights, or a double-door unit. But it's only recommended for doors 7 feet and under; over that, use their 8-foot door finishing kit. Colors available are Natural Oak, Cherry, Walnut, English Walnut, Light Oak, Cedar, and Antique White.

Before you can apply the top coat, allow the stain to dry at least 24 to 48 hours. Check for dryness by affixing a piece of masking tape to the stained skin sample surface. Rub the tape firmly with a hard object and remove. If the tape comes off clean, then the stain is completely dry.

When brushing on the top coat, apply even coats (following the grain) and allow 18 to 24 hours before applying a second coat. (That second coat is strongly recommended.) Personally, I'd go for three coats. Be sure to coat all wooden edges. Because you're working on a fiberglass door, the manufacturer recommends that you wipe with a tack cloth instead of sanding between coats.

These coats carry a five-year satisfaction warranty. Their top coat will last three times longer than exterior polyurethane or varnish. But explain to your customer that the exterior finish will eventually be affected by exposure and weathering from the sun, moisture, and air pollutants. Once the gloss fades, the top coat will need to be redone. Are you going to take care of it when that time comes (for a charge, of course), or will you leave this responsibility to the customer? Discuss this maintenance issue with your customer before the contract is signed.

Manhours — The following manhours are based on one experienced carpenter for new construction. On the double-wide units, consider using a helper. The Therma-Tru units come with optional brick molding.

Hardware:

▮ Deadbolt: 0.800 manhours per unit

▮ Handlesets: 1.00 manhour per unit

▮ Oak threshold with vinyl weather seal ($^{5}/_{8}$" or $^{3}/_{4}$" × 37"): 0.381 manhours per unit

▮ Hinges ($3^{1}/_{2}$" × $3^{1}/_{2}$"): 0.18 manhours per hinge, which includes both door and jamb

▮ Weatherstripping: 1.53 manhours per 3'0" door

Finish:

- ▮ Staining: 0.006 manhours per square foot per coat

- ▮ Finish: 0.007 manhours per square foot per coat

- ▮ Light sanding: 0.011 manhours per square foot per 3'0" door

- ▮ Painting (handwork): 0.020 manhours per square foot for three coats for both door and frame

Sidelights:

- ▮ Installation (12-inch and 14-inch × 6'8"): 1.36 manhours per unit

- ▮ Lights: 0.131 manhours per square foot for both sidelights and door

Door:

- ▮ Installation: 1.32 manhours up to 3'0" to fit; this doesn't include frame, trim, or finish

- ▮ Frame: 0.850 manhours per unit up to 3'0" × 6'8" and up to jamb width of a 2 × 6 framing member

- ▮ Trim: 0.250 manhours per side, unfinished up to a 3'0" door

- ▮ Prehung steel, solid core, or molded door: 0.750 manhours up to 3'6"

- ▮ Prehung: 1.00 manhour for 5'0" to 6'0" double-wide units

Windows

Windows are just as important as doors, if not more so. Besides trying to get the maximum amount of R-value out of an insulated unit, a window also has to provide daylight, view, sound control, and style. My biggest concern about windows is style; for me, they just have to have class. I'm talking metal-clad wood windows here, with their nice price tags. It's very hard to sell a customer on windows with such high quality. It can be done, however. I know — I did it for years.

Vinyl-clad windows came along a few years back. They're affordable, available in colors, maintenance free, and less conductive than metal. But my experience with the earlier models was that they weren't particularly pretty.

To help increase energy efficiency, thermal break frames were introduced for metal (aluminum) frames, which gave them a big efficiency improvement over the earlier models. Metal is metal, but these windows certainly have their place, especially in the bathroom or in other moist areas.

Today, window manufacturers go to great lengths to design the ultimate window that delivers both superior overall R-value and great eye appeal. Some windows will fool you the first time you see them; you won't believe they're made from composition materials, fiberglass, and even metal. Your customer will be surprised as well. These new products have opened up a whole new outlook on the new designs and energy-efficient windows coming to the market.

Traco

How about a company that makes history while it makes windows? It all started in Pittsburgh back in 1943 when Mae and E. R. Randall founded the Three Rivers Aluminum Company, a small window facility. Today, Robert Randall runs the company that offers some very interesting and innovative products for both residential and commercial markets.

TRAWOOD

This wood-like finish dresses up solid vinyl or aluminum windows. And it isn't a laminate or foil tape, but rather a multiple heat-cured finish that permanently adheres to the frame. This process gives it the feel, warmth, and look of real wood. The two shades available are light and medium oak. The advantage of this system is that it makes the overall window maintenance free: no warping, no swelling, and no need to restain.

Power Two

A composite window system, Power Two combines the thermal resistance of vinyl (interior) and the strength and durability of finished aluminum (exterior) in one product. This results in a corrosion-resistant window able to stand up to Mother Nature's

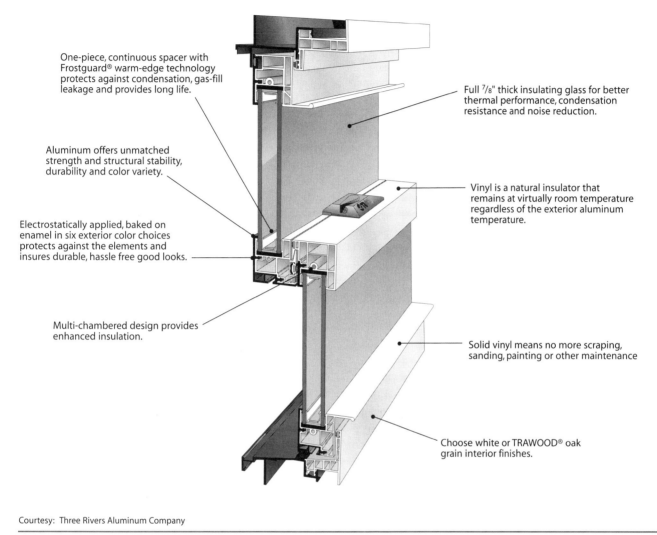

One-piece, continuous spacer with Frostguard® warm-edge technology protects against condensation, gas-fill leakage and provides long life.

Aluminum offers unmatched strength and structural stability, durability and color variety.

Electrostatically applied, baked on enamel in six exterior color choices protects against the elements and insures durable, hassle free good looks.

Multi-chambered design provides enhanced insulation.

Full ⁷/₈" thick insulating glass for better thermal performance, condensation resistance and noise reduction.

Vinyl is a natural insulator that remains at virtually room temperature regardless of the exterior aluminum temperature.

Solid vinyl means no more scraping, sanding, painting or other maintenance

Choose white or TRAWOOD® oak grain interior finishes.

Courtesy: Three Rivers Aluminum Company

Figure 8-3
Power Two window system of vinyl and aluminum

harsh challenges — and it's available in six exterior colors. As you can see in Figure 8-3, the Power Two Window System offers a few other features that make this an attractive addition to your repertoire of regular building materials, including:

▌ A mechanical interlock at sash meeting rails to resist exterior forced entry for security and increased window airtightness.

▌ Triple weatherstripping at sash meeting rails to increase window airtightness.

▌ Narrower window frame to permit more sunlight and viewing area per window.

▌ Electrostatically applied, baked-on enamel in six exterior colors to protect and beautify the windows. Or choose white or oak grain interior finishes.

▌ Two locks for added security.

They also offer "low-E" (low-emissivity) insulated glass. This gives a 30 percent improvement over regular insulating glass in reducing heat transmission. The ultrathin pyrolitic coating (metallic film) on the glass allows the light to pass through, while reducing harmful ultraviolet rays. It performs day and night to either reflect heat into the building or away from the building. It can be tempered and combined with tinted or reflective glass as required and can also be enhanced with argon gas. For customers who live in colder climates, this could represent a 40 percent improvement in the reduction of heat transmission over regular glass.

You can also choose the Heat Mirror insulating glass. This product offers a 58 percent improvement over regular insulating glass in reducing heat transmission, as well as enhanced acoustic performance. It actually exceeds the benefits of triple glazing without the added weight, and reduces interior cold drafts and condensation by raising the inboard glass temperature.

What about tempered glass? Traco offers View-Safe tempered glass that looks and feels exactly like ordinary glass yet is only $\frac{1}{8}$ inch thick. It's been tested to withstand impacts of up to 24,000 pounds (that's 12 tons per inch). This is four times the strength of regular glass. Since it's hard to break and nearly impossible to cut with a glass cutter, View-Safe discourages unwanted intruders or the errant baseball. If the glass should break (which is a possibility), it's much safer than regular glass because it crumbles into small square blunt-edged nuggets which are harmless and easy to clean up. This is definitely something to consider for a child's room.

Manhours — Installing these windows is no different from installing any other windows on the market. While some windows can be installed by one person, depending on the size of the window, I think it's a good idea to use two. You should have a helper hold the product in place while the carpenter makes adjustments and secures it. These manhours are per opening for setting the window only — no carpentry:

Casement:

▌ Up to 8 square feet: 1.00 manhour

▌ Over 8 square feet to 16 square feet: 1.50 manhours

Double-hung:

▌ Up to 8 square feet: 1.00 manhour

▌ 8 square feet to 16 square feet: 1.50 manhours

▌ 16 square feet to 32 square feet: 2.00 manhours

▌ Over 32 square feet: 3.00 manhours

Double or sliding:

▌ Up to 8 square feet: 1.00 manhour

▌ 8 square feet to 16 square feet: 1.50 manhours

▌ 16 square feet to 32 square feet: 2.00 manhours

▌ Over 32 square feet: 3.00 manhours

Comfort Line

It takes a lot for me to get excited about a new product, but when a sample of the Comfort Line Fiber Frame window arrived, I knew I'd seen the ultimate in window construction and design. This window system provides the best of both worlds: a frame constructed of pultruded fiberglass with real oak veneer applied to the interior (Figure 8-4). It's just a good-looking window! And it gives you the flexibility to match the surrounding environment.

What makes this window special is the pultrusion technology. *Pultrusion* is a process in which several thousand high-strength glass fibers are saturated with a specially-blended resin formulation and pulled through a heated forming die. This proprietary process produces fiberglass lineals that won't split, splinter, shrink, warp, swell, or bow. The Fiber Frame lineals are thermally nonconductive and have relatively the same coefficients of expansion and contraction as insulated glass. That is to say, it's

Courtesy: Comfort Line Inc.

Figure 8-4
Fiber Frame window with oak veneer

virtually nil. The result is an energy-efficient, virtually indestructible framing material. Built around double-insulated glass with low-E and argon gas, this is the ultimate window for year-round comfort and efficiency. Optional solid oak grilles add to the overall quality.

The entire system is available in double-hung, casement, awning or picture window styles. If you need dynamite patio doors, they're available as well. I always look for an almost maintenance-free window for my customer. It looks to me like this is the answer. A fiberglass system is the next step up from a vinyl window. In general, a fiberglass window will cost about 40 percent more than a good quality vinyl window. Your customer will see the value once you point out the benefits:

▮ Superior structural strength

▮ No flexing or bending

▮ No thermally inefficient (metal) structural reinforcement required

▮ Superior dimensional stability

▮ Virtually unaffected by heat, cold, or moisture

▮ Low maintenance

▮ Paintable and stainable

Comfort Line has manufactured and supplied windows and patio doors for the building products industry since 1959. In 1964 they pioneered the use of vinyl for storm window and storm door fabrication. Fiber Frame technology revolutionized the window industry again in 1988. How did I let such a window slip by me for so long? For you professionals who haven't seen this system yet, call Comfort Line (see Appendix) for a sample to check it out for yourself.

Installing the product is the same as any other window with a nailing fin. Use the following manhours, but keep in mind that this is for setting the window and/or patio door per opening only — no carpentry:

Casement and/or awning:

▮ Up to 8 square feet: 1.00 manhour

▮ Over 8 square feet to 18 square feet: 1.50 manhours

Picture:

▮ Up to 8 square feet: 1.00 manhour

▮ Over 8 square feet to 18 square feet: 1.50 manhours

▮ Over 18 square feet to 32 square feet: 2.00 manhours

▮ Over 32 square feet: 2.50 manhours

Double-hung:

▮ Up to 8 square feet: 1.00 manhour

▮ 8 square feet to 16 square feet: 1.50 manhours

▮ 16 square feet to 28 square feet: 2.00 manhours

Patio doors (1-inch glass):

▮ 5 foot: 2.00 manhours

▮ 6 foot: 2.75 manhours

▮ 8 foot: 3.25 manhours

▮ 12 foot: 3.75 manhours

▮ 16 foot: 4.25 manhours

Wenco

Wenco Windows, a division of JELD-WEN, has found a way to conserve timber resources by constructing frames and sills for the Eliminator-PF double-hung windows out of Werzalit. That's a tough composite that won't split, warp, or rot. Werzalit is composed of wood fibers and a resin composite that's molded under high heat and pressure. The sash (frame) is made of a cellular PVC, which the manufacturer claims has twice the insulation value of wood. But what's unique is that the sashes are prefinished inside and out, with the interior finished to resemble Ponderosa pine. It's available in a fruitwood stain with a clear satin topcoat. The exterior is available in white, beige, and earthtone.

The entire interior is protected during the manufacturing process by a peel-off protective film that's easily removed after the window is completely installed.

WENCO's insulated glass is cushioned in an extruded vinyl marine glazing gasket that's an integral element of the unit. It carries a limited lifetime warranty. The window also features a flexible weather seal. Upper and lower sashes tilt in so customers can easily clean both sides of the window from the inside. You can get an optional grille that's permanently sealed between the insulating glass. Windows are prefinished on the interior side and capped on the exterior side in the three colors previously mentioned. A round top window is available for the top of the Eliminator.

A similar technology goes into their Eliminator swing patio door. Stiles and rails are made of laminated strand lumber (LSL) core material and encased with an aluminum exterior cladding and rigid urethane interior with an embossed oak grain finish ready for stain (Figure 8-5). It has four commercial-grade hinges and the deluxe "Eclipse" lockset and deadbolt are standard. But the casing isn't included.

Courtesy: Wenco Windows/A Division of JELD-WEN, Inc.

Figure 8-5
Eliminator patio door with embossed oak grain finish

The door sill is thermally broken to minimize condensation and freezing. Insulated tempered safety glass is available in clear or high-performance. High-performance insulated glass uses a specially-coated low-E glass which incorporates a thin — essentially invisible — metallic film that controls the full spectrum of light. It helps reduce heating bills in the winter and cooling bills in the summer. In the winter, high-performance glass reflects long-wave heat energy back into the home, keeping the house warmer. On the other hand, in the summer the glass reflects long-wave heat energy away from the home, keeping the house cooler.

Nailing fins and a drip cap are included for standard $4^1/_2$-inch wall thickness, with exterior extension jambs available for $6^1/_2$-inch walls. Use the following manhours, but keep in mind that this is for setting the window and/or swinging patio door only per opening — no carpentry:

Double-hung:

▌ $1'1/_2"$ × up to 4'9": 1.00 manhour

▌ $2'2^1/_2"$ up to $3'5^1/_2"$ × 5'5" up to 6'5": 1.50 manhours

Patio doors ($^7/_8"$ glass):

▌ $4'10^1/_2"$ × $6'8^3/_{16}"$: 2.00 manhours

▌ $5'10^1/_2"$ × $6'8^3/_{16}"$: 2.75 manhours

DecoRoof

One area I find difficult is building a roof system for a bay window. I've never installed a bow window, so I can only guess it presents similar difficulties. Since I've installed a few bays, I know there must be an easier way, and I believe I found it in DecoRoof. DecoRoof is a copper roof system that's very appealing in design, function, and installation. Your customers will like it as well.

The DecoRoof Collection by Stillwater Products includes an assortment of copper roofs and accessories that complement any bay or bow window. But you can also use them just as well above flat windows, patio doors, and even entries. They're available in pure copper, aged copper, primed steel, or painted aluminum. Copper roofs, including all the

Figure 8-6
Installing copper DecoRoof

parts, come protected by a polyethylene sheeting, which the manufacturer recommends leaving on until installation is completed. Then, remove it to expose that "new penny" shine (Figure 8-6), which ages to a natural green patina in time.

If your customer wants a natural aged look now, the company offers a true patina roof system. Your customer doesn't have to wait years for it to develop naturally.

The primed steel, on the other hand, allows you to select specific colors to match the customer's home. The painted aluminum unit is available in high gloss black and white and low gloss white (also used as a primer). If you want custom colors, contact the manufacturer directly before bidding to make sure of the pricing.

These units save installation time because they require no sub-framing and are self-flashed. Each unit is custom built to fit any manufactured bay or bow window of any brand or size. All you have to supply is the model number of the window. If you build your own bays, the manufacturer will send you

a worksheet so you can provide them with the measurements necessary for them to build a custom roof.

These systems are kits with instructions and all the necessary weatherproofing materials. Accessory items include a brick soffit and overhang kit for a brick or masonry installation, which are frequently difficult work areas. The company normally has a two-week lead time, but large custom units could take three weeks or more.

Construction Notes — In a wood structure, the installation is basically the same for most of their units. The installation differs when installing their Canopy system, or when working in brick and/or masonry. To learn more, contact the company for the installation manual. Here's some information to give you a better insight into how this system is assembled and applied in a standard wood wall application.

1. All panels and parts are identified on the back, and read from left to right as seen from the front. Most roof systems have three or four panels. See Figure 8-7.

2. Place the left-hand mounting strip (marked #1) on the top outer edge of the window frame above the left-hand flanker window. If your window uses brickmold or casing, mount the strip to its top outer edge. Carefully predrill through the vinyl into the window frame, and screw that strip into place using the $1/2$-inch stainless-steel screws provided.

3. Repeat this step for all other mounting strips, left to right. The tips of the strips should just about touch one another, but not overlap. Caulk the joint with silicone wherever two pieces from the mounting assembly come together.

4. Apply any insulation before installing any of the roof panels. Be careful to position the insulation so the roof panels sit snugly on top of the window and flat up against the wall. If you're using support cables to hold the window in place, make sure they're installed low enough to clear the angle of the roof.

5. When installing the roof panels, start with the left (#1) panel and insert the bottom panel flange into the slot, making sure the flange sits solidly in the bottom of the slot and the panel flashing falls flat up against the wall.

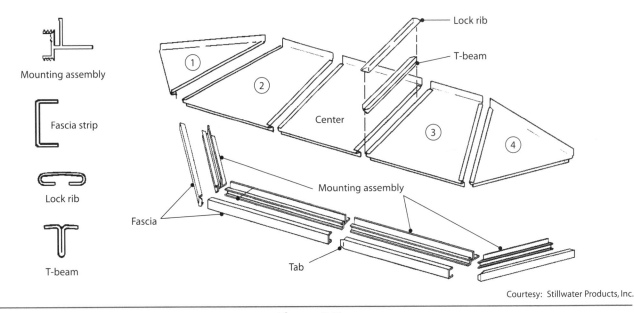

Figure 8-7
Parts for the Capri design

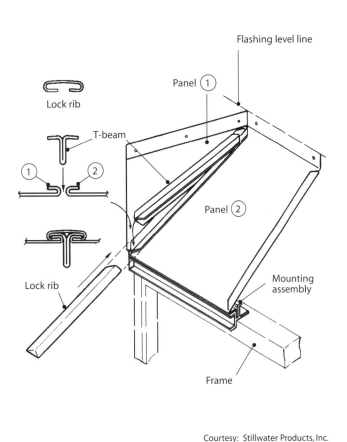

Courtesy: Stillwater Products, Inc.

Figure 8-8
Temporary T-beam

6. Working from left to right, install the remaining roof panels, leaving a gap of about $1/8$ inch between the panels. Insert the appropriate T-beam into each panel junction, then slide the temporary lock-ribs up the panel flanges all the way (Figure 8-8).

7. Make sure the roof panels are lined up with the window, then fasten the panel flashing to the wall using screws or nails. Screw the panel flanges to the vinyl mounting strips with the $1/2$-inch stainless-steel screws provided. Pilot-drill the screw holes through the "V"-shaped screw-guide on the vinyl strip. Use at least two screws per panel.

8. Remove the lock-rib and T-beam from the panel, one junction at a time. Apply a small bead of silicone down the top of each panel flange. Reinsert the T-beam so it's sealed to the panel flanges by the silicone. Run another bead of silicone into the bottom fold of the lock-rib, where it'll clasp the roof overhang. Slide the lock-rib back in place, making sure its flanges slide up under the panel flanges. With the lock-rib about $1/2$ inch from closing, seal the upper end of the T-beam and the top section of the panel seam where it meets the panel flashing. Close the lock-rib snugly. Finally, seal the flashing seams and the nails. Repeat this procedure for each remaining panel junction.

9. Now snap the fascia strips into place. Before you start, take the strip without any tabs (marked #2) and hold it to the end. Start by running a small bead of silicone (this is optional) along each of the two inside angles of the fascia marked #1. Carefully snap it into place over the corresponding mounting strip. Repeat this procedure for all remaining fascia strips, fitting each strip over the tab of its neighbors and ending with the strip marked #2.

10. Finally, add the siding over the flashing. When completed, remove all protective polyethylene. One word of caution. Don't expose the film to more than about a week of sunshine. More than that can make the film extremely difficult and costly to remove.

Manhours— The manufacturer point outs that installation time is 30 to 90 minutes (0.50 to 1.50 manhours) on the average for all roofs. For first time around, use the following manhours:

▌ Up to 48-inch unit: 1.86 manhours

▌ 60-inch unit: 2.07 manhours

▌ 72-inch unit: 2.27 manhours

▌ 84-inch unit: 2.48 manhours

▌ 96-inch unit: 2.69 manhours

▌ 108-inch unit: 2.90 manhours

▌ 120-inch unit: 3.11 manhours

▌ 132-inch unit: 3.32 manhours

▌ 144-inch unit: 3.54 manhours

Skylights

Keep in mind that you're not just a contractor any more, but a businessman who stays abreast of all the wonderful highly-efficient functional building products coming to the market. They certainly can make a difference in your customers' lives. Whatever you do to make their lives more comfortable means you're on the right path to being an all-around professional. So what are you waiting for? If you haven't installed any of the following products, you should — introduce your customers to the next century!

CeeFlow

The biggest problem with any skylight, it seems, is water leakage. The CeeFlow skylight is guaranteed not to leak. That's because of its unusual design; the skylight has no curb. This eliminates drainage obstruction and provides for uninterrupted eave-to-ridge ventilation. Your customers get a skylight that maintains the visual integrity of the roofline, an uninterrupted airflow, *and no water leakage.*

The system's design isolates the roof deck from the warm (or cool) air from the living space, minimizing winter's freezing and thawing cycles which create ice backup, a major cause of leakage. Summer heat penetration is also retarded through double-glazed, low-E insulating glass for both upper and lower panels. Laminated safety glass is used for the interior panel.

CeeFlow technology addresses other problems traditional skylights ignore. For example, the lower panel opens, using a latch and hinge system. When the panel is open, it draws fresh air into the room from the eave edge, and it permits easy cleaning. While the system is designed to insulate best in a closed position, the panel can still be left open and unattended without any concern for the weather. After installation, the skylight is flush to the roof's surface at the upper edge, and gracefully makes the transition for shingling down the side rails to a slightly raised lower edge. This tapered shape prevents any interruption of the roof's normal ability to shed weather and debris (Figure 8-9). This curbless design eliminates leaking, caulking and complicated flashing. Independent testing shows that the system carries an R-factor of 4.97.

CeeFlow is working with the concept of low-wattage roof heating to relieve drainage problems caused by water production of skylights and other insulation interruptions in otherwise highly insulated roof systems. These systems can be either passive or active. They can relieve the destructive freeze-thaw cycle by allowing just enough heat to penetrate to the snowpack. A small void is left between the snow and the roof shingles. This heat extends down-roof from a warm spot in a cold roof and it contacts the eave drip so that it won't hold ice.

Skylights can be used in single, double, or gang installations to create whatever effect the customer desires. Interior frames are made of selected clear

Courtesy: CeeFlow, Inc.

Figure 8-9
This curbless skylight eliminates leaking, caulking and complicated flashing

hardwood, basswood, poplar, and various plywoods (usually mahogany). The system is designed to accept an optional, interior mounted, translucent shade which can be either manual or electric (not provided by the manufacturer). The skylights are available in two standard widths to fit within framing members spaced 24 or 32 inches on center and three standard lengths (29, 47, or 59 inches). They also offer custom-made units. All units carry a full five-year transferable warranty.

Construction Notes — The system is designed to be installed within a ventilated roof system. The manufacturer recommends venting procedures, but they're not required for the skylight to function perfectly. CeeFlow's structural case fits into the rafter system and has a narrow waist which allows air to flow around it. Holes are drilled in the upper header and lower cross-member. You could reduce the heights of the headers and cross-members to assure air flow. Just make sure the air can pass above or through the framing members to an outside vented area. CeeFlow isn't recommend for installation in roofs pitched 3:12 or below. To learn more about the installation of these units, which includes the use of membrane material, contact the manufacturer for installation instructions.

Manhours — Labor costs include installation of the unit, flashing, and (manual) sun shade. These manhours don't include carpentry, roof work or interior trim and finishing. Figure on a crew of two. Keep in mind that these sizes make up the multiple units. The manhours listed are based on each individual unit:

- $21^1/2''$ or $29^1/2'' \times 29''$: 3.42 manhours
- $21^1/2''$ or $29^1/2'' \times 47''$: 3.52 manhours
- $21^1/2''$ or $29^1/2'' \times 59''$: 3.62 manhours

Sun Tunnel

There's a new breed of tubular skylights that can bring in more natural light than traditional box skylights. They're especially helpful to bring natural light into those areas that have a clammy dark feeling, like a hallway or dark corners. The beauty is in their design. They fit between framing members spaced either 16 or 24 inches on center without any additional framework! So they can be installed where ordinary skylights can't.

There are a few products like this on the market, but the one I want to zero in on is the Sun Tunnel. It makes it easy to bring natural lighting into areas like hallways, bathrooms, kitchens, even walk-in closets — just about anywhere the customer needs a little sunshine. Its flexible tubing curves around obstructions so it can be installed wherever it's wanted (Figure 8-10).

The tube is made of Sola-Film, a highly reflective ultraviolet-proof quadruple laminate, consisting of a double outer layer of metalized polyester fill with a double inner layer of reinforced glass fiber filament and yarn mesh aluminum metal foil. The entire unit fits within framing members spaced at 16 inches on center. The dome is a transparent UV-stabilized acrylic. You can also get a double dome.

The system includes a built-in double-sealed roof flashing that prevents water buildup, essentially guaranteeing a watertight installation on any roof design. The system is very simple: the mirrored ring in the dome captures light from all angles, reflecting it through a reflective tube down to the diffusion panel which disperses the light. Light can come from the sun or the moon. This particular product originated in Australia in the late 1980s, and was introduced to the U.S. market in 1993. Since the pipe fits between framing members, it also means the hole cut in the ceiling wallboard doesn't require additional finish work for the diffusion panel.

They have systems that will fit 16 inches or 24 inches on center. When you purchase a unit, you get everything necessary to complete the project. The maximum run on the tubing is 10 feet for the 14-inch unit and 12 feet for the 20-inch. Installation on a tile roof requires a special kit, available at no additional cost. The diffuser's round ceiling ring is white. Depending on the exposure and the length of the tube, the 14-inch unit will normally light up a 10 × 10-foot area; the 20-inch unit will handle a 15 × 15-foot area. The product also carries a seven-year warranty.

Construction Notes — Figure 8-11 shows all the components of the Sun Tunnel. It's pretty easy to install, but you need to be careful with the flashing and reinstalling the roofing material.

Manhours — The manufacturer claims their system can be installed in less than 2 hours. The only problem I can see is the amount of insulation that could be in the attic. Personally (and especially the first time around), I would figure on 3.5 to 4.5 manhours.

Courtesy: The Sun Tunnel

Figure 8-10
Sun Tunnel's flexible tube skirts obstructions

The type of roofing could affect the total time required. One person should be able to install this unit with no problem.

Moldings

Plaster (or a form of it) has been used extensively to produce beautiful effects in cornice, chair rail, and panel moldings, ceiling medallions, door and window surrounds, base moldings and all the other traditional trim. In fact, the early use of plaster provided many of the patterns for today's reproductions. Now these moldings are being manufactured (plain or detailed embossed design) in wood, metal-clad wood, and lightweight "user friendly" materials made from furniture-grade polymer, fiberglass, reinforced polyester and high-density polyurethane.

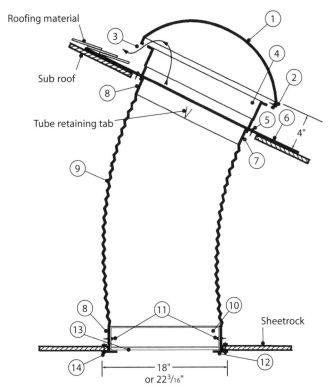

1. Transparent acrylic dome
2. Dome fixed with four noncorrosive screws
3. Vent tabs (vented or solid), interchangeable
4. Black skylight base
5. Base attachment using screws and two continuous caulk beads to seal the frame to the roof flashing base
6. Prepainted flat metal flashing is also available in prebent form to take lead strips for tile roofs
7. Mirror-finished stainless steel ring, 28-gauge with four bendable tabs for tubing attachment
8. Three continuous wraps of duct tape
9. Highly reflective flexible tube
10. 28-gauge mirror-finish stainless steel ring
11. Four sheet metal screws attach stainless steel ring to ceiling frame
12. White ABS injection-molded ceiling frame
13. Prismatic acrylic diffusion panel
14. Screws attach ring to ceiling with retainer blocks

Courtesy: The Sun Tunnel

Figure 8-11
Sun Tunnel components

These materials allow the moldings to be flexible to conform to surfaces like stairwells, bay windows, and arch radii.

Personally, I prefer the free, graceful, natural look of wood. The unlimited combinations available today give the opportunity to create unique effects, to add a personal touch to a room to suit a customer's individual preferences. Wood molding adds warmth and charm to any room — but it can take months to achieve a full and luxurious finished look. It's also costly, so the question is, "Can your customer afford it?" If so, then go for that masterpiece! If not, suggest some alternatives to your customer, like a nonwood molding that reproduces the finest in historical and classical designs.

Nonwood moldings eliminate all the steps it takes to build a cornice piece in wood. Now it's available all in one piece. Because of the process and materials used in manufacturing, some moldings are extremely flexible, which allows them to conform to most wall irregularities. Because of near-perfect reproductions, these moldings could be used in

historical preservation projects. However, you'll want to check with your local Historic Preservation Office and/or Historic Landmarks Commission before recommending their use.

Because the molds used to produce nonwood moldings are made from actual wood patterns, some manufacturers are able to produce a wood grain effect. This makes it a little easier to reproduce a natural look using stains. However, I've found that it takes time, practice, and special finishes to achieve the same effect. Personally, as I've mentioned before, I prefer to use wood for the natural look, and reserve the nonwood molding for areas I plan to paint. They're factory primed, requiring no sealing. And, of course, painting can create the most beautiful effects imaginable.

The following products are only a small sample of the variety you can find on the market today. I suggest you contact the manufacturers and request samples and catalogs. That's the easiest way to stay abreast of what's available so you'll have just the right product for any project that comes along.

Focal Point

I first encountered Focal Point in early 1994 when I was introduced to their St. James Woodgrain cornice molding. This particular profile, molded in high-density polymer, has an actual wood grain emboss molded into the surface. It's lightweight, yet has the workability and density of white pine. The real beauty and heart of Focal Point's moldings is in the process that allows accurate reproduction of a complete line of architectural and historic designs. Figure 8-12A shows Stanton Hall from the Historic Natchez Foundation collection. (Natchez has the largest collection of pre-Civil War homes in the country.)

Focal Point is authorized to reproduce architectural details from many historic properties around the world, including the Colonial Williamsburg Foundation, the National Trust for Historic Preservation, and the Victorian Society in America. You can provide your customers with architecturally-correct products from classic buildings around the world — at a price they can afford; a custom look without the custom price normally associated with hand-crafted wood or heavy plaster moldings. Wood moldings, of course, are more labor intensive because of all the component pieces required for complete depth of detail. With Focal Point, a complex design can be molded as a completed unit, making it easier to handle and to install.

Whatever your molding needs, Focal Point has something to satisfy you in their collection. If you can't find what you need, contact their Custom Services Department. They can provide a quote within 48 hours once they've received a sample or drawing.

Besides the polymer molding, they also offer:

❚ Contour-all material — a flexible polymer for curved applications

❚ FocalFlex material — manufactured to a specific radius

❚ Focal Point domes — made of fiberglass or high-density polyurethane

❚ Focal Point's Classicast Columns — made of marble and fiberglass. They're load-bearing, and they wear like stone (Figure 8-12B)

❚ Other architectural details such as door surrounds, medallions, niches, and stair brackets.

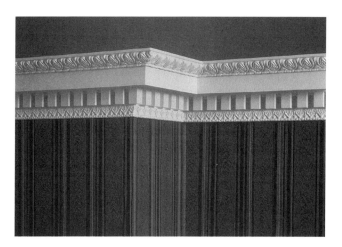

A Stanton Hall polymer molding

B Tuscan columns of marble and fiberglass

Courtesy: Focal Point Architectural Products, Inc.

Figure 8-12
Focal Point molding and columns

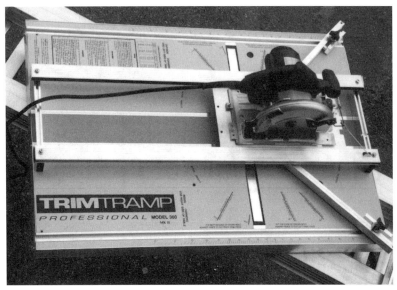

Courtesy: TRIMTRAMP Ltd.

Figure 8-13

TRIMTRAMP Model 300 — known as the Three-Fence-Miter-Table

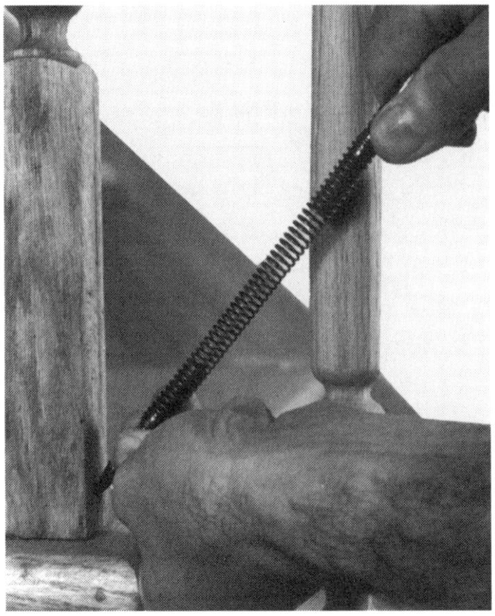

Courtesy: Noxon Inc.

Figure 8-14

The Two Bit Snapper nailset

Construction Notes — Focal Point provides a chart in each box of molding to help you determine the proper degree of cut for cornice and/or crown moldings. Pay attention. This chart makes the job go a whole lot easier. If you don't buy an entire box, ask your dealer for a copy of this important chart. For some of their wider moldings, you may have to build your own miter box or use a specialized horizontal panel saw designed to cut wider moldings.

Figure 8-13 shows the TRIMTRAMP Model 300 mounted on a Model 301 extension table. (The circular saw shown in the photo isn't included.) It's all-aluminum except the kerf board, which is high-density board. Clearance between guidebars and the deck is adjustable ($1^9/_{16}$ to $2^1/_{16}$ inches) to suit a $7^1/_4$-inch or $8^1/_4$-inch saw. The two variable miter fences are stopped at exactly 45 degrees in either direction. A crosscut fence flips up at the touch of a button, allowing cuts up to 20 inches with an accuracy of 0.002 inch. Other miter angles can be set to about 0.015-inch accuracy. You get a chart that shows how to set the miter fences to cut crown molding on the flat.

These moldings are held in place with Focal Point Adhesive applied to the bedding edges and nails or screws at the framing members. Outside and inside corners are also glued and fastened with nails or screws. Remove any excess adhesive with mineral spirits while it's still wet. Nails and screws should be recessed and filled with a sandable putty, then spot primed. When countersinking nails, you may find it easier to use the tool shown in Figure 8-14. Noxon Tools offers a full line of SpringLine (hammerless) tools. The Two Bit Snapper, a nailset containing two tip sizes in the same tool, lets you get to those hard-to-reach areas. With one hand, hold the end on top of the nail head, while you pull back on the plunger and then release it with the other hand. The sudden snap sets the nail with one motion. Because you're working with composite material, you want to protect the surface from an accidental hammer blow. This tool helps to eliminate that risk.

To achieve a good solid tight joint, cut your piece at least $1/_8$ to $1/_4$ inch long to create pressure on the joint by springing the piece into place. For the finishing touch, because ceilings and walls aren't always smooth and flat, use caulk between the molding and the ceiling and/or wall. When painting, be sure to use a nonalcohol-based paint.

Manhours — Because Focal Point offers so many products, I'll only give manhours for cornice crown, cove moldings, and accent blocks. When installing any molding above your head, you'll find the job goes a lot smoother and faster with a second person. These manhours include applying adhesive and fasteners, and preparing the molding for finish. They assume you've got a couple of similar jobs under your belt:

▊ Cornice up to 8 × 8¼-inch: 0.067 manhours per linear foot

▊ Crown up to 10½ × 13½-inch: 0.067 manhours per linear foot

▊ Cove up to 14½ × 14½-inch: 0.067 manhours per linear foot

▊ Add additional manhours for anything above these measurements: 0.034 manhours per linear foot

▊ Accent blocks: 0.333 manhours each

Style-Mark

Style-Mark Inc. has produced high-density urethane decorative trim for over 20 years. Their catalog is full of decorative millwork, door and window trims, louvers, and moldings. They also have the classic and shell niches available in various sizes, as well as 29 different ceiling medallion designs.

Figure 8-15A shows the shell niche with a fan-shaped seashell with the top curvature, easily installed in a 2 × 4 wall cavity between 16-inch on-center studs. The depth is approximately 9 inches, including the shelf.

Their variety of dentil and crown moldings includes three bracket designs in oak grain, and their exterior louvers come in a redwood grain-like pattern. They're coated with a neutral color that accepts non-penetrating stains, or paint if your customer prefers.

The products I want to focus on are the trim pieces that can highlight any door or window. Just consider the possibilities! These component parts can easily customize an entrance to near perfection, either side or both. Pilasters, crossheads, pediments, and accessories can be combined to create a masterpiece, like the door shown in Figure 8-15B. It consists of a 10-inch decorative arch with an inside diameter of 64 inches, a 64-inch half-round sunburst window pediment, a 10-inch keystone designed to fit over the arch, and two 10-inch pilasters.

Windows, on the other hand, are too often forgotten during the design phase. Trim such as this can make any window a standout in the neighborhood. The installer in Figure 8-15C is using a pneumatic

A Shell niche

B Entrance trim

C Half-round window pediment

Courtesy: Style-Mark Inc.

Figure 8-15
Style-Mark line of urethane decorative trim

nailer to attach a half-round sunburst window pediment. The keystone at the top of the arch sets off the window.

Maintenance and upkeep are minimal for Style-Mark trim. It's factory sealed with two protective coatings suitable for exterior elements, so no additional finish is required. You won't find chipping or peeling normally associated with painted wood trim, and it's resistant to decay and insect damage. Another feature is that if you've cut the trim for any reason, the exposed surface can be sealed with automotive body filler. After it's set and sanded, it'll have a surface as smooth as the original.

Construction Notes — This product is very simple to work with and install. But I learned, after installing a few systems, how important it is to order the correct style and size of components. That will help keep your job on schedule. And, since you're dealing with a finish item, take your time and double-check the measurement before making that first cut.

For a door treatment similar to that shown in Figure 8-15B, follow these steps:

1. To determine the required height of the pilasters, measure the distance from the landing to the top of the door or overhead doorlight, including the brick molding. Then cut the pilasters from the base. Before installation, select the width of the sunburst and/or arch, then find the desired width for pilaster placement. Install the pilasters by placing urethane-based construction adhesive on the backside and fastening mechanically. Whether you use a screw or nail, be sure to countersink. Try to stay on a flat surface, which makes it easier to sand once the filler has been applied.

2. If you plan to use a crosshead on top of the pilasters, install it slightly wider (up to 2 inches) than the opening as long as the overhang is even on both sides. Now mount the pediment on top of the crosshead. Normally the pediment is even with, or a little larger than, the crosshead.

3. The arch and/or sunburst will be flush to the outside edge of the pilasters or have a slight overhang.

4. Know what you want to do before installing the pilasters. If the door will have a screen door, then the brick molding should stay in place; if not, remove it.

5. Before painting, caulk any gap areas where the trim meets the wall surface.

To learn more about applications in different wall surfaces, contact Style-Mark for their Technical Catalog.

Manhours — While one person can handle the project, a second body would be handy, especially when installing the pediment. Consider a two-man crew for these types of projects. The manhours listed include applying adhesive and fasteners and preparing for paint. There's a learning curve: these figures are based on the third time around. Knowing that, you'll want to add 25 percent to your overall manhour figure until you're comfortable with the product. Also, you may need scaffolding, depending on window heights. That's not included in these figures:

- Pilaster: 0.800 manhours each ($3^1/2$ to 11 inches × 52 to 144 inches). This includes plinth blocks located at the base of the pilaster. For length at 168 or 192 inches, use 1.00 manhour for each.

- Arched/combination pediment: 1.45 manhours per unit (47 to 85 inches).

- Crosshead: 0.800 manhours each (47 to 94 inches).

ABTco

While interior plastic moldings have been on the market for some time, I've never been a very big fan of them even though they worked out just fine when I did use them. It wasn't until I received a few samples from ABTco that I got a new perspective on these alternative products. ABTco produces a wide variety of moldings from a polystyrene substrate with a density similar to pine. This substrate has the feel of the real thing, regardless of the prefinish (wood grain or paint). They have no imperfections and there's no worry of warping, splitting, or cracking. These features can only save in material and labor costs. Because they're prefinished and ready to install, sanding, priming, and/or sealing aren't necessary. The company offers the following:

▌ *PrimeMolding* — The surface of the molding is treated with a light-quality enamel primer and is ready to finish with any premium latex or oil-based paint. There are 24 different profiles available. It costs about the same as finger-jointed pine.

▌ *Canterbury Architectural Molding System* — It's prefinished in crystal white in a classic designer style. It includes six profiles and six inside and outside corners, making this system "miterless." The corner pieces eliminate the need for difficult angle cuts; only 90 degree cuts are required.

▌ *PinePlus* — This is a paintable and stainable molding with a finish that resembles the grain of natural pine. The wood grain is great looking, and the customer gets the choice of having it painted or stained to complement the room. There are 29 different profiles.

▌ *UltraOak* — This is a prefinished molding with a wood-like oak grain, yet the product costs less and is lighter than the real McCoy. There are 13 different profiles and they're designed to be combined to create custom profiles.

▌ *Affinity* — It's about as close as you can come to the real thing — the look, feel, and texture of wood. It has a smooth satin finish and the warm hues have been carefully chosen to match virtually any home decor. Available in three wood grain colors, Natural, Amber, and Autumn, that range in appearance from bleached wood to dark honey. Nine profiles are available.

Construction Notes — Here are some tips for installing these moldings:

1. Prior to installation, give the molding at least 24 hours to adjust to room temperature.

2. On prefinished wood-like grain moldings, use a putty stick; on preprimed moldings, use a water-based filler.

3. Make cuts the exact size. Don't "spring" or force the molding into place.

4. Glue all joints and splices using white or woodworking type glues.

5. For staining, don't use polyurethane stains or one-step products. Light stains work best. Consider painting or staining prior to installation.

6. Molding can be installed using a pneumatic nailer set at 95 psi or less, or simply hand-nailed. Predrilling isn't necessary

7. Splicing should be done on a 45-degree angle.

8. Coping can produce a perfect fit.

Manhours — Most applications can be completed with a crew of one, but there'll be times (especially when working overhead) when a second set of hands will help speed production. Don't forget to factor in that second body. The manhour figures included are based on the most commonly-used profiles:

▌ Base: 0.012 manhours per linear foot

▌ Casing: 0.016 manhours per linear foot

▌ Cove: 0.025 manhours per linear foot

▌ Chair rail: 0.015 manhours per linear foot

▌ Crown: 0.025 manhours per linear foot

▌ Casing corner block: 0.025 manhours each

▌ Plinth block: 0.025 manhours each

▌ Base corner block (inside and outside): 0.025 manhours each

▌ Ceiling corner block (inside and outside): 0.025 manhours each

ResinArt

Have you avoided half-round windows or other radius situations because you couldn't find a molding to follow the radius? I know I did. But I shouldn't have and neither should you. ResinArt East, Inc., has produced flexible moldings since 1968. First developed for the furniture industry, the DuraFlex brand has been tested in residential and commercial markets for some time now with excellent results. DuraFlex is a flexible composite trim molding that encourages creativity. Whether the molding is placed in a concave curve, convex curve, or radius, it holds

Courtesy: ResinArt East, Inc.

Figure 8-16
DuraFlex flexible molding trims a round window

its profile with no distortion whatsoever (Figure 8-16). This dense product has the weight of oak and it cuts like wood, but it smells like fiberglass resin.

The color is consistent all the way through and there's no need for a primer coat. It can be painted as well as stained. The manufacturer claims the product is impervious to insects, and once installed, can withstand any harsh environment, including salt air, extreme heat, high humidity, and severe cold without the warping, cracking, and rot frequently associated with wood. ResinArt East offers a variety of profiles, and they'll make you custom profiles if you want a particular design they don't offer. In addition to moldings, they also sell architectural accessories like corbels, ceiling medallions, stair brackets, capitals, and keystones.

Construction Notes — Here is some additional information that may help you to better understand the product:

1. A pneumatic pin nailer is recommended, but for best results use a construction-type adhesive as well. Don't overnail. Tape or clamps may come in handy to secure the material while the adhesive sets.

2. Space nails at least 6 inches apart and $^3/_8$ inch from any edge. Predrilling isn't necessary with #6 or smaller finish nails.

3. DuraFlex can be cut, shaped, or sanded using any woodworking tool.

4. The product may not absorb stain as rapidly as some woods and may require additional coats. Artist oils and paint pigments also work well for staining. Finish with a clear-coat product. You get the best results by spraying.

5. The material is less flexible when installed at temperatures below 55 degrees F. Before installing, allow the material to reach room temperatures of between 55 and 90 degrees F.

Manhours — There's a learning curve with these moldings. Personally, I'd make an actual trial installation to get a feel for the product and its characteristics before estimating the job. Use the following manhours for the basic profiles:

▌ Casing: 0.050 manhours per linear foot

▌ Base (up to 6 inches): 0.050 manhours per linear foot

▌ Base shoe and quarter round: 0.030 manhours per linear foot

Fen-Tech

Jamb extensions always seem to consume a lot of wood, especially around windows. One company that provides the best of both worlds is Fen-Tech, Inc. They produce a composite jamb with veneer wood and vinyl finishes. What's unique about the product is that the substrate is made from refined wheat stalks — yes, the same wheat plant that we eat. The wheat plant stalks are milled into fine particles which are compressed into an engineered composite board called Wheatboard. It's been on the market since late 1994.

The binding agent used in Wheatboard is formaldehyde-free MDI (Methylenediphenyl Diisocyanate), a very stable product that's chemically inert after curing. Every year, American farmers grow acres and acres of wheat. After harvest, the usual practice is to till the wheat stem by-product back into the ground or to burn it, which produces

Courtesy: Fen-Tech, Inc.

Figure 8-17
Wheatboard veneered jamb extensions

airborne pollution. That's a major concern in our area. Now this agricultural by-product can be used for worthwhile building products.

You can't tell the difference between this and particleboard, for instance, unless you smell it and notice its unique straw aroma. Once pine, oak, or cherrywood has been laminated to it and you look at its rolled edge, you'd take it for a solid piece of wood. Figure 8-17 shows how wood veneer jamb extensions enhance the vinyl window.

Fen-Tech also offers a vinyl universal adapter (receiver) which fits over the jamb and is held in place against the window frame with double-sided glazing tape. On those remodeling projects where the wallboard got wet and needed replacement, you could use their vinyl jamb. Both products come in a variety of widths and lengths.

Construction Notes — This is a system to frame out the rough opening to a window. From experience, I prefer to build a framed box. The nice thing about this system is that once the box has been fitted to the window, you can remove the box, place the receiver over the edges, remove the tape, and slide the box back into place. The receiver sticks to the window and you can remove the box again to install the casing.

Once completed, slide the unit back into the rough opening and into the receiver.

The only nails required are in the outside casing edge. The receiver protects against any moisture entering the exposed edge of the jamb extension, and the glazing tape creates an airtight seal between the window and jamb, one of the more common areas for air infiltration.

Manhours — In most cases this is a one-man operation. These manhour figures are to cut extension jambs, assemble, shim, case, and install the receiver. However, they don't include finishing. Consider having the material prefinished before cutting and/or assembling.

▌ Up to 5 feet wide on any side: 0.950 manhours per opening

▌ Add for any side over 5 feet long: 0.090 manhours per side

I encourage you to not let the opportunities these products represent slip through your fingers. Be aware of the products that are out on the market. One day you'll have to dig deep for that special product — and since you're a well-informed professional, it won't be a problem at all! In Chapter 9 we'll look at products to finish off the interior of any home — so let's continue.

Interior Products

We've spent the first eight chapters building a sound, efficient structure. It's tightly buttoned up to protect it from Mother Nature and it's time to finish the interior. Now where do you go?

This is where your creativity and artistry really come into play. If you can customize the home to reflect the owners' taste — a difficult task for any professional — you'll have satisfied customers and some excellent recommendations. These last chapters will showcase products to help you produce just what your customers want. In this chapter, I'll zero in on wallboard, cabinetwork products, and finish flooring.

Wallboard

Personally, I have a love/hate relationship with wallboard and finishing. Depending on job size, I prefer to hang and finish my own wallboard instead of using a subcontractor. Because I hang the wallboard on the majority of my own jobs, I'm always on the lookout for new and better products to make the job go a little smoother and faster. These next few products are designed to do just that.

But first, I want to bring up the issue of gypsum product disposal. What do you do with all the wallboard scraps? Do you haul it to a landfill, or do you hire "scrappers" (or "recyclers") to take away construction or renovation debris? Some local landfills won't accept the wallboard. If you use scrappers, you may have another problem — there's no guarantee

that they'll handle the debris in an appropriate manner. A company I contacted, New West Gypsum Recycling (NWG), Inc., told me there are scrappers running scams throughout the country. They also clued me in on a second group of recyclers called "gatherers." They rent a property, put in a scale and either pick up the debris or charge to have it dropped off. After accumulating the waste, they pocket the cash, and before you know it, they've left the area, leaving the debris behind. Then the taxpayers (and that includes you) have to pay to have the mess properly recycled. It's up to you, as a professional, to check out anyone you use. The bottom line here is to know who you're dealing with so situations like this don't come up.

Up in my neck of the woods we're fortunate to have NWG. Obviously, most of you reading this book aren't going to ship your scrap materials to NWG, but there's probably a company doing a similar job, probably for similar prices, in your area. This particular company will provide a 32-cubic-yard bin at no drop-off charge, no daily rental, no hauling charges, and assess only one fee — $60 to $80 per ton — for collection, hauling, and disposal. If you can find a company that's this good, and this reasonable, you're in luck. Most companies charge a whole lot more.

New West Gypsum Recycling, Inc. publishes an interesting newsletter, *Full Circle*, and a video that describes the gypsum recycling process. You can contact them at 800-929-1817, or e-mail them at: nwgypsum@direct.ca.

The Nailer

Once in a great while, a product comes along that can really make a difference out on the job site. The Nailer is one of those. It can replace virtually all wood backers used to support wallboard at inside corners and along the ceiling line at the double top plates. These are the places where wood backing should be installed — but too often it's not. You'll find it's missing after the floor or roof have been completed. This only makes the job more difficult when you have to install the backings. Of course, these are areas where you can't use a nail gun, swing a hammer, or get a screw gun into the space allowed.

The Nailer can save a lot of headaches *and* knuckle bashing. It was designed for use in areas without backing, but builders rapidly started using it to help prevent joint cracking. Joints tend to crack where the wall meets the ceiling, due to the truss expansion and contraction near the top plates.

You can also use The Nailer to do this. Attach The Nailer to the top plates on either side of the truss, then hold the first ceiling wallboard fasteners in from the wall 16 inches. Fasten the ceiling wallboard to the clips (not within 16 inches of the trusses) to allow the wallboard to float independently of the trusses. That helps eliminate cracks.

Courtesy: The Millennium Group, Inc.

Figure 9-1
You can see The Nailer in place

You can attach The Nailer with nails, screws or staples. The $1^{1}/_{2}$-inch textured square face (made of 100 percent recycled plastic) prevents skating; screws zip right into the plastic tab. If you use a staple gun, be sure the staples are *at least* $^{1}/_{2}$ inch long. Place two staples through the beveled face of the stem, one near the top and the other near the bottom. Using a hand, pneumatic, or electric stapler, you can attach the plastic tabs (stem) to the studs or plates in about a third of the time it would take to install solid wood backing. You may not find this necessary, but if extra holding power is needed, use one or two nails (wallboard or roofing) or screws for added strength. In Figure 9-1, you can see Nailers fastened to the double top plates and to the last framing member on the partition wall.

One drawback I see is that you have to take time to understand the layout. You've got to ensure that The Nailer lands on a joint between two sheets. Of course, you can try to install The Nailer at the same time as the wallboard. Following the layout of the vertical framing members speeds up installation at the ceiling line. For inside corners, I recommend using a template, so you transfer the correct layout markings to the stud the plastic tabs will attach to.

Using five Nailers 16 inches on center eliminates one 8-foot stud, cutting your material costs and helping to conserve our natural resources. And there's one additional benefit. Using The Nailer in corners improves energy efficiency, because with wooden backers out of the way, you can fit the insulation properly to stop air flow.

The manufacturer estimates that it takes 150 to 200 Nailers to cover 1,000 square feet of living space. That's after you're familiar with the product — probably after three installations. When figuring manhours, calculate one hour per 1,000 square feet. Remember, it takes five Nailers for each 8-foot vertical corner.

FIBEROCK

FIBEROCK is a fiber-reinforced gypsum panel produced by United States Gypsum Company. It's designed for two types of application: underlayment and wallboard. Both are manufactured with cellulose fibers, obtained from recycled newspaper, blended with gypsum and perlite to provide strength throughout the panel. FIBEROCK isn't made like standard gypsum wallboard, with its paper surface. Instead, it's fabricated on an assembly line in an all-in-one process so the bottom, middle and top layers create a solid uniform panel. This process eliminates the use of paper, which can delaminate. For now I want to zero in on their regular and VHI Abuse Resistant Gypsum Fiber Panels. We'll cover their underlayment later in the chapter.

You can get FIBEROCK wallboard in lengths up to 12 feet. FIBEROCK VHI wallboard is available in lengths up to 10 feet, and it has a fiberglass mesh reinforcing layer on the back. Figure 9-2 shows VHI wallboard being fastened to a steel stud. On the left you can see the reinforced fiberglass mesh on the back of another sheet. This added reinforcement is great for use in areas where *very high impact* (VHI) resistance is required, like hallways or entryways that get a lot of traffic, and in dorms and schools. Choose VHI wallboard where you expect high rates of abuse. That's why it's mostly used in the commercial market.

Both products are fire-resistant, but aren't recommended for use as a base for tile or wall panels in wet areas. As with other gypsum products, you can use them in wall, partition, and ceiling applications. What I like about these wallboards is that they're fabricated all in one process, so the three individual layers (bottom, middle, and top) create a solid uniform panel. This makes them stronger and more resistant to dents, breaking, and puncturing. I see this as a viable product to eliminate callbacks from a door handle puncturing the wall or dents from the kids wrestling.

Construction Notes — You can fasten FIBEROCK with screws, nails, approved adhesives, or (what I find most interesting) staples. Use the same tools for installation, and score and snap it just like you would for traditional gypsum wallboard. Be sure to have the proper tools and make sure they're sharp and in good running condition to speed production and make the job go smoothly. Finish FIBEROCK just like gypsum wallboard.

Speaking of tools, the *Rockeater* by Takagi Tools makes cutting through wallboard a real pleasure (Figure 9-3). The Rockeater is no ordinary keyhole saw; it packs colossal punching power with great cutting speed. It's great for cutting holes for water lines as small as ¹/₂ inch diameter.

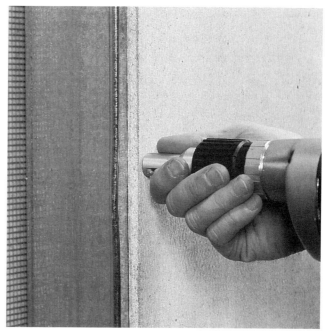

Courtesy: United States Gypsum Company

Figure 9-2
Installing FIBERLOCK VHI wallboard

Manhours — Plan on a two- or three-man crew. Here are typical manhours for interior applications on walls 8 feet high when you're fastening to wood framing. They include either nails, staples or screws, but no taping or finishing:

- ¹/₂-inch FIBEROCK: 0.010 manhours per square foot.
- ⁵/₈-inch FIBEROCK: 0.013 manhours per square foot.

Courtesy: Takagi Tools, Inc.

Figure 9-3
The Rockeater keyhole saw

When fastening to metal furring or studs:

■ $1/2$-inch FIBEROCK: 0.012 manhours per square foot.

■ $5/8$-inch FIBEROCK: 0.015 manhours per square foot.

Add an additional 0.001 manhours per square foot for wall heights between 8 and 12 feet.

Wallboard Repair Clips

Repairing holes in wallboard is nothing new to anyone who's spent time out in the field. And I suppose each of us has a unique way of approaching it. But, depending on the damage, this simple task can turn into a major project. I prefer not to get involved in these projects because customers have a hard time accepting the costs for what seems to them like a simple job. If you could just eliminate a step or two to cut the time down, you could charge less and still make a profit.

Well, I believe I found a product that can speed production and cure the "damaged wall blues." United States Gypsum Company (USG) has developed Sheetrock Drywall Repair Clips designed for both $1/2$- and $5/8$-inch wallboard. They meet building code requirements for repairing one-hour fire-rated wallboard partitions (providing you use fire-rated wallboard, of course).

The kit comes with 6 clips and 12 screws to handle a damaged area larger than 2 inches. Follow along while looking at Figure 9-4. If the damage lands near a stud, use a utility knife to cut the wallboard centered on the stud. Then square out the hole as shown in section A. I find my Rockeater keyhole saw handy here to cut the remaining three sides (or four if it lands between the studs). Slip the drywall repair clips onto the edge of the squared hole, centering them on the length of the cut. Now fasten the clips with wallboard screws about $3/4$ inch in from the edge and centered between the tabs.

Cut a new section of wallboard to fit the hole and install onto the clips, as shown in section B of Figure 9-4. Attach the remaining screws in the wallboard in line with the first set of screws and again hold back about $3/4$ inch from the edge. Just remove the tabs from each clip and you're ready to finish out the wall.

This is one product you should always have on hand — you just never know when you'll be called out for that midnight repair!

A Installing the clips

B Inserting the wallboard patch

Courtesy: United States Gypsum Company

Figure 9-4
Using USG drywall repair clips

Manhours — Manhours are pretty hard to predict because there are so many variables. Probably most cases will fall under a standard minimum fee you need to set for your company. Since your travel time will likely exceed the repair time, I recommend setting a standard flat fee for small jobs like these. Then when a customer calls, you'll be prepared to quote a price and you'll know immediately if the customer thinks the job is worth the cost. It's better to know now than to waste time running out to look at the damage. I charged $128 plus materials, if I had another job nearby. If not, I charged $225 plus materials, depending on how much travel time was involved. The bottom line is that you have to decide if it's worthwhile to go out on such small projects.

Flexible Wallboard

Who said a standard wall has to be flat and straight? Why not add a little dimension by adding architectural details like contours or niches that give the wall its own personality? And don't stop there; how about rounded soffits or archways? Did you know you can accomplish all this on your own without bringing in a specialist? USG has developed Sheetrock brand ¼-inch Flexible Gypsum Panels that make it easy to create attractive smooth curved surfaces. The sheets are only available in 4 × 8 feet.

What has kept you from tackling these architectural details? Perhaps it was the prospect of wetting down ½-inch wallboard just to make a simple radius. With USG's ¼-inch panel, you can make a radius as tight as 11 inches with a dry panel. If you wet it, you can make it 7 inches. The wallboard is made of a fire-resistant gypsum core, encased in heavy natural-finish paper on the face side and strong liner paper on the backside. The face paper is folded around the long edges to reinforce and protect the core. The ends are square-cut and finished smooth, with the long panel edges tapered, allowing normal taping for the joints.

Construction Notes — To achieve a superb-looking job, pay close attention to the alignment of the framing members. The ¼-inch flexible board is so thin that it'll follow the surface of the framing exactly. It's a whole lot easier and less expensive to fix any framing areas that might pose a problem now than later on down the road. Callbacks are a pain. One way to check stud alignment, in either in a convex or concave design, is with a long piece of ½-inch PVC piping. The PVC is flexible enough to bend along the framing, allowing you to spot any studs which are out of alignment. It's a little like "measuring twice and cutting once." Recheck the surface of the framing twice before installing the wallboard.

While we're on the subject of framing, would you still frame at 16 inches on center for radius walls? Nope. Instead, space your studs 3 to 8 inches apart, depending on the radius size. Also, you should know that for best results you'll need to install two layers of wallboard. Consider the first sheet as the base and the second as the face. Here are a few other things to consider when working with the wallboard:

▮ This is a two-man operation, and you need to take care when installing the first sheet so it retains its shape.

▮ Secure the base sheet to the studs with 1-inch fasteners 16 inches on center, with a minimum of four screws per stud.

▮ Always center the face sheet over the joints of the base sheet.

▮ For the face, use 1⅝-inch fasteners spaced 12 inches on center, with a minimum of five screws per stud.

▮ When applying the wallboard to an outside radius of the frame, attach the sheet to the last stud on either tangent wing wall, then screw-attach the panel to the studs as you carefully work the panel around the frame.

▮ On the inside of the arc, support the sheets at each end. You'll need two people, each supporting and pulling in on the sheet end with one hand and using the other hand to carefully push the panel in at the center. Fasten panels to studs beginning with the center stud and working outward.

Figure 9-5 shows three stages of a typical installation. In section A, notice how tight the framing is, especially in the niche area on the right-hand side. In this area alone there are a niche, an archway, and an entry rotunda (circular area with a dome ceiling). In section C you see the final job — very clean and smooth. Use the chart in Figure 9-6 for minimum bending radii.

A Completed framing

B Installing the panel

C The finished job

Courtesy: United States Gypsum Company

Figure 9-5
Installing Sheetrock flexible panels

Manhours — This is a tough one, but use $25 to $45 to review the framing for each special detail design. For interior applications, fastening to wood using screws with no surface taping or finishing, estimate 0.050 manhours per square foot for double $^1/_4$-inch panels. This includes a two-man crew. For a third person, add 0.015 manhours per square foot.

Cornerbeads

USG recently introduced a line of paper-faced metal tape-on bead. In most cases, it's easier to handle and install than standard nail-bead. I've used both and found that you really need both kinds of bead, depending on the project. What I like about the new bead is the soft round design for outside and inside corners. With standard 90-degree corners, you can count on damage, especially in high-traffic areas.

Figure 9-7 shows the tape-on corners: bullnose outside corner on the top and 90-degree outside corner on the bottom. On commercial projects they're great around all corners of an entryway. It gives a wonderful free-flow feeling. These beads can help to eliminate the edge cracking and chipping normally caused by wood framing members shrinking, normal

Application	Condition	Lengthwise		Widthwise	
		Bend radii	Maximum stud spacing	Bend radii	Maximum stud spacing
Inside (concave)	Dry	32"	9" o.c.	20"	9" o.c.
Inside (concave)	Wet	20"	9" o.c.	10"	6" o.c.
Outside (convex)	Dry	32"	9" o.c.	11"	6" o.c.
Outside (convex)	Wet	15"	9" o.c.	7"	6" o.c.

Courtesy: United States Gypsum Company

Figure 9-6
Maximum bending radii for Sheetrock flexible panels

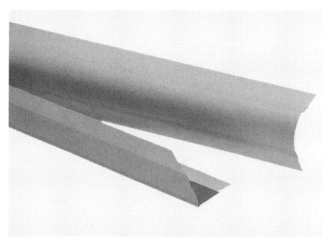

Courtesy: United States Gypsum Company

Figure 9-7
Tape-on corner bead

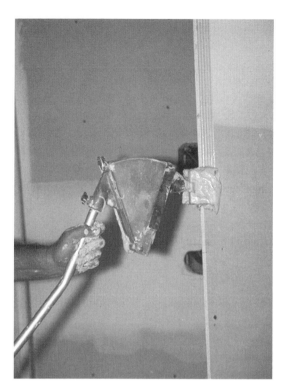

A Tool applies right amount of mud

building movement, and, of course, everyday wear and tear. The rounded corners also make it easier to apply wallpaper, and eliminate damage to the paper at the corners.

The difference between paper-faced and standard galvanized beads is the application: paper-faced bead is applied with joint compound, not nails. This eliminates the possibility of nail popping or hairline cracks that seem to appear at about 1 foot to $3^{1}/_{2}$ feet off the finish floor. Tape-on beads could reduce callbacks. The tape-on bead does cost a bit more per piece than the standard, but you might make it up in labor. When you're doing repair jobs, the tape-on bead is easier to remove.

I noticed that the tape-on takes a while to learn to use efficiently, especially in mud application. You have some choices to make. You can put the mud on the corners by hand, running the bead through a hopper filled with joint compound and then pressing it in place using a roller tool. Or you can use a corner-box applicator to apply the mud directly to the corners (Figure 9-8). On smaller jobs, I'd use the standard method — apply the mud by hand. For larger jobs, the corner-box tool is a better choice, both for speed and to apply the right amount of mud. Use a 6-inch taping knife to embed the corner bead firmly into the mud by running the knife at a 45-degree angle over the corner, applying an even pressure. Make sure there are no air bubbles under the paper.

B Embedding the bead in the mud

Courtesy: Stan Fellerman/United States Gypsum Company

Figure 9-8
Using a corner-box tool to apply the bead

The tape-on style is also ideal for installing wallboard over metal framing, since no fasteners are required. For those who aren't convinced or are unwilling to try something new, USG also manufactures paper-faced metal nail-on corner bead. The nail-on style uses the same paper-faced metal technology to provide good bond adhesion for the mud, spray texture finishes, and paints. You can fasten it to the wallboard using nails, staples, or screws. Pick up one of USG's catalogs to check out the variety of inside and outside corner styles and sizes they offer.

Manhours — After scaling the learning curve and using the proper equipment, you'll find the tape-on applies faster than nail-on. For bidding purposes, figure on 0.030 manhours per linear foot for inside or outside tape-on corners. This includes the initial bonding coat. For nail-on use 0.025 manhours per linear foot. (Keep in mind these figures are for installing the corner beads only — no taping.) Both products include lengths from 8 to 12 feet.

Easy Corner

The biggest problem associated with corner beads themselves is coming to a three-way outside corner. It's very time-consuming, to say the least, and to achieve a good-looking corner is a challenge in itself. At least, that's what I always thought. One day, out of left field, a funny-looking piece of plastic hardware landed on my desk. It looks like a pyramid or a spaceship ready to take off. After a closer look, I realized it was a molded outside corner designed to work with corner beads, either tape-on or nail-on. It's called *Easy Corner*, distributed by Unimast Incorporated. And it creates a good-looking three-way corner.

Easy Corner is designed with recessed flanges so the corner bead ends can hold it in place while remaining flush with the finish surface of the corner itself. The holes in this area help to lock the mud into place. Even though it's held in place by the corner beads, you can staple it in place to speed production. If you decide to staple, be sure to staple into the recessed flange (Figure 9-9). I see this not only as a timesaver, but as an excellent way to create perfect and consistent corners every time.

Unfortunately, Easy Corner is so simple and easy to install that it's hard to attach any manhours. With all the linear foot prices I've given so far for installing corner beads, just include this as part of those manhour figures. You can't go wrong because you'll save time in the long run.

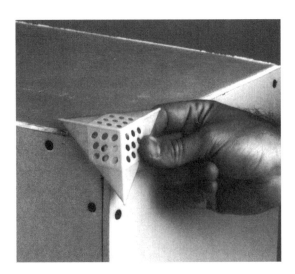

A Staple into the recessed flange

B Finishing the corner

Courtesy: Unimast Incorporated

Figure 9-9
Using Easy Corner

NO COAT

As I mentioned earlier, I believe wallboard finishing should be left to professionals. In my experience, more mud got on me than on the wall — or I put so much mud on the wall that I spent way too much time sanding. A good taper does very little sanding. The biggest problem I've noticed is that joint tape is just a little too thin to work with. It's hard to maintain a uniform line in an inside corner, the taping knife easily cuts the paper, paper snakes emerge from nowhere, and I can't float more than $1/4$ inch. Of course, when the mud dries, hairline cracks develop on both sides of the outside edge of the tape. So why hasn't someone designed a better product?

Someone has, and I just heard about it. Tim Smythe, who has been in the wallboard business for 20 years, has designed a couple that came to the market this year under the company name of NO COAT Products. The two I want to deal with are Ultraflex and QUICK'N'EASY CORNER.

Products — Both of these are designed as prefinished joint tape, which means they require very little joint compound to finish them off. No mud is applied directly to the surface, only on the edges. You only need to feather the edges for a quick and smooth transition to the wallboard. The overall quality of these products is something you'll notice the first time you handle them. Ultraflex is $4^3/16$ inches wide and laminated with a high-grade white paper board for the top finish coat. It has a plastic center, and the back is standard joint tape (off-white in color) that's $3^5/16$ inches wide and separated in the center (lengthwise). It's very rigid, but remains flexible because a built-in hinge allows it to be used in just about any angle for inside and outside corners. It's also great for those off-angles found in tray, coffer, and vaulted ceilings.

Ultraflex's rigidity allows you to make one inside or outside corner out of a continuous length — no more worry about two or three pieces (Figure 9-10). After you cut Ultraflex to length with scissors, fold and put it into place. It'll maintain its form regardless of length. It's self-straightening and will withstand movement and extreme impacts.

QUICK'N'EASY CORNER is the little brother to Ultraflex. It, too, is a joint tape laminated in two layers (the topside is white and the back is off-white) and it's $3^5/8$ inches wide. The edges are torn to taper

Courtesy: NO COAT Products

Figure 9-10
Installing Ultraflex prefinished joint tape

toward the back and it's also self-straightening. QUICK'N'EASY CORNER appears to be about five times the thickness of standard joint tape currently on the market, yet it's still pliable enough to work with (Figure 9-11). Once it's installed into the first coat of mud, smooth out the joint tape surface as you normally would, using a 6-inch taping knife. Finish by feathering the edge. QUICK'N'EASY CORNER is distributed by both NO COAT and GRABBER Construction Products.

Remember, both prefinished products were designed to eliminate the heavy costs of mud, days of drying time, heavy sanding, and — most of all — labor costs.

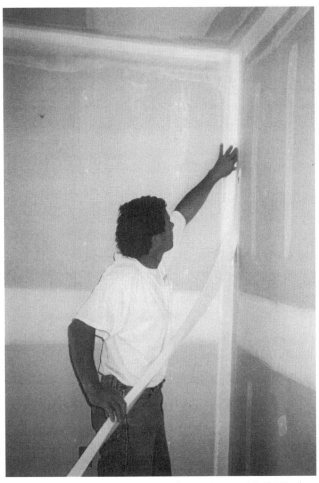

Courtesy: NO COAT Products

Figure 9-11
QUICK 'N' EASY Corner goes into place

Manhours — You can buy both Ultraflex and QUICK'N'EASY CORNER for between $30 and $45 for a 100-foot roll, which is high. But you could save up to 75 percent on your labor cost, plus savings on the amount of mud required. That's after you've learned to use the product, of course. With that in mind, remember that I've already established 0.030 manhours per linear foot as a working figure for standard tape-on corners. Since you'll be saving up to 75 percent, use 0.0075 manhours per linear foot as your new figure for inside and outside corners using Ultraflex and QUICK'N'EASY CORNERS. Don't use this figure the first time around; instead use 0.030 manhours per linear foot. For all other seams use 0.007 manhours per linear foot for wall joints and 0.009 manhours per linear foot for ceiling joints using standard joint tape. These figures are based on a one-man crew.

Plaster Wall Liners

There are times when a wall isn't bad enough to need replacement, but it has enough surface cracks that the labor costs to repair it exceed the cost of installing new wallboard. What do you do in cases like this? What if you're working in an historical building or need to encapsulate lead-based paint or asbestos? Flexi-Wall Systems manufactures veneer plaster liners in different weights and mil thicknesses that can help renew old walls. Basically, they're a gypsum-impregnated jute fabric, installed like a wallcovering, that comes in 48-inch-wide rolls.

The plaster liner can be applied to most rigid surfaces, hiding cracks, patches, mortar joints, or minor imperfections in surface walls. These surfaces could include cinder blocks (new or old), wallboard, plaster, glass, and (I've been told) even metal. In block walls, you don't even have to fill mortar joints before refinishing the wall. Just remove all excess grout burrs or other protrusions. Existing paint shouldn't affect the bonding. Once applied using the manufacturer's water-based adhesive, the liner will crystallize and set the plaster, giving a permanently finished one-step decorated wall with all the properties of conventional plaster.

Products — The company has a full line of products, but I want to zero in on two. *Plaster In A Roll* is a one-step, crack-proof wallcovering available in three decorative textured weaves. It covers walls and ceilings in one easy step, with little or no preparation, and can bring a low-stress, sound-absorbing atmosphere to new or existing interior spaces (Figure 9-12).

The second, *Faster Plaster*, is a two-step upgrade to most finishes. It's available in heavy duty (42 ounces per linear yard) and medium weight (34 ounces per linear yard). Install this versatile underliner on walls or ceiling, then apply the customer's preferred final finish: paint, vinyl, plaster, paper. You can upgrade cinder blocks, wallboard, paneling, or tile, and give it a fresh, modern look.

Construction Notes — In general, applying Flexi-Wall finishes is similar to installing conventional wall covering material. Here are some specific suggestions:

▌ Use only Flexi-Wall Adhesive #500. Adjust the amount of adhesive used according to the porosity and condition of the substrate.

▐ Don't paste too far ahead or you'll get dry pockets.

▐ Use a medium nap roller to spread adhesive on the wallcovering, not on walls.

▐ This type of material rapidly dulls blades, so have plenty on hand to change regularly.

▐ For *Plaster In A Roll*, don't double-cut on the wall. Trim on a table and cut back from the edge about 1¹/₂ inches to assure color uniformity and pattern match at the seams.

▐ Keep all seams away from external corners by at least 4 inches. On internal corners, overlap the corner by 1 inch and install the adjacent wallcovering by applying over the 1-inch lap and into the corner.

▐ Wipe off any excess adhesive immediately with a clean, slightly damp (not wet) cloth.

Courtesy: Flexi-Wall Systems

Figure 9-12
Plaster In A Roll applied to a block wall

Manhours — Use 360 square feet as minimum job size. If wall preparation (filling holes, and so on) is required, use 0.010 manhours per square foot for the preparation. The following figures include the use of the #500 adhesive:

1. Plaster In A Roll: Classics, Contemporary, and Images line: 0.012 manhours per square foot.

2. Faster Plaster: medium-weight plaster wall liner #605 and heavy-duty plaster wall liner #609: 0.010 manhours per square foot.

Wallcovering/Cabinetry Products

There are so many new products in this area that I can't claim to have used them all. But I have watched them being used by others, and previewed them at The National Hardware Show a couple of years ago. All of them add some type of flair or provide additional flexibility. You can use them either to improve the overall appearance or just to bring back some nostalgia, like embossed wallcoverings or wainscoting in molded plastic. You can offer your customer timeless elegance without the major expense of designing and building from scratch.

Wainscoting

I've done a few wainscoting projects in my time. They are, without a doubt, complex and expensive projects, but worth it when finished. A well-done job will add warmth and elegance to any room. As I said, the overall project is both time-consuming and expensive, especially depending on the type of wood and pattern/design chosen.

Outwater Plastics Industries, Inc., Architectural Division, has entered the market with a plastic wainscoting panel that's virtually indistinguishable from real hardwood paneling. Only the termites will know the difference! You can paint or stain it, and create the effect for a fraction of the time and cost of real hardwood paneling (Figure 9-13).

The panel is made of high impact polystyrene, and patterned after the finest oak wood grains to ensure perfect blending with bordering panels. A special lip on each panel overlaps adjacent panels for a seamless installation without face nailing or exposed nail holes to fill. The ¹/₁₆-inch thick unfinished

Courtesy: Outwater Plastic Industries, Inc.

Figure 9-13
This plastic wainscoting panel looks like the real thing

Courtesy: Outwater Plastic Industries, Inc.

Figure 9-14
Wallcovering from the Lincrusta collection

pecan-colored 12 × 36-inch and 18 × 36-inch wainscoting panels and 9 × 36-inch filler panels easily adhere to walls using cove base adhesive. It's impervious to moisture and insects, and because it's dimensionally stable, it requires little maintenance.

Manhours — The manhours listed here don't include painting or staining. Figure 0.015 manhours per square foot, which includes adhesive troweling. To install the wainscot cap, use 0.016 manhours per linear foot. Depending on the length of the molding, you may need a helper. You could use the same 0.016 manhour figure per linear foot for installing base molding.

Embossed

First impressions seem to be the most lasting. And a good way to capitalize on this perception is by installing embossed wallcoverings. They give the impression of a classic home, and most of your customers would love to have one. The inventors, Frederick Walton and Thomas Palmer, who created high-relief embossed wallcoverings over 100 years ago, probably had no idea their wallcoverings would still be used today. Even now, some of their designs, including Lincrusta (Figure 9-14) and Anaglypta, are still produced on the same brass rollers using the same techniques.

Outwater Plastic Industries distributes an affordable product that you can finish to resemble more expensive materials such as oak paneling, carved wood, plaster and metalwork, and Cordovan leather. You can now, quite easily and cheaply, provide your customers with lavish interiors, once the hallmark of the wealthy. Both Lincrusta and Anaglypta are manufactured of natural, raw, recycled paper, virgin pulp from managed forests, cotton linters or linoleum-like materials to suit the needs of the application.

The detailed embossed patterns are great for hiding old, uneven, or otherwise unsightly walls and ceilings. Install them using traditional wallpaper hanging methods. Or (and I recommend this) hire a subcontractor!

Manhours — Every order is packed with its own detailed instructions, so use the following figures when assembling a bid. The manhour figures assume a two-person crew, and are figured on material 20.5 inches wide by 11 yards long (approximately 55 square feet) for both the Lincrusta and Anaglypta wallcoverings. Borders are 10.5 to 20.875 inches wide by 33 feet long for Lincrusta, and 3.375 to 7 inches wide by 16.5 feet long for Anaglypta. These figures don't include finishing (painting or staining).

1. Scrape off old wallpaper, up to two layers: 0.015 manhours per square foot.

2. Patch holes and scrape marks: 0.010 manhours per square foot.

3. Blank stock (underliner 18 inches × 8 yards): 0.668 manhours per roll.

4. Embossed wallcovering (20.5 inches × 11 yards): 1.68 manhours per roll.

5. Coordinating borders (Lincrusta): 0.020 manhours per linear foot.

6. Coordinating borders (Anaglypta): 0.017 manhours per linear foot.

Pliable MDF Board

There seems to be no end to the products offered by Outwater Plastic Industries. I recommend you pick up one of their catalogs so you can see for yourself. Their latest is the Pliable MDF (medium-density fiber) Board. It allows professionals like furniture and cabinetmakers, woodworkers, builders and remodelers to produce all types of circular and wave-like curves to create unconventional and unusually-shaped furniture, walls, and columns.

It's smooth on one side, while the other side has $1/16$-inch relief cuts spaced $1/4$ inch apart that start in from the sides at $1^1/8$ inch (leaders). The relief cuts run in the long direction of the panel, which is offered in two sizes: two 8 mm boards create a $5/8$-inch thickness and two 9.5 mm boards create a $3/4$-inch thickness. The overall finish size is $40^7/8$ inches wide by 103.10 inches long. It takes shape and comes to life when the two panels are glued together on the grooved sides and molded into shape before the glue sets.

Look at Figure 9-15. It's an unusual shape, but just imagine what you can create — perhaps an unusual coffee or conference table? Notice that the relief cuts on both panels are glued together, relief cuts to relief cuts. To achieve and maintain a uniform consistency for curved cabinet doors or furniture components,

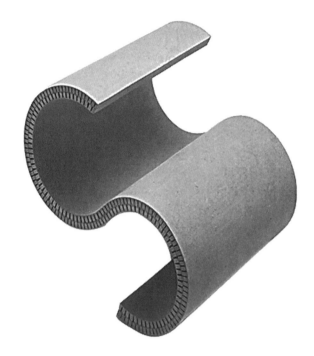

Courtesy: Outwater Plastic Industries, Inc.

Figure 9-15
S-shape formed from two sheets of Pliable MDF Board

use a template and hold the work in place with clamps and straps. Let it cure for at least five hours.

Construction Notes — The surface is suitable for veneer, paint, or plastic laminate. The manufacturer recommends that you apply all surfacing material *before* shaping the Pliable MDF Board, with the exception of paint. Paint it *after* the forming is done. When using a plastic laminate, be sure it's a post-forming product (as its flexibility allows it to conform to the shape of the surface). Use yellow or white glue for wood veneers. Remember to bend and apply clamps as you follow the template. Proceed slowly so you don't break the Pliable MDF Board.

Manhours — Figuring manhours for custom work is difficult because there are so many variables involved and because you have to consider setup time and making the templates. Here are some working manhour estimates:

▮ Installing post-forming plastic laminates and/or wood veneers, with a two-man crew: 0.25 to 0.40 manhours per square foot.

▌ Building templates or jigs for one person: 0.50 manhours per square foot.

Before using these figures, go through the mental process and size up the overall project to see if they seem realistic. If not, adjust them to meet your particular manhour requirements.

Pole-Wrap

I first saw Pole-Wrap at the 1996 National Hardware Show. It's similar in design and construction to Outwater's Pliable MDF Board, with two major exceptions: its relief cuts point to the outside, and the finished surface is laminated with an oak veneer. It was designed to wrap basement support posts (Figure 9-16). I've always tried to eliminate such supports, if it was possible, to avoid having to build an unattractive square box around the post. Pole-Wrap adds a little class to the post, especially if you stain it. Painted, it would just blend in with the rest of the environment.

Courtesy: Pole-Wrap

Figure 9-16
Basement support post covered with Pole-Wrap

Pole-Wrap is fabricated of $1/32$-inch thick unfinished red oak veneer bonded to $1/8$-inch medium-density fiberboard with a resin-impregnated paper backing, for a total thickness of $3/16$ inch. The slats running the 8-foot length are designed with a 30-degree V-groove that measures $1/2$ inch at the bottom and $3/8$ inch at the finish surface. It's available in three sizes:

▌ 12 inches × 8 feet for post circumferences up to $11^1/2$ inches.

▌ 16 inches × 8 feet for post circumferences up to $15^1/2$ inches.

▌ 4 × 8-foot sheets — enough to do four standard $11^1/2$-inch posts. You can also use the full 4 × 8-foot sheet as wallcovering, or even around a built-in bar to match the post, for example.

Construction Notes — Once you've cut Pole-Wrap to length, fit the sheet around the post. Using a utility knife, cut between the slats, leaving from $1/8$ to $1/2$ inch of extra width. In other words, make it fit loosely around the pole. The amount of adhesive you apply will take up any slack.

Once you're satisfied with the fit, remove it and apply construction adhesive in a full $1/4$-inch bead around the pole. Space the beads 6 inches apart, starting from the top and working toward the bottom. Then promptly place Pole-Wrap around the pole. Start by affixing the inside middle and then wrap to the outside edges. This is when a second pair of hands will come in handy. Make sure the ends meet evenly and hold the ends together with masking tape every foot. When the adhesive has dried, be careful removing the tape and any glue that may have oozed out from the seam. Trim any excess and you're ready to finish the surface.

Manhours — Depending on the diameter of the post, figure you could spend at least $1^1/2$ manhours prepping the post and installing Pole-Wrap, using a two-man crew.

Dimensional Wall and Ceiling Systems

A sample from Natural Impressions by Holz Dammers Moers (HDM) USA came across my desk as I wrote this chapter. What is it? Essentially, it's a complete system that lets you decorate walls and the ceiling surface in a mix-or-match of woodgrain planks to create your own pattern. Natural Impressions is a multifunctional $^1/_2$-inch MDF core with a laminated woodgrain decorative plank available in six finishes: Coastal Ash, Laredo Oak, Vintage Oak, Kingston Cherry, Concord Maple, and Northern Beech. When I picked up the sample of Concord Maple, I thought it was the real thing. You'll probably also do a double-take.

All four sides of the planks are dadoed to accept the aluminum clipping system. It's the clips (which are fastened to the wall) that join different-sized planks together (Figure 9-17). The planks come in widths of $6^5/_8$ and $11^3/_4$ inches and lengths of $31^1/_2$ and 96 inches. Spaces between the planks are covered with multicolored battens for a unique look. The battens, $^7/_8$ inches wide, are available in a variety of lengths, both long and short, to cover the intersections at the end grains. The company also offers matching molding, including base, casing, crown, chair rail, inside and outside corners, and cap, all 102 inches in length.

Installation is suitable for any room or ceiling in both commercial and residential projects, except the high humidity areas of a bath or laundry room. The advantages of the system are that you can remove individual planks and battens for easy repair, or just to change the theme.

Because this product is made from medium-density fiberboard, it contains a urea formaldehyde resin which may release low concentrations of formaldehyde vapors. Formaldehyde can be irritating to the eyes and upper respiratory system, especially in people with allergies or respiratory ailments. You should check with your customer before purchasing this or any building material made of MDF. Your customer may have an allergic reaction to it. Find out before you install it, or you could end up removing it all, at your own expense. And when you're installing it, make sure you have adequate ventilation. Look in Chapter 13 for more information about formaldehyde.

Construction Notes — HDM offers two packages, either in the 96- or $31^1/_2$-inch lengths, which include five planks, five long matching battens, and five short battens. Of course, as mentioned before, you can combine these with different woodgrain blanks,

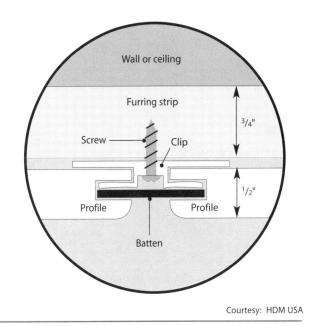

Courtesy: HDM USA

Figure 9-17
Top section view of Natural Impressions planks

lengths, and widths, and contrasting battens to achieve that just-right look (Figure 9-18). Here's what you need to know before installing these planks:

▌ The material needs to be conditioned to room temperature (minimum 64 degrees F) for 48 hours before installation.

▌ When mounting clips into wallboard, make sure the wallboard screws penetrate the framing members. If not, attach furring strips 16 inches on center. If you're installing it over metal studs, it's a good idea to use furring strips.

▌ Over masonry walls, attach a 4-mil vapor barrier before you attach furring strips, then install the clips. Place two clips at the bottom of every vertical plank to prevent moisture wicking from slab to plank.

▌ Leave at least a $^1/_4$-inch space between the ceiling and the top of planks, and between the wall and the side of planks. You can surface-nail these areas and cover them with the prefinished moldings.

▌ For ceiling installation, attach to the bottom of the floor joist or bottom chord of the truss. If you're not sure of the penetration, use furring strips 16 inches on center.

Courtesy: HDM USA

Figure 9-18
A finished Natural Impressions installation

Manhours — When figuring your manhours, remember two things: You won't meet maximum efficiency the first time you install it, and the complexity of design will also affect your installation time. These manhour estimates are for a one-man crew, except you'll probably need two people to install the furring strips and trim. Use these figures for a straightforward project. Consider upping your manhour estimate up to 50 percent for installing planks, depending on the complexity of design.

1. Installing furring strips over masonry 16 inches on center: 0.020 manhours per square foot.

2. Installing furring strips over wallboard and into wood or steel framing members 16 inches on center: 0.013 manhours per square foot.

3. Installing 4-mil film tack stapled: 0.003 manhours per square foot.

4. Installing planks on wall surface: 0.050 manhours per square foot.

5. Installing planks on ceiling surface: 0.100 manhours per square foot.

6. Installing all molding except for crown: 0.030 manhours per linear foot

7. Installing crown molding: 0.044 manhours per linear foot.

Cedarline

The fresh aroma of Eastern Red Cedar is strong, distinctive and very pleasant. Have you ever had the opportunity to line a closet with it? It could start a whole new side business for you. The advantage of cedar is that the natural forest-fresh fragrance eliminates smoke and other odors and keeps linens and clothing smelling fresh. It repels moths without that mothball smell, as well as silverfish, roaches, and other insects. And it resists mildew. Do I have your attention now?

Cedarline by Giles & Kendall is a 4 × 8-foot cedar pressboard panel made from 100 percent aromatic Eastern Red Cedar flakes. The cedar logs are select cut by small 2- to 4-man crews. I've been told the manufacturer doesn't clear cut, but harvests only the cedar trees they need. They still do much of the logging with mules to minimize damage. First-growth trees are few and far between, so second- and third-growth trees (and even smaller) are used to manufacture Cedarline. That makes the cost of the panels about half the price per square foot of cedar planks. The logs are debarked and chipped into large flakes. Then the flakes are sprayed with adhesive and wax and steam-pressed into panels. Once the panels have cooled for four days, they're trimmed to size and sanded on one side for dimensional consistency. They're available in two thicknesses: 3/16 and 1/4 inch.

Cedarline is made by an exclusive low-temperature process that allows the cedar oils to release their fragrance and pest repellant properties. Other manufacturers use a high-temperature OSB process that seals in the oils and destroys the cedar benefits. Look for the "100% Aromatic Cedar" Seal as your assurance that it's made of Eastern Red Cedar by a process that will deliver the natural cedar results you want.

Over time, the aromatic cedar will lose its aroma. Slowly, cedar oil crystallizes on the surface, sealing the panel. Fortunately, it can be regenerated by lightly sanding the surface with fine sandpaper or steel wool. Obviously, don't apply a finish over the cedar.

Just think of the many places where you could use Cedarline: pantry, sink cabinets, attic storage areas, garage storage, pet areas, chests, and drawers — you get the idea. Cut the panels using any woodworking tools. The manufacturer doesn't recommend scoring and snapping. Install the panels over solid backing or on open framing members. Use $1^1/_4$-inch panel nails over bare studs, and 2-inch nails over wallboard. Nail the panels every 12 inches along the studs. If you're using adhesive, run a bead along the panel's top and bottom and along each stud. Of course, it's a good idea to check with your local building department when attaching it to open framing members. Over concrete or cinder blocks, first seal and fur the wall.

Manhours — One person should be able to handle lining a standard-sized closet. These manhour figures are per square foot of floor, wall, ceiling, or door(s) of the closet, using compressed cedar chip 4 × 8-foot panels.

1. $^3/_{16}$ inch thick: 0.016 manhours per square foot.

2. $^1/_4$ inch thick: 0.016 manhours per square foot.

Flooring

There's one area of a home that often doesn't receive the attention it deserves — and that's the floor, both its structure and the final finish. With so many different types of floor coverings available, it's difficult to know what to recommend to your customers. Sometimes customers have definite ideas. That makes it easy to choose the material, but how do you stay current with proper installation techniques? One way is to attend classes and seminars sponsored by manufacturers when new products are introduced. It's one sure way to keep abreast of new techniques, learn helpful tips, receive guidance on proper handling, and acquire some hands-on experience. You'll also learn how to sell the flooring and understand the manufacturers' warranties. For whatever flooring you and/or your customer select, be sure to check with the manufacturer about the availability of instructions and training opportunities.

Knowing how to install the flooring is mandatory — but it's not enough by itself. You also need to understand the underlying floor structure you need to build on to get the flooring's full warranty protection. Neither you nor your customer want callbacks! The best way to be certain of the structure (where it's possible, as in a new addition) is to begin with the floor joists. One of the more annoying problems in any house is a squeaky or bouncy floor. It may be caused either by a poorly-built floor system, or the materials used (or both!). And you don't have any control over either — or do you? Be sure to review Chapter 4, especially the products under "Framing Members." The next step to achieve a solid flooring system is the subfloor. Again, I refer you back to Chapter 4 for the section on sheathing. It could make a difference!

The final phase in the flooring system is the underlayment, probably the most crucial step in the system. There are many underlayments on the market, each designed to work best with a particular floor covering. You probably think that if you use an APA-trademarked plywood with an Exposure Durability Classification marked "Exterior" A-C (a fully sanded face), you'll be okay. But are you? It's widely recommended, and fine for most installations. But it's still best to contact the flooring manufacturer (carpet, tile, vinyl, or wood) to ask what they recommend for use under their specific product.

In this section I'll cover some of the new underlayments. You can find others in Chapter 10. I'll also cover some of the finish flooring products, with the exception of tile (also covered in the next chapter). It's important to follow the manufacturer's recommendations both for the underlayment and the specific floor covering. That protects you by protecting that manufacturers' warranty. It just might save your hide if something should go wrong later on.

Sound Barrier Floor Systems

Homasote Company's ComfortBase and 440 CarpetBoard are high-density fiberboard available either $^1/_2$ or $^5/_8$ inch thick, with an overall size of 4 × 4 or 4 × 8 feet. They both provide a permanent effective sound system in residential and commercial structures. And the insulating quality (R-value 1.2 for $^1/_2$ inch and 1.33 for $^5/_8$ inch) of ComfortBase and CarpetBoard increases the floor surface temperature. They're also a whole lot softer than a concrete slab. Those are all qualities that your customers can appreciate!

ComfortBase is recommended for application over concrete slabs or concrete floors. The $^1/_2$-inch 440 CarpetBoard is best for wooden subfloors as an underlayment (Figure 9-19). Under carpet, it's all the underlayment you need. They're both suitable for use with laminated or floating floor systems.

Courtesy: Homasote Company

Figure 9-19
CarpetBoard for installation under carpet and pad

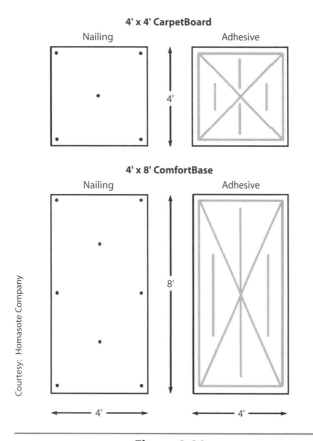

Courtesy: Homasote Company

Figure 9-20
Glue-nail pattern for CarpetBoard and ComfortBase

If you use CarpetBoard as an insulator and plan to install vinyl, wood parquet, wood strip flooring or ceramic tile, install an appropriate underlayment at least ¼ inch thick over the CarpetBoard. Be sure to allow 48 hours of drying time for the adhesive to set. CarpetBoard isn't recommended for areas where moisture is present because it'll expand. But it's environmentally safe and contains no formaldehyde or asbestos additives. It's made from recycled papers, like the rest of their products that we covered in Chapter 4.

Construction Notes — There are a couple of different ways you can install ComfortBase and CarpetBoard, but here's what the manufacturer recommends. Before installing, store the material flat indoors (uncovered) and allow it to adjust to room temperature for at least 24 hours. If the building is excessively cold, hot, or damp, don't install it. In other words, try to install it when the indoor temperature is about the same as it'll be when the structure is occupied.

Stagger panel joints about like you would for a brick pattern. Space all edges ³/₁₆ inch from adjoining panels and ³/₈ inch from walls, partitions, and cabinets to allow for panel expansion and contraction.

Here's how to install ComfortBase over concrete for use under carpet and pad:

▌ The concrete subfloor must be dry and waterproofed.

▌ Glue wood nailers (¹/₂ or ⁵/₈ inch thick by 1¹/₂ inches wide) around the perimeter of the room where you'll install carpet strips.

▌ Use only 4 × 4-foot panels to allow more joints for expansion control.

▌ Apply a ³/₈-inch bead of subfloor adhesive to the back of the ComfortBase, holding back ³/₄ inch from the panel edges, as shown in Figure 9-20. Turn it over with the adhesive side down and apply it to the subfloor. Nail from the center out, using concrete nails and following the glue-nail pattern. Be sure to stagger joints and leave a ³/₁₆-inch expansion space between sheets.

▌ Let it cure for a minimum of 48 hours before carpet installation.

For installation of 440 CarpetBoard over wood for use under carpet and pad:

▮ Glue ^{1}/$_2$- or ^{5}/$_8$-inch-thick by 1^1/$_2$-inch-wide wood nailers around the perimeter of the room where carpet strips will be installed.

▮ Use 4 × 8-foot size panels.

▮ Apply a ^{3}/$_8$-inch bead of subfloor adhesive to CarpetBoard, holding back ^{3}/$_4$ inches from panel edge. Turn it over and place adhesive side down to the subfloor. Nail from the center out, using ring shank nails, as shown in Figure 9-20. Be sure to stagger the joints and leave a ^{3}/$_{16}$-inch expansion space between sheets.

▮ Let it cure for a minimum of 48 hours before carpet installation.

If you decide to use ComfortBase as an insulator under vinyl, wood parquet or strip flooring, or ceramic tile, then install an appropriate ^{1}/$_4$-inch thick underlayment over the ComfortBase. Be sure to allow 48 hours of drying time for the adhesive to set.

Manhours — The difficulty of installing ComfortBase may arise when installing fasteners to a concrete subfloor. Check with the manufacturer on alternatives, like the loose lay or floating methods. Otherwise, use the following manhours:

1. Wood nailers: cut, adhesive/nail, and install on wooden subfloor: 0.037 manhours per linear foot for a crew of one.

2. Wood nailers: cut, adhesive/nail, and install on concrete subfloor: 0.074 manhours per linear foot for a crew of one.

3. Installing ^{1}/$_2$- or ^{5}/$_8$-inch CarpetBoard on wooden subfloor: 0.012 manhours per square foot for a crew of two.

4. Installing ^{1}/$_2$- or ^{5}/$_8$-inch ComfortBase on concrete subfloor: 0.036 manhours per square foot for a crew of two.

5. Applying adhesive to CarpetBoard: 0.042 manhours per 100 square foot for a crew of one.

Fiberock Underlayment

United States Gypsum Company manufactures Fiberock, fiber-reinforced gypsum panels designed specifically for floor underlayment in residential construction. The panels feature a smooth, uniform, indentation-resistant surface you can use as a substrate for resilient flooring, carpet, ceramic tile, and hardwood flooring. According to the manufacturer, it's approved by major resilient floor covering and adhesive manufacturers. The surface is coated with moisture-resistant sealer that enhances floor covering adhesive bonding and workability.

Fiberock panels are available in three sizes: ^{1}/$_4$ inch × 4 × 4 feet, and ^{3}/$_8$ inch × 4 × 4 foot or 4 × 8 feet. The ^{1}/$_4$-inch panels have up to 52 percent more indentation resistance than traditional wooden underlayments (75 pounds per square foot). But they're not recommended for use in areas subject to prolonged exposure to standing water, like shower stalls, saunas, and hot tub decks. Also, don't install them directly over concrete subfloors, either below grade, above grade, or suspended.

Construction Notes — A nailing pattern is printed on the top surface of each 4 × 4-foot square. Always carry the panels upright, not in a horizontal, flat position, or you could crack them. Follow these additional guidelines:

1. Store panels flat and at room temperature for at least 48 hours before installing.

2. For installation over crawl space areas, provide a minimum of 18 inches of well-ventilated space and cover the ground with a minimum 4-mil vapor barrier recommended for below-grade application.

3. Inspect the subfloor for an even, flat surface. Check end and edge joints for uneven edges and variations in panel thickness which could show through the underlayment. If necessary, sand uneven joints and/or fill voids and large gaps with an approved floor leveler.

4. Cut panels using the score-and-snap method and install smooth side up, maintaining a ^{1}/$_4$-inch space between perimeter walls. Butt edges and ends lightly together (maximum of ^{1}/$_{32}$-inch gap allowed).

Courtesy: United States Gypsum Company

Figure 9-21
Installing Fiberock with pneumatic tools

5. Use a fastener with a leg length the combined thickness of the underlayment and subfloor. Space fasteners ¹/₂ inch in from the panel edge, every 3 inches or less around the perimeter, and 6 inches in the field. Fasten one panel at a time (Figure 9-21). If you use pneumatic tools, apply pressure on the gun to hold the panel tight to the subfloor before squeezing the trigger. Don't hold the trigger and bounce the tool.

6. If you staple ¹/₄-inch panels, a combined glue/staple application is required. Avoid adhesives that can stain vinyl flooring, like solvent-based adhesives or adhesives that contain dark processing oils.

7. Apply the adhesive with a ¹/₈-inch notched trowel in a 6-inch-wide strip around the perimeter and lengthwise down the middle of each panel. If you're using a cartridge-style adhesive, apply in a zigzag pattern spanning the 6-inch-wide strip, then level it with the ¹/₈-inch trowel. Don't allow adhesive to fill panel joints.

8. Finish joints by lightly sanding uneven panel joints and/or fill all end and edge joints and any other imperfections using a portland cement-based floor patching compound.

Manhours — These figures are based on a crew of two installing 4 × 8-foot panels using either the nailing or adhesive/stapling method. If the floor joist and subfloor surface aren't level, you need to add in additional time for prep work. Be sure to check the subfloor before giving a bid.

▌ Installation of underlayment: 0.008 manhours per square foot.

▌ Finishing end and edge joints: 0.014 manhours per linear foot for a crew of one.

Cork Underlayment

Why cork? It's comfortable, warm, quiet, aesthetically pleasing and a renewable resource. And what is cork? It's the outer skin (bark) of the cork oak tree found in Spain and Portugal, which takes 25 years to mature and can live to be 500 years old. In the summer months the tree dehydrates, loosening the bark so it can be harvested. Harvesting doesn't harm the tree because the cork bark is self-replenishing. The newly-stripped area is protected by a thin inner bark. It grows back and the process can be repeated every nine years (Figure 9-22).

Courtesy: Wicander Enterprises, Inc.

Figure 9-22
Harvesting the bark from a cork oak

Cork Underlayment on Wood Joist — Lightweight Concrete Floor System					
Subfloor	**Suspended ceiling**	**WECU cork thickness**	**Floor covering**	**STC (Sound Transmission Class)**	**IIC (Impact Insulation Class)**
1½" lightweight concrete on ⅝" plywood	No	½" Soundless+	Tile	58	52
1½" lightweight concrete on ⅝" plywood	Yes	½" Soundless+	Tile	59	56
Cork Underlayment on 6" Concrete Slab					
Subfloor	**Suspended ceiling**	**WECU cork thickness**	**Floor covering**	**STC (Sound Transmission Class)**	**IIC (Impact Insulation Class)**
6" concrete	No	½" Soundless+	Tile	56	53
6" concrete	Yes	½" Soundless+	Tile	56	63
6" concrete	Yes	¼" Soundless	Tile	56	56
6" concrete	No	¼" Soundless	Hardwood	56	52
6" concrete	Yes	¼" Soundless	Hardwood	56	59

For results on independent testing, contact Wicander Enterprises. In the last two columns the higher the number above 50 (Standard Building Code), the better the rating.

Courtesy: Wicander Enterprises, Inc.

Figure 9-23
Sound ratings for WECU cork underlayment

Cork is 50 percent air (that's 200 million air cells per cubic inch) which provides natural thermal insulation, excellent acoustics, and contributes to a healthier home because it doesn't trap dirt and fungi.

Wicander Enterprises, Inc., has promoted cork underlayments since 1979 for application under ceramic tile and marble. Their special high-density underlayments contain a resin binder to prevent the cracking and moisture problems normally associated with lower-density and polyurethane-binder underlayments. Today, along with their distributor, W.R. Bonsal Company, they offer WECU-Crackless, a $3/32$-inch underlayment that provides stress protection for ceramic tiles by preventing existing or future stress cracks in the subfloor from telegraphing through the underlayment. For construction in areas that need to meet sound control requirements (like condominiums in California or Florida), they offer WECU-Soundless ¼ inch and WECU Soundless+ ½ inch underlayments. They offer efficient sound control over concrete or wood subfloors for use with ceramic tile or hardwood floors. The chart in Figure 9-23 gives sound ratings for ¼-inch and ½-inch cork underlayment.

Construction Notes — To lay tile over cork underlayment, set with sanded thinset mixed with minimum acrylic latex solids of 20 percent to insure a strong bond. For marble, install using a marble-set or mud-set method. For hardwood floors, follow the floor manufacturer's recommended adhesive for application on cork underlayment. Don't use nails or screws. Follow these guidelines:

1. $3/32$-inch (WECU-Crackless) for stress crack suppression (Figure 9-24). Available in a 4-foot-wide roll 200 feet long. Be certain to follow the manufacturer's recommended curing time before permitting traffic on the finished floor.

2. ¼-inch (WECU-Soundless) for sound control for concrete structures with suspended ceilings. Available in 100 feet by 4 feet wide.

Courtesy: Hoobs© 93/W. R. Bonsal Company

Figure 9-24
Installing tile over WECU-Crackless underlayment

3. $^1/_2$-inch (WECU-Soundless+) for optimum sound control for concrete and wood-joist structures with or without a suspended ceiling. Available in widths of 3 and 4 feet and lengths of 4 to 10 feet.

4. Spread mastic using a $^1/_{16}$-inch V-notched trowel (Type 1 mastic).

5. Install it up to walls snugly, and with butting and staggered seams.

Courtesy: Natural Cork, Ltd., Co.

Figure 9-25
The cork pattern on the floor continues on the doors

6. Starting from the center, use a 75- to 100-pound roller and roll to the outer edge to prevent air bubbles.

7. Wait 24 hours for the adhesive to cure before installing floor covering.

8. For sound control, leave a $^1/_4$-inch gap between the perimeter wall and floor covering.

Manhours — If the floor joist and subfloor surface aren't level, you need to add in some time for prep work. Be sure to check the subfloor before giving a bid. These figures are based on a crew of two:

▌ Installing $^3/_{32}$-, $^1/_4$- or $^1/_2$-inch cork underlayment over concrete or wooden subfloor: 0.014 manhours per square foot.

▌ Applying adhesive to subfloor: 0.0215 manhours per 100 square feet.

Cork Flooring

Farmers who grow cork oaks generally contract their harvest to the manufacturer. In the case of cork flooring, the manufacturer is Applicork. In the past 15 years they've turned their attention from cork stoppers to a more profitable area — interior furnishings. Now they have a vested interest in Natural Cork Ltd., Co.'s multiple lines of cork products ranging from floor and wall coverings, underlayment and acoustical insulation, and textile fabrics distributed throughout United States and Canada.

Interestingly enough, natural cork is a fire inhibitor. It doesn't encourage flame-spread or release any toxic gases on combustion. The bark is made up of dead cells filled with air and lined with alternating layers of cellulose and a waxy substance called *suberin*. Suberin is a natural insect repellent — even against termites. This unique cellular structure gives cork acoustic and impact noise resistance, and it's a moisture barrier, impervious to water (good for bathroom floors). Its resiliency and elasticity give cork flooring a comfortable, cushioned feeling.

One of Applicork's products is Natural Cork Planks Plus. Prefinished in six design/finishes, these $^1/_2 \times 12 \times 36$-inch planks simply look good. Look at the handsome office in Figure 9-25. The Merida pattern on the floor has been incorporated into the doors. The planks include three laminations: a $5/32$-inch cork top, a $3/32$-inch cork bottom, and a $1/4$-inch thick core made of fiberboard with tongue-and-groove detailing. They're ideal for remodeling projects because you can install them over virtually any existing floor that's smooth and level. The planks don't require underlayment, only a 6-mil vapor barrier over the substrate, according to the manufacturer.

Construction Notes — To learn more about the installation and care of Natural Cork Planks Plus, contact the manufacturer for installation guidelines. Basically, it's a floating floor system held down and in place by its own weight, not attached to the substrate but glued together only in the T&G. The manufacturer recommends Franklin's Titebond II wood glue at approximately 8 ounces per 100 square feet of flooring. Here are a few installation highlights:

1. Over concrete above, below, or on grade, first apply 6-mil film, overlapping seams by 8 inches and taping the sheets together.

2. For wooden subfloors over basements, cover the ground with 6-mil film.

3. For homes with an in-floor radiant heating system, contact the manufacturer.

4. Acclimate the planks to room temperature (between 60 and 85 degrees F with relative humidity between 50 and 70 percent) for at least 72 hours prior to installation.

5. Try to begin with the longest wall that runs parallel to the plank floor, and leave $1/2$ inch around the perimeter of the room. You'll need spacers at the end and sides (two per plank) when installing. Start on the left-hand side with the groove facing the wall.

6. Apply a $1/8$-inch bead of glue to the groove (not the tongue) on both the sides and ends.

7. Stagger joints by at least 10 inches and use a knocking block to protect the tongue of the plank.

8. Once the floor has cured for at least 24 hours, remove spacers and install the base moldings.

Manhours — These manhours are based on a one-man crew and don't allow for a learning curve. To install the complete system, including base molding, use the following:

▌ To install planks: 0.042 manhours per square foot.

▌ To install base moldings up to $3^{1}/_{2}$ inches: 0.016 manhours per linear foot.

Natural Bamboo Flooring

Bamboo is a plant (genus Bambusa) of the grass family which normally grows in warm or tropical regions — China, for instance. Bamboo stalks are hollow, usually round and jointed, and can reach heights of 100 feet and diameters exceeding 6 inches. If you know anything about bamboo, you know it doesn't matter how many times you pull it out of the ground — it still grows. That's because mature bamboo has extensive root systems that continue to send up new shoots for decades. Here in the States we call it *cane*. When bamboo is harvested in managed forests, it replenishes itself. Since harvesting is done by hand, it minimizes the impact on the local environment.

In 1989, Smith & Fong Company began to develop new uses and applications for bamboo. They developed a process called *plyboo*, which simply means "laminating bamboo." Because bamboo has to be milled into flat narrow strips from the core of its wall, it has to be laminated to manufacture products such as flooring, paneling, and plywood. Prior to lamination, the strips have to be boiled in a bath of boric acid and lime solution to extract the starch that attracts termites or powder post beetles. This process assures the finished flooring won't be affected by pests.

After kiln drying, the bamboo strips are sanded and laid side by side to create a single-ply panel. These panels are then laminated to each other again to form multi-ply products. The manufacturer offers two colors, amber and natural (the natural color of bamboo). Amber, a color similar to tea stain, is naturally created by a hot steam treatment that cooks the bamboo, creating a permanent coloration that won't fade or sand out. Both colors look great, especially in their T&G plank bamboo flooring. The natural joints of the bamboo provide a distinctive look

Courtesy: Smith & Fong Company

Figure 9-26
Natural bamboo flooring has a distinctive pattern

throughout the finished flooring. The planks shown in Figure 9-26 are available in $^5/_8 \times 3^5/_8 \times 72$ inch. Notice they're "end-matching" as well (T&G ends).

Construction Notes — This flooring installs just like fir or oak T&G floor planks, so use the same techniques and guidelines recommended by NOFMA (National Oak Flooring Manufacturers Association

— see Appendix). Use a $^1/_2$-inch nail gun, either manual or pneumatic. Space fasteners 8 to 10 inches apart with a minimum of two per board near the ends.

The QuikJack by Cepco Tool Company (Figure 9-27) is specially designed to pull in the last flooring board. Remember to protect the wall with a bearing plate long enough to span at least three or four framing members. A bearing plate that falls between framing members may push through the finish wall as you use the tool. (Use caution when pressing against the wall, as the jack may push the wall out.) Because of the versatility of this tool, it can push, pull, spread, or join virtually any type of material. You can work with it anywhere within the room because the accessory gripper and retaining strap allow the tool to straddle a 2 × 4 (providing it doesn't exceed 10 feet unless a temporary brace is used). Here are some other things to keep in mind:

1. Bamboo is about 90 percent the hardness of red oak (oak is 1,290 psi; bamboo, 1,130 psi).

2. You can change or match the color with dyes and stains.

3. Water- or oil-modified finishes work well on bamboo flooring.

4. The flooring carries a 10-year warranty from date of purchase for delamination under normal conditions.

Courtesy: Cepco Tool Company

Figure 9-27
The QuickJack helps pull in the last flooring board

Manhours — The figures include installing felt and nailing flooring for a crew of one, but don't include installation of base moldings. I recommend a subcontractor to install and finish the floor, but if you plan to finish the floor yourself, use the following manhour figures:

▌ Installing $^5/_8$ × $3^5/_8$ × 72-inch bamboo plank flooring: 0.062 manhours per square foot.

▌ Sanding, three passes (60/80/100 grit): 0.023 manhours per square foot.

▌ Two coats of stain/sealer: 0.010 manhours per square foot.

▌ Two coats of urethane: 0.012 manhours per square foot.

▌ Sanding and 2 coats of lacquer: 0.047 manhours per square foot.

Courtesy: Wilsonart International Inc.

Figure 9-28
Use spacers when installing Wilsonart flooring

Laminate Flooring

I wouldn't have believed it until I saw the sample — laminate in a T&G plank flooring. But that's just what Wilsonart International created. It's Wilsonart Flooring, new to the market in 1996. It's available in 20 woodgrain patterns and colors with a size of $^5/_{16}$ × $7^3/_4$ × $46^1/_2$ inches. The T&G is milled into the ends as well as the sides. This floating floor system is manufactured in a three-part lamination. The *top* is the finish surface, made of high-pressure laminate material similar to countertop material but ten times more durable. The *core*, which is the bulk of the finished flooring, is made of 50- to 55-pound HDF. The *bottom* is a high-pressure backer laminate to help balance the plank and add strength. Where would you use flooring like this? In 1996, Wilsonart conducted a survey and received nearly 40,000 responses. About a third of those considering a laminate floor planned to use it in a kitchen.

From a remodeler's standpoint, the floating floor can be a great problem-solver, since you can install it directly over most existing floors. This might include tile, vinyl, wood, or even concrete, without the worry of removing any flooring. This alone could make the difference between getting the job or not getting the job.

Construction Notes — This is a prefinished flooring, which means you never want to sand, lacquer, or wax it. Wilsonart offers an impressive warranty (to the original purchaser) that its laminate flooring won't wear through, fade, or stain under normal use and service for 16 years for residential use, and five years for commercial applications. This is a limited warranty. Contact Wilsonart (see Appendix) to get your copy. Wilsonart also offers prefinished moldings and other products to complete the job.

Here's a quick outline so you can get an idea of how Wilsonart flooring is installed. To get the full scope of the installation, request their guide.

1. Leave $^1/_4$ inch around the perimeter of the room and the plank flooring for expansion. Use spacers to maintain a consistent $^1/_4$ inch (Figure 9-28). Wilsonart sells spacers specially designed for use with floating floor systems.

2. When installing over concrete, install a vapor barrier and overlap the seams by 16 inches, but don't tape them. Since this is a hard substrate, consider using Wilsonart's foam padding to absorb noise and impact before installing finish flooring.

3. Store flooring at room temperature in unopened cartons for at least 48 hours prior to installation.

4. Try to begin in a corner of the room where there will be no traffic for at least one hour.

5. Always check the starting wall for straightness. For an uneven wall, use a straightedge that can be nailed to the subfloor. Place the groove side of the plank facing the wall on the first course.

6. Never apply glue to the tongue, only the grooves. Wipe off any excess glue as you go.

7. Never tap on the tongue. Use a scrap piece or their tapping block.

8. After the first four rows of planks have been completed, allow the floor to set for 45 minutes before continuing the rest of the installation.

9. To install planks underneath door jambs and casings, they'll need to be cut. The easiest way is to use a piece of foam, a flooring scrap, and a Takagi double-edge saw (Model #10-2440). Place the saw flat on the scrap's surface with the 17 point side toward the cutting surface. Make sure most of the teeth are on the face before pulling to make the first cut. Take it slowly and make easy strokes.

Manhours — Each box contains eight planks or enough to cover approximately 20 square feet. Wilsonart suggests that 300 square feet of flooring can be installed in less than eight hours. If this is the case, it works out to around 0.02666 manhours per square foot. But don't expect to match that estimate the first time you install a floating floor system. Consider using the following figure the first time around for a crew of one:

▌ To install planks: 0.042 manhours per square foot.

▌ To install base moldings up to 3½ inches: 0.016 manhours per linear foot.

Enviro-Tech Carpet

Roughly 3 million plastic soft drink containers are discarded each hour. While they only constitute 6.5 percent of all landfill waste by weight, they take up eight times the space per pound of paper or glass. Containers like soft drink and ketchup bottles are produced from polyethylene terephthalate (PET), a thermoplastic material which can be melted and reformed with virtually no changes in its physical properties. Image Carpets has been successfully manufacturing carpets using PET for over 15 years. They help to keep more than 180 million pounds of recyclable high-grade plastic bottles out of our landfills each year.

The process they use is interesting. The bottles that come into the plant are sorted, the heavy plastic bottoms and caps are removed, and the remainder is ground up into irregular ¼-inch chips along with the paper labels, glue from bottles and labels, and other impurities. Any paper is blown out of the mixture by air jets, then it moves along the assembly line to a chemical bath that strips the glue. From there the chips go to a two-step water/electrostatic process that isolates PET chips from other impurities. Figure 9-29A shows clean PET chips ready for the final process. These chips end up in about 35 percent of all polyester carpet made in the United States.

These PET chips are melted to the consistency of honey and forced through molds that create fibers about the thickness of a human hair. Finally, the fiber is spun into yarn and tufted into carpet (Figure 9-29B). PET polyester is one of the strongest man-made fibers. Its versatility, durability, and inherent resistance to stains make PET the popular choice for a variety of consumer products ranging from seat belts to tire cords to carpet.

Image Carpets handles their own recycling, processing for the polyester fibers, and manufacturing of the carpet. They also provide other manufacturers with PET and fibers. Two yarn systems made from the Enviro-Tech process are *Resistron* and *Duratron*. Both are made from the high-molecular-weight PET fibers. Resistron, a hydrophobic PET fiber, resists wet and dry soil and diffuses static build-up. Duratron blends nylon and PET for rich, lustrous colors with increased static control, as well as soil-, stain-, and wear-resistance. These are environmentally-friendly carpets that you need to feel and see to believe. They give you something new to offer your customers.

Manhours — This carpet installs just like any other carpet, with this one exception: it's available in widths of 15 feet (as well as the standard 12-foot width). Just think, you could be installing with no seam! Normally I recommend using a subcontractor because it's so hard on the knees to install carpet, but depending on circumstances or the size of the job, you may end up installing it yourself. If you're

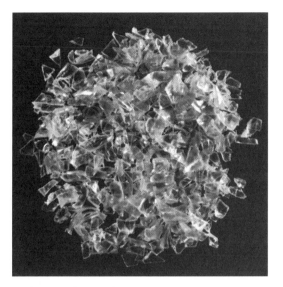

A Clean PET chips ready to use

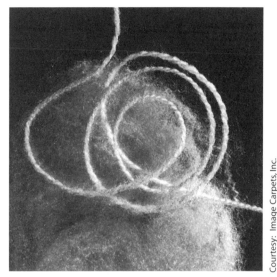

Courtesy: Image Carpets, Inc.

B The fibers and yarn made from the chips

Figure 9-29
The raw ingredients of Enviro-Tech carpet

installing Resistron and Duratron carpet with rebond pad, use these manhours figures based on a crew of two:

- 21 to 36 ounces: 0.114 manhours per square yard.
- 36 to 50 ounces: 0.119 manhours per square yard.
- 50 ounces plus: 0.125 manhours per square yard.
- Add for box steps: 0.076 manhours per riser.
- Add for wrapped steps, open riser, sewn: 0.121 manhours per step.
- Add for sewn edge treatment: 0.114 manhours per riser.
- Add for circular stair steps: 0.243 manhours per step.

Problem Solvers

These "problem solvers" don't fit into a specific category except they're interior products that can make life easier. Some of them would make a good gift from you to your customer at the end of the job. Keep them in mind for that finishing touch in any remodel.

Hinge-It

In every home there are at least $3^1/_2$ inches of unusable space behind every door, between the adjacent wall and the open door. This is space that most of your customers don't realize exists. So what can you do with such a small depth of space? Plenty! With the help of Hinge-It Corporation, you have products at your fingertips that can easily convert this space so they can hang pants in the bedroom or towels in the bathroom. They manufacture over 75 different *behind the door* hangers that can help people gain control over what my wife calls "organized clutter."

Two of their spacesavers have really solved a couple of problem areas in our home, in the boys' bedroom and in the bathroom. The Hinge-It Clutterbuster is made of five 18-inch long birch arms with $1^1/_4$-inch solid birch ball end caps, finished naturally. It's great for the kids' room for hanging miscellaneous clothes or pants (Figure 9-30A). The Hinge-It Eurorack is a one-piece $^3/_4$-inch diameter tubular pipe available in white, polished brass, and stainless steel. It really gives bathroom towels a chance to dry (Figure 9-30B).

A Clutterbuster

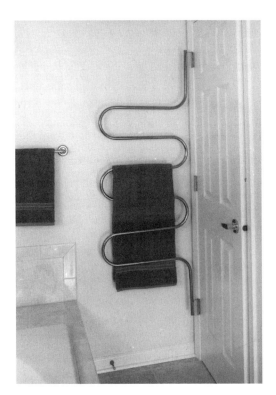

B Eurorack

Figure 9-30
Hinge-It space savers

Hinge-It products are held in place by the pins from the door hinges, except Jam-It, which attaches to the door jamb with screws. Because they're reasonably priced, consider installing one or two at the end of a job as your way of saying "thanks" to the customer for choosing you to do the work.

I'll give you the manhours for the two products I've installed (from the time I took them out of their boxes to hanging them on the hinges), which will give you a good idea how to bid. Of course, in some cases, you might spend more time pulling pins out of the hinges than it takes to build or hang the unit, especially on older homes where the hinges might sport ten years' worth of paint. Also, if you want to install a Heated Eurorack and there's no outlet close by, you'll have to add the cost of an electrician to your overall figure. In this case, forget the freebie.

▎ Hinge-It Clutterbuster (64 inches × 20 inches), which has to be assembled: 1.0 manhour per unit.

▎ Eurorack (polished brass), a one-piece unit: 0.25 manhours per unit.

Quick Connect

How many remodeling projects have you done where installing a dryer vent was part of the job? Perhaps you've replaced the floor covering in a laundry room, only to move the dryer and discover the hose was connected incorrectly, the flexible plastic pipe was filled with lint, or it was even crushed and restricting air flow? Customers have no idea that roughly 13,000 fires are caused each year because of built-up lint. You can do your customer a service by correcting the situation. Quick Connect by Best Dressed Homes Company is a product that's recommended by major dryer manufacturers. It's simple to install, allows easy access behind the dryer, and is clean and safe for the home.

As you can see in Figure 9-31, Quick Connect can be appropriate for many applications. Before running out to the job site, be sure you have all the correct components, such as additional elbows and/or vent pipe, to complete the project in one trip. For all the installations shown, you'll need to purchase or use the existing outside 4-inch dryer hood. The secret to the kit is the specially-designed donut-shaped connector that allows the pipe from the dryer to hit the target and slide in without any hassles. Estimate 0.50

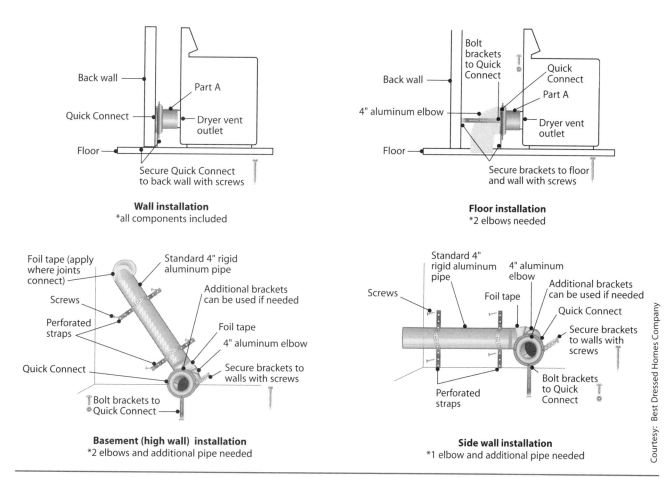

Back wall

Part A

Quick Connect

Dryer vent outlet

Floor

Secure Quick Connect to back wall with screws

Wall installation
*all components included

Bolt brackets to Quick Connect

Back wall

Quick Connect

Part A

4" aluminum elbow

Dryer vent outlet

Floor

Secure brackets to floor and wall with screws

Floor installation
*2 elbows needed

Foil tape (apply where joints connect)

Standard 4" rigid aluminum pipe

Screws

Additional brackets can be used if needed

Perforated straps

Foil tape

4" aluminum elbow

Quick Connect

Secure brackets to walls with screws

Bolt brackets to Quick Connect

Basement (high wall) installation
*2 elbows and additional pipe needed

Standard 4" rigid aluminum pipe

4" aluminum elbow

Screws

Additional brackets can be used if needed

Foil tape

Quick Connect

Secure brackets to walls with screws

Perforated straps

Bolt brackets to Quick Connect

Side wall installation
*1 elbow and additional pipe needed

Courtesy: Best Dressed Homes Company

Figure 9-31
Installing Quick Connect

manhours for installing Quick Connect on an existing system. If you need to start from scratch or install it new, then figure on spending at least an hour.

Organizers

Here's one of those ideas that provides a finishing touch to a kitchen project. Vance Industries offers organizers for the kitchen, bath, office, shop, and many other areas that just need a helping hand. Their Trim-Fit, made of high-impact polystyrene, is just what it says: an organizer that can be trimmed for a custom fit. It's available in white, almond, or the Perfect-Fit premium line that comes in white or gray granite (Figure 9-32). The unique design, combined with extra deep (2-inch) compartments, provides improved organization and increased storage capacity. Again, this is one gift you can give to your customers as your way of saying "thank you." It can certainly create good public relations at a small cost.

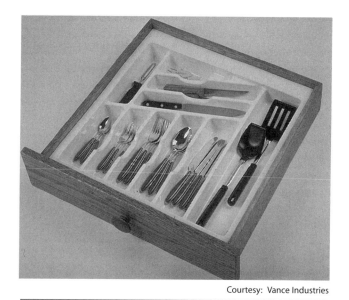

Courtesy: Vance Industries

Figure 9-32
Perfect-Fit Premium drawer organizer

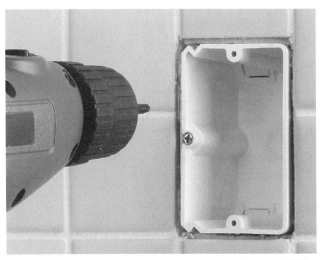

Courtesy: RACO Inc.

Figure 9-33

Just loosen the screw to bring the box flush with the tile

Courtesy: RACO Inc.

Figure 9-34

The Adjust-A-Brace to attach outlet boxes to metal studs

Organizers are available in different sizes and configurations to fit just about any drawer size and any type of application your customer might have in mind. It's easy to trim all of Vance's drawer organizers for a custom fit using a score and snap method (with a straightedge and utility knife). Their cabinet and universal shelf organizers may require a bandsaw or table saw. On the job site, figure at least 0.25 manhours per drawer.

Adjust-A-Box

Of all the problem-solvers currently on the market, this one is a remodeler's dream come true. Adjust-A-Box by RACO/Hubbell Electrical Products is an electrical box that actually clamps onto a stud. The universal 16-gauge steel bracket has tangs that bite in and grip the stud, securing it in place without nails. The coined holes on the bracket's face are used to attach to metal studs with two sheet metal screws. This clip-on feature saves time, the box mounts squarely, and it allows for easy repositioning. It's UL approved to attach the single-gang box without a fastener. But I recommend using a wire staple to attach the bracket to the side of the stud. The double-gang box has to be secured with a fastener.

Its ease of attachment isn't the main feature of this outlet box. The icing on the cake is that the box is completely adjustable for any wall thickness up to 2 inches. You can remove a single gang box and after making just three cuts to enlarge the hole, install a

double gang box on the same bracket. (This is where Takagi's Rockeater Drywall Saw will come in handy.) Just think of the expense you can save, not to mention wall repairs. This box fits flush to any new addition to the wall (Figure 9-33), can be removed without opening a wall, and provides easy access to any inner wall cavity.

RACO/Hubbell Electrical Products have one other product I want to mention. Their Adjust-A-Brace is a simple solution to installing outlet boxes to metal studs (Figure 9-34). There's no need to make your own brace; save that valuable time for something more constructive. It also gives you the option of hanging boxes at any point between the studs. You can do back-to-back and side-by-side installations, mount gang boxes, and combine boxes of different depths on the same hanger. Keep the electrical boxes by RACO in mind during the framing stage, covered in Chapter 4 under "Framing Materials."

Figure your manhours at 0.025 per unit for the Adjust-A-Box. For Adjust-A-Brace, use 0.012 manhours per unit.

The next chapter, Bathrooms and Kitchens, will introduce several worthwhile products that will help keep down the maintenance in two of the most heavily-used rooms in a home. Some of them will also help boost your creative juices into overdrive. I don't know about you, but the bathroom is my favorite place in the home to remodel. So let's get started.

10

Bathrooms and Kitchens

What's your favorite area of a home to work on? Personally, I enjoy bathroom projects, both remodeling and building new. Perhaps you prefer working on kitchens. These are both areas that present unique problems because they receive constant use and take considerable abuse. The upside is that they're a good way to showcase your professional skills.

Bathrooms started out as a basic utility, but they've gone way beyond that. Today's customers want luxurious comfort, including more space, steam rooms, exercise equipment, or larger (even social) bathing facilities. And kitchens frequently serve as the focal point for family gatherings. So what's your role in all this? You'll want to suggest installations and products to enhance your customers' bathrooms and kitchens and meet their very personal needs. As a skilled contractor, you can create bathrooms that are important — and yet very private — rooms. You can design and build kitchens filled with warmth and personality.

Bathrooms and kitchens are big business. Both can generate large sales and big profits. How profitable they are depends in large part on your knowledge of new products on the market, and how well your company can sell and install those products.

But here's a word of caution: Don't overlook the terms of the Americans with Disabilities Act (ADA). Familiarize yourself with barrier-free issues and codes as defined by your local building department. Right now, barrier-free codes generally only apply to commercial installations, but I think it's only a matter of time before they become mandatory in residential applications. Just look at what's happening in people's lives. More and more people remain in their homes longer than previous generations did, and many parents are now moving back in with their children instead of going to commercial facilities like nursing homes. The time will come (if it hasn't already) when you'll be asked to create barrier-free living space, particularly in bathrooms and kitchens. Being abreast of the law and current trends in the field, as well as available products and their proper installation, can give you a major edge over the competition. For detailed guidance on this, order *Accessible Housing,* (www.asktooltalk.com). It's packed full of useful information.

With that in mind, I want to introduce you to some state-of-the-art products. Some may lengthen the life expectancy of some areas of the bathroom or kitchen. And their sheer good looks will ensure that you leave behind a satisfied customer, the kind that provides you with many referral customers!

Repair Instead of Replace

Most major problems in any bathroom appear in the bath/shower enclosure (fixture or wall). It may be a lack of proper maintenance by the customer, but problems can also arise because a product was improperly installed or the wrong product was used in this very specialized environment.

Courtesy: Re-Bath Corporation

Figure 10-1
Trial fitting a liner over a bathtub

Courtesy: Re-Bath Corporation

Figure 10-2
Trial fitting over an existing tile wall

Re-Bath

What do you do if you're lucky enough to get a job remodeling multiple bathrooms, such as in a hotel or motel? Or if there's nothing wrong with an existing residential bathroom except for the surface of the bathtub? Maybe the porcelain has chipped, lost its shine, or the new owners just want a color change.

A bathtub is a pretty expensive item to replace because the project always requires opening three walls. If your customer is on a budget, you may want to suggest Re-Bath Corporation. They manufacture a bathtub liner that completely covers existing cast iron or steel (not fiberglass) bathtubs. Figure 10-1 shows a Re-Bath installer trial-fitting a liner over an existing tub. The customer already had an enclosure installed before deciding she wanted a tub to match the enclosure's color. Normally, the enclosure would be installed on top of the liner to create a better seal. The end of this tub has a small pony wall with a finished ledge that will require some attention to make it match or blend in with the finish product.

Re-Bath also offers shower base liners and wall surrounds designed to fit over existing shower pans and tile walls (Figure 10-2). Just think of the mess you and your customer avoid by using a custom-made product for the job!

Re-Bath's products are made of smooth, lustrous, nonporous ABS (acrylonitrile butadiene styrene) acrylic, available in a wide range of colors to fit any decorating scheme. They're installed by factory-trained applicators (subcontractors). The dealer will measure the existing bathtub, take photos, and then match those measurements and photos to one of over 550 molds stocked in their manufacturing facility. Then the acrylic is vacuum-formed and the liner is shipped ready for installation.

Construction Notes — Installation doesn't take long. First they clean the existing tub of foreign material, rust, or any other surface imperfections, and remove the trim from the overflow and drain. Then they precisely fit the liner to the tub, an important step since the caulk between wall/floor and liner must provide a smooth transition. The gap on the liner must be as small as possible to achieve a good bond. Next they apply special adhesives to the existing tub, and position the liner in place. After installing the overflow and drain trim (also supplied by the manufacturer), they caulk the liner's joint for a final seal and a professional finished look. Re-Bath can be a cost-effective (and great looking) solution to bathtub problems.

Manhours — Since this is a dealer-installed product, the likelihood of your installing it is slim unless you become a dealer. I'll give you some estimated manhours so if you do subcontract this out, you can at least provide your customer with an estimated time frame:

▌ To install bathtub liners: 4.0 to 5.0 manhours per unit.

▌ To install shower base liners: 2.0 to 3.0 manhours per unit.

▌ To install wall surround: 6.0 manhours per unit.

Unique Refinishers

Have you ever bid on a project where you have to freshen up a bathroom without removing the fixtures or walls? How about when the customer just wants a different color scheme, but the fixtures are in good shape? What should you do? You could tell the customer "No problem. I'd be glad to bid the job but it'll be expensive. Removing a tub is costly and can inconvenience the household for a week or two." Or you could consider alternative products that can help you get the job done *and* save money.

I've never believed in reglazing bathtubs, ceramic tile, plastic laminate, cultured marble, and other surfaces — until now. But procedures and products have vastly improved over the last ten years, so don't rule it out of the bidding process. It may be the best solution for some jobs.

Unique Refinishers has been doing it for more than 30 years. They offer a one-day process at about 20 percent of the cost of replacement. This is definitely something to consider, especially if your customer is on a tight budget. Sure, it's not as good as ripping everything out and starting afresh, but you'll find a lot of customers have no idea how expensive a bathroom job can be. When you tell them, they'll politely show you the door. If you can offer them reglazing, you're offering them an option they can afford.

Construction Notes — Factory-trained applicators clean the surface using their exclusive etching process, an important step because it assures a strong bond between the old and new surfaces. Chips and cracks are individually repaired before the glazing is applied. Then they apply the glazing, available to match any color sample, with a spray gun. Incidentally, the applicator uses an efficient ventilating system to remove odors and dust from the area. What a great way to give a tired bathroom a fresh new look, at an affordable price!

Manhours — Again, this process requires a factory-trained applicator. However, so you know the time involved, it takes approximately 2.0 to 3.0 manhours for either a tub or wall surround. A two-man crew can do a complete bathroom in 8.0 to 9.0 manhours.

Swan

What do you do when a tub/shower valve springs a leak? If the shower walls are still good, what alternatives can you offer the customer? There are many choices, but which one is most economical? In some cases, you can repair or replace the plumbing by accessing it from the backside (an option discussed in more detail later). If not, your only option is to enter the front wall to make repairs.

Depending on the wall covering, entering the front wall can be costly, but Swan's fiberglass Shower Tower makes this job almost a snap. The unit permits you to replace a valve while maintaining the surface of the existing plumbing wall, whether it's solid, fiberglass, or tile. It comes as a single piece, preplumbed and in a full range of colors to match or complement any color scheme or existing walls. A standard shower head or hand-held shower finishes it off.

Figure 10-3 gives you a bird's eye view of the entire system and its components. The unit is $12^1/4$ inches wide and 58 inches tall, with foam strips installed on the back to prevent water seepage. The model shown, TM1-S, is a standard shower head for tubs only. They also make a unit for shower stalls, called SM1-S.

The Shower Tower in Figure 10-4 was installed over a fiberglass tub enclosure. Notice how well it blends in with the rest of the enclosure. Just think of the time and money savings if you had to do an entire apartment house.

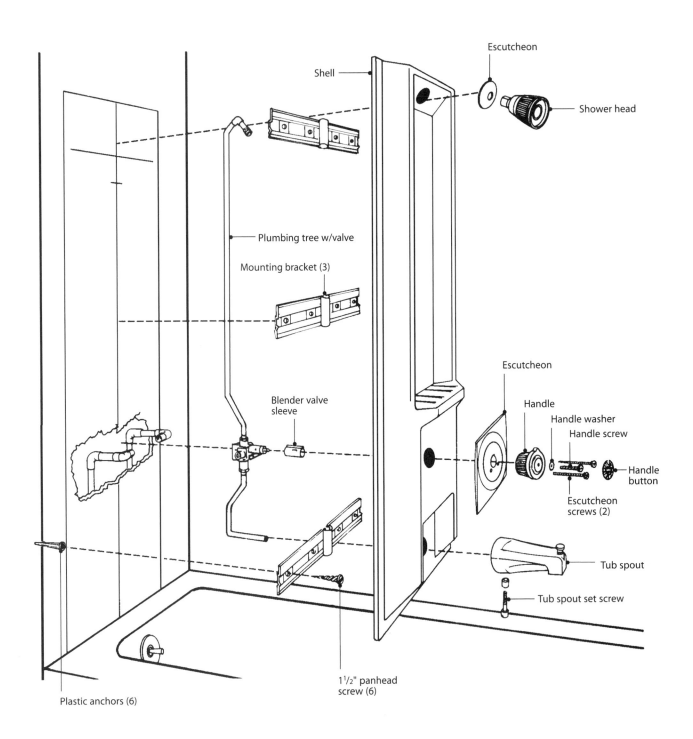

Figure 10-3
Shower Tower standard shower/tub system and components

Courtesy: The Swan Corporation

Figure 10-4
The Shower Tower installed

But what about many customers' favorite: the ceramic tile bathroom? It gives a rich feeling and allows the customer to express individual tastes with one-of-a-kind designs. Unfortunately, tile enclosures are high maintenance. Even with improved grout, the best solution to many tile problems is a disciplined customer wiping down the shower walls after every use to get rid of most of the moisture. This isn't going to happen — you know it, I know it, and the customer knows it. So, as soon as the grout discolors or cracks or mold starts to appear, your customer calls to inquire about options.

Introduce them to SwanTile (TI-5), a unique enclosure that resembles both the look and feel of an expensive grade of tile, without the maintenance headaches. The unit includes heavy-gauge molded fiberglass panels in coordinated colors, and includes two shampoo shelves and a built-in soap dish (Figure 10-5). Extension panels are available that add an additional 24 inches of height to get you closer to the ceiling. Swan's five-piece kit also includes two apron strips for installation down the side walls up against the front of the tub.

Courtesy: The Swan Corporation

Figure 10-5
SwanTile bathtub enclosure

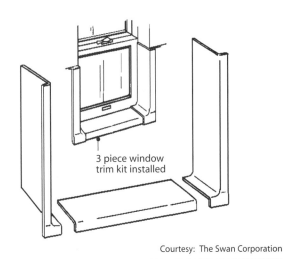

3 piece window
trim kit installed

Courtesy: The Swan Corporation

Figure 10-6
Swan's window trim kit

Another favorite of mine is Swan's Tubwal Model TW-32, a smooth-wall five-piece unit. If there's a window in the shower, use the three-piece matching window trim kit which fits most windows and completes the enclosure's overall appearance (Figure 10-6). Personally, I recommend removing the window from the shower area altogether. You could replace it with a horizontal custom-made window just above the enclosure and below the ceiling.

Panels for both kits can be adjusted to fit all standard tub sizes (29 to 32 inches deep and 57 to 62 inches wide with 53 × 60 inches wide for the SwanTile). Kits are also available for shower units.

Construction Notes — I've installed a lot of Swan's products and find them generally easy to work with, but I can share a few tricks to help your installation go smoothly. First, before you even consider a kit installation, take time to protect the surface of the tub or shower pan. Then follow these helpful hints:

1. Inspect all pieces right from the box. Grind off any imperfections around the edge on the back side of the panels with a small disk sander that can be used in a drill. Do a quick trial-fit of each piece.

2. Make all cuts to the *bottom* edges of the panels, not the top edges. When using a jigsaw, mask the bottom of the footplate as well as the area to be cut on the panel to help prevent any damage to the enclosure surface.

3. Start with the back wall and trial-fit the two edge panels so they fit into the corners vertically. Use a level on the sides of the panels to check for plumb. If the panels lean into the corners, level the panels at the bottom where they meet at the tub. To accomplish this, scribe the bottom of each panel and cut with a jigsaw so the panel drops squarely to the tub, and yet remains level. This isn't always possible, but try to get as close as possible. Your goal is a level enclosure all the way around the tub, with the panels matching evenly at both ends of the tub. You can't be sloppy here. Any variance will be clearly visible once the shower door is installed because the tub enclosure will be higher at one end of the tub than the other, and the level shower door will only emphasize that fact. If you install the panel plumb and the tub is crooked, you'll have a noticeable gap at one side of a panel when you apply caulk. You might have to fudge your level a bit to ensure that the panels match evenly at both ends of the tub.

4. Now for the tricky part. After you apply the adhesive, remove the strip off the double-sided tape that's located on the back of the outside edges. Then rest the panel on the edge of the tub, keeping the top of the panel away from the wall. Start at the bottom, pressing firmly as you work your way toward the top. Do this carefully for each piece. What's so tricky about this? Once the glue's on the wall, you only have one chance to install the panels. If the panel should go on the wall crooked, it's extremely hard to remove it for adjustment because of the double-sided tape. If it does go crooked, it will take two people to remove it — one to pull (starting at a corner) while the other one uses a stiff putty knife to break the seal of the tape.

5. Every place where one panel overlaps another, as well as at the top and down the sides where the enclosure meets the wall, apply latex caulk. Select caulk the same color as the enclosure to unite the installation visually. Finish the job with silicone caulk where the panels meet at the tub.

Manhours — All of these products can be installed with a one-man crew. A second pair of hands will only slow down the overall project. With that in mind, consider the following manhours:

▌ To install the Shower Tower in new construction: 1.5 to 2.0 manhours per unit.

▌ To install the Shower Tower in remodeling: 2.0 to 4.0 manhours per unit (depending on the existing plumbing and whether you need to bring in a plumber).

▌ To install tub/shower enclosure: 4.0 to 5.0 manhours per unit.

▌ To install window trim kit: 1.0 to 1.5 manhours per unit.

Access Panels

As mentioned earlier, there's a second choice when it comes to replacing or repairing tub/shower valves. Access can often be achieved from the back with a little preplanning, eliminating costly repairs to the wall and/or plumbing wall. Sachwin Products markets *Access Able*, an access panel versatile enough for three different applications:

1. Mount to rough framing during new construction. The flange is nailed face down so you can apply plaster or stucco or install wallboard.

2. Reverse the outer frame so the unit can be used on any finished wall at any thickness using construction adhesive.

3. Replace rusted or damaged access panels. Depending on the situation, you might need to fasten through the frame, but consider construction adhesive as your first choice.

Access Able comes in three sizes, offering 6 × 9 inches, 14 × 14 inches, or 14 × 29 inches of full access. It's designed to be cosmetically acceptable in cases where the panel must be located in full view. You can paint or wallpaper the lid. Consider using the unit on the ceiling to access hidden electrical and plumbing pipes.

The panel shown in Figure 10-7 provides access to a drain for a custom built-in aquarium. Even though the customer has full access to the drain's shutoff valve, the 2 × 4s make it pretty restrictive.

Courtesy: Sachwin Products

Figure 10-7
Using an Access Able panel

More attention to the design of the tank supports (4 × 4s, 4 × 6s, or even a header) would have left more room for the customer to operate the valve comfortably.

Consider a minimum of 1 hour to install this product.

Solid Surfacing

When you admire a bathroom or kitchen you just completed, what stands out the most? For me, it's the countertop. The countertop frequently gets lost in all the details of the project, but it really needs some pretty careful consideration. Whether it's a new surface design or a unique and contrasting edge, something beyond the normal plastic laminate can really give the room some personality. Solid surface material makes great-looking countertops, especially when it's paired up with a molded sink. Hopefully the following products will provide some fresh and innovative ideas you can incorporate into your next project.

Some companies offer one-piece vanity tops and bowls, kitchen sinks, ready-to-install tub/shower wall kits, bath floor tile systems, and accessory items in a wide array of colors that can put your creative juices into overdrive. But nothing beats starting with the raw stock, and milling it to create a unique design according to the customer's taste.

My experience with solid surface products is that they can be a hard sell because they're expensive, both in materials and in labor. But there *are* customers out there who want the high-end look, and who are willing to pay for it. Those are the customers you want.

Solid surfacing materials require sharp carbide-tipped blades and a handful of good tools. The fine dust created from cutting is hard on power tools, so be sure to blow out the dust with an air compressor. The manufacturers of these products require you to be trained to work with and install their materials. If you decide this isn't your particular cup of tea, then hire a factory-trained fabricator as a subcontractor.

There's a lot of information available regarding the installation of solid surface materials, so don't take what I'm giving you as the final word on the subject. This overview won't qualify you as a fabricator of any product. I'll just give you the basics to familiarize you with the product.

Swanstone

Swanstone, developed by The Swan Corporation, is a unique nonporous material with integral reinforcements. It has solid color molded throughout its entire thickness, and a wonderful silky feel. It's not subject to stress or heat cracking. For bathroom installations, Swanstone is available in vanity top blanks and with molded bowls. For kitchens, you can make your countertop and then install an undercounter sink.

Installing a Sink

You have several options when installing a sink (Figure 10-8). For a bathroom with a $1/2$-inch preformed top, you can mount the sink in a mechanical undermount, as shown in section A. The rounded edge of the countertop slightly overhangs the bowl. A seamed undermount is shown in section B. The countertop is flush with the bowl, which has a square

top edge. In both cases, the seams are secured with #14 × $3/4$-inch panhead sheet metal screws and seam adhesive.

For $1/4$-inch solid surface material installed over wood substrate (either in a bathroom or kitchen), the sink requires a little different handling. Working in an environment with a temperature of at least 65 degrees F, cut the substrate to the outside perimeter of the sink. Then glue the solid material to the substrate with contact cement. Drill a 1-inch hole in the center of the solid material. With 60-grit sandpaper, rough up the top ledge of the sink (which is made of the same solid surface material). Clean the surface thoroughly with denatured alcohol and a white towel. *Don't use lacquer thinner, acetone, or other solvents* because they leave an oily film that can interfere with adhesion and discolor seams or caulk lines.

After mixing the adhesive, pour the contents around the inside cut of the substrate and spread evenly in a 2-inch band around this perimeter. Puncture any air bubbles and then press the sink onto the adhesive, working it back and forth to ensure total surface contact. Use five or six deep throat clamps to secure the sink firmly to the top, allowing the adhesive to cure for one hour. Once the adhesive has cured, remove the clamps and position and install the sink clamps using #10 × $3/4$-inch wood screws. Now you can turn the countertop right side up and, beginning at the 1-inch hole, make the sink cutout with a router. The edge can be tapered, rounded, sanded (220-grit sandpaper) and buffed using an abrasive finishing pad.

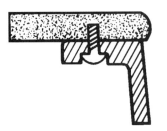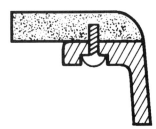

A Mechanical undermount   **B** Seamed undermount

Courtesy: The Swan Corporation

Figure 10-8

Installing a sink in a Swanstone countertop

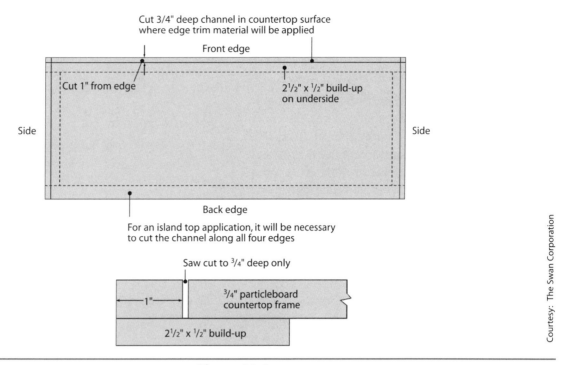

Cut 3/4" deep channel in countertop surface
where edge trim material will be applied

Front edge

Cut 1" from edge

$2^{1}/_{2}$" x $^{1}/_{2}$" build-up
on underside

Side

Side

Back edge

For an island top application, it will be necessary
to cut the channel along all four edges

Saw cut to $^{3}/_{4}$" deep only

1"

$^{3}/_{4}$" particleboard
countertop frame

$2^{1}/_{2}$" x $^{1}/_{2}$" build-up

Courtesy: The Swan Corporation

Figure 10-9
Expansion cut in substrate under Swanstone

Forming the Front Edge

Because the substrate will expand and contract, make expansion cuts 1 inch in around the perimeter to prevent damage to any finish edges (Figure 10-9). The best way to do this is to cut the substrate to the desired size. Then use wood glue and attach a $2^{1}/_{2}$ × $^{1}/_{2}$-inch build-up to the underside and flush to the front edge. Install wood screws about $1^{7}/_{8}$ inches back from the front edge. After it has dried, turn the substrate right-side up and, using a saw, make a $^{3}/_{4}$-inch-deep cut 1 inch in from the edge.

Figure 10-10 shows two popular finish edges to consider, the fabricated and the "V" groove. Traditional fabrication methods are necessary for some jobs (section A). But for most installations, the "V" groove provides a quick and easy countertop edge. Here's how to do it. When the substrate is $1^{1}/_{4}$ inches thick, make sure at least 2 inches of the material overhangs the substrate. Mark the underside of the material and tape this area as well as the front edges of the substrate to protect both areas from contact cement. When the finish surface has been glued to the substrate, trim the 2-inch overhang to $1^{1}/_{2}$ inches (thickness of substrate plus the thickness of the solid surface material). Now apply 2-inch-wide clear

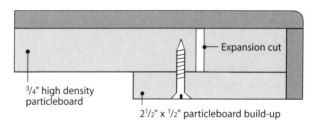

$^{3}/_{4}$" high density
particleboard

Expansion cut

$2^{1}/_{2}$" x $^{1}/_{2}$" particleboard build-up

A Fabricated front edge & build-up

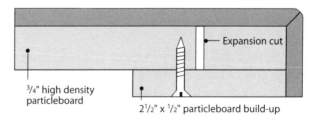

$^{3}/_{4}$" high density
particleboard

Expansion cut

$2^{1}/_{2}$" x $^{1}/_{2}$" particleboard build-up

B "V" groove fabricated front edge

Courtesy: The Swan Corporation

Figure 10-10
Installing the front edge of the countertop

tape on the top side of the finish surface, centered over the edge of the substrate. The tape will act as a hinge once you cut the "V" groove and fold the finish edge flush to the substrate's edge.

Use a specially-designed router base and router to cut a "V" groove along the underside of the finish sheet. (You can buy the complete tool from The Swan Corporation.) The substrate edge serves as a guide. The router won't cut the tape and gives a perfect mitered edge you can fold and glue into place. Then remove and discard the corners and rough up the "V" groove with 60-grit sandpaper. After cleaning the groove with denatured alcohol, apply a generous bead of color-matched seam adhesive along the groove. Don't forget to apply a bead of adhesive to the vertical seams.

With the countertop still upside down, fold the finish edge up against the substrate's edge and hold it in place with spring clamps. After an hour, flip the countertop over and remove the tape. Now rout the finish edge with a $^3/_8$-inch radius bit. Finish by sanding with 220-grit and then 320-grit sandpaper. Finally, wipe the surface with denatured alcohol.

Making Seams

If there's one thing a customer will complain about, it's a visible seam. I'm sure that at one time or another you've installed plastic laminate and couldn't avoid a seam or two. That's one beautiful advantage these materials have over laminates — their special adhesive makes seams almost invisible. Think of the flexibility this gives you, simply because seams aren't an issue! The key here is to make sure the two joining surfaces have clean and straight edges. Once the seamed surface has been sanded and buffed, you have one solid surface with no visible seams.

There are a couple of ways to make seams. One way is to fabricate a solid wooden substrate. Cut a dado $1^7/_8$ inches wide and (in this case) $^1/_4$ inch deep in the substrate just under the center of the two pieces to be joined. A spline $1^7/_8$ inches wide and the depth of the substrate, made of the same solid surfacing, will fit into this dadoed-out space.

Alternatively, you can split the substrate in the same area where the countertop material will meet. Figure 10-11 shows the finished seam. The substrate

is held together and supported by a 12-inch block the length of the seam. The trick is to completely glue up half of the countertop — the surface material, substrate, and support block. The surface material should overhang the substrate by $^1/_{16}$ inch to help achieve a tight glue seam and allow for substrate expansion. Then glue the spline and put it into place on the first half. Add wood glue to the top of the support block, then pull the second half (which has the surface material glued to the substrate) to within $^1/_8$ inch of the first half. Apply tape to both sides of the joint, then use two special suction surface clamps. Fill the seam with the adhesive, tighten the surface clamps, then scrape off any squeeze-out with a wooden scraper. Secure the other half of the support block with wood screws. Use 1-inch screws when the 12-inch block is $^1/_2$ inch thick.

If the substrate is supported by the base cabinets, you can add the support block once the two surfaces have been pulled together with the surface clamps. It's also a good idea to use biscuits along the edge of the substrate to help align the two substrates. Allow one hour for the seam adhesive to dry, then you can sand the seam to perfection.

Manhours — I've installed a lot of solid surface products, and I've come to the conclusion that it's more economical to use factory-trained installers. They do these fabrications day in and day out, with most of the work done in a controlled shop environment. Here are the manhour estimates that some installers use:

▌ To install $^1/_4$-inch solid surface material: 0.181 manhours per linear foot for a crew of two.

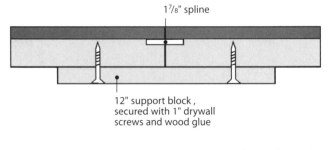

1$^7/_8$" spline

12" support block, secured with 1" drywall screws and wood glue

Courtesy: The Swan Corporation

Figure 10-11

Making the seam in Swanstone

■ Working of seams: 2.5 manhours for the first seam (includes 1 hour of adhesive curing time).

■ To install a sink in a mechanical undermount: 1.5 manhours per unit.

■ To install a sink in a seamed undermount: 2.0 to 2.5 manhours per unit (includes 1 hour set-up time for adhesive).

Wilsonart

Wilsonart Gibraltar Solid Surfacing is a blend of resins, made in flat panels that you can apply to horizontal and vertical surfaces. It's available in six solid colors and eighteen granite-look designs that you can use inside or out, for decorative or functional applications. The panels are $1/2$ inch thick and come in widths of 30 and 36 inches and lengths of 96 and 144 inches. You can also get strips that are $1^3/4$ or $5^3/4$ inches wide by 12 feet long. And you can bond the panels to each other to create blocks of matching or coordinating/contrasting colors.

This is a great product for vision-impaired individuals because a different-colored edge makes it easier to tell where the countertop starts and stops. In Figure 10-12, notice how two different colors highlight the edges of the countertop. From the Wilsonart Custom Edges collection (available in plastic laminate, color-through laminate, and wood), this edge from the XP series features a tongue you can fit into a groove cut into the countertop's edge. The low sheen of the product is also easy on the eyes. And you can taper or round off the edges and outside corners to help prevent a serious accident if someone should slip and fall.

These countertops have to be installed by a factory-trained fabricator. The $1/2$-inch panels don't have to be bonded to a substrate, although Wilsonart also makes a $1/8$-inch solid surfacing veneer that does have to be bonded. Even though Gibraltar is self-supporting, it may require additional support, depending on the overall design.

A complete line of sinks and vanity bowls completes the collection. What's really great is that these products are color-matched to the laminate line, creating even more options and possibilities.

Courtesy: Wilsonart International Inc.

Figure 10-12
Two-color edge created from Wilsonart materials

Manhours — The manufacturer provided detailed manhours for installing their material, even though one of their trained fabricators must do the installation. The manhours may help you decide if you're interested in trying to get certification to install Gibraltar Solid Surfacing. These labor estimates are based on a standard fabrication, working with the material at least three times and using a semi-automated shop. Except where noted otherwise, they assume a one-man crew. For other manhours associated with detailed custom fabrication, contact the manufacturer.

■ To install $1/2$-inch solid surface material: 0.181 manhours per linear foot for a crew of two.

■ To install an undermount sink: 0.335 manhours (this includes cutting the hole), plus 0.50 manhours of adhesive curing time.

■ Seaming: 0.25 manhours per 30 to 36 linear inches, including leveling and sanding seam. Add 0.50 manhours of adhesive curing time.

■ To fabricate edging (monolithic drop edge, two layer buildup): 0.25 manhours per 6 linear feet, plus 0.50 manhours of adhesive curing time.

■ Sanding: 1.666 manhours per 100 square feet.

Other Countertop Material

Of course, not all of your customers want (or can afford) solid countertops made of Swanstone or Wilsonart. What can you offer them to dress up their kitchen at a more moderate price? Here are a couple of products to consider.

Kuehn Bevel

This company has taken bevel molding to a higher standard. You can mix and match any brands of plastic laminate in any combination of colors or finishes to create a designer look, then finish it off with this molding. This alternative countertop edge gives the custom look of a solid surface, but at a laminate price. There's no minimum order and 95 percent of all custom orders are ready to ship within three to five working days. The molding is available up to 12 feet long in many different profiles, including single and double bevel and single and double round. It's manufactured with laminate bonded with an industrial grade contact adhesive to medium-density fiberboard. And now Kuehn has decorative solid surface edges made with Corian, designed to enhance laminate countertops with a solid surface edge.

Construction Notes — Two of the nicest things about this molding are that it's really enjoyable to install, and you don't need to be certified to do it. Figure 10-13 shows the steps you follow. After trimming the laminate deck, you precut the molding,

Laminate deck and finish trim with router and straightedge.

Precut Kuehn Bevel Molding longer than edge to be covered, using a 60 tooth carbide tipped blade or solid surface blade in trim saw or table saw.

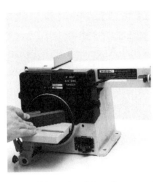
Fit molding. Sand in mitres and joints with disc sander equipped with mitre fence.

Cut biscuit slots (minimum of 12" apart). Insert "O" biscuit. If preferred, a $5/32$" slotter cutter can be used.

Prefit all pieces prior to gluing.

Apply adhesive sealant over entire back surface of molding to assure a good watetight seal.

Affix molding using filament tape every 2" - 3". Align top edge of molding precisely with edge of the laminate while taping. Scrape off excess adhesive sealant. Let adhesive sealant dry 24 hours before removing tape.

The finished project.

Courtesy: Kuehn Bevels

Figure 10-13
Installing Kuehn Bevel solid surface edges

check the fit, glue and align. It's easy to create an expensive, custom look at a budget price. With seven different profiles and a wide variety of colors, you can color-coordinate the edge to the laminate countertop to create a dramatic effect that's affordable.

Manhours — When installing Kuehn Bevel edge profiles, either laminate or solid surface, keep in mind that complex angles in your overall design can affect labor costs for cutting inside and outside miters. After the third time around, consider using 0.50 manhours for every 12 feet. This figure is based on a one-man crew. But my experience is that it's nice to have a second pair of hands when installing such edges.

TerraGreen Ceramics

Nothing is more appealing than a tile countertop in a kitchen or bathroom. It just conveys a certain element of class. Take, for instance, the ceramic tile collection by TerraGreen Ceramics, Inc. They create their products using a sophisticated system called *glass fusion*. This process combines recycled glass and minerals to create an entirely new ceramic material with a distinctive look and feel. The body of the tile contains over 55 percent waste glass (windows, mirrors, and post-consumer glass like bottles and jars). The rest of the tile is made of nonmetallic minerals such as special clays, feldspar, sand, and silica. The manufacturing process is designed to have no negative impact on the environment, or the employees.

Because of the handwork involved during the manufacturing process, no two tiles are exactly alike. This is especially noticeable in the TerraClassic line, due to the hand-rubbing each tile gets before firing. In the TerraTraffic line, the color isn't applied to the surface. It's consistent through the entire body, and the surface isn't as glossy. They also offer custom colors for both lines. There are textured surfaces available for skid-resistant applications. The only way to really appreciate the beauty of these tiles is to see and handle them yourself.

TerraGreen offers a full line of tiles for just about any type of application, residential or commercial, for countertops, flooring or walls. Install the tiles as you would any other tile installation. Note, however, that this tile is dense, so it's important to use the

Courtesy: Makita U.S.A., Inc.

Figure 10-14
Makita cordless saw Model 4190D

proper cutting tools. Use a tile saw with water and a diamond blade for a cleaner cut. If you have a minimal number of cuts, then the Makita Cordless Cutter will come in handy (Figure 10-14). It's a 9.6V $3^3/_8$-inch cordless saw designed to cut glass and ceramic, available with a diamond blade and refillable water coolant bottle. If you plan to use the tool a lot, consider having an extra battery on hand.

The following manhours assume a crew of one installing 4 × 4, 4 × 8, and 6 × 6 inch tiles. They don't include labor for surface preparation.

Allow for installing the tile with organic adhesive and grout:

- For countertop installation: 0.180 manhours per square foot.

- For floor installation: 0.110 manhours per square foot.

- For wall installation: 0.131 manhours per square foot.

- Trim pieces, including bullnose, sink rail or cap, quarter round or bead, and base: 0.0385 manhours per linear foot.

- Trim pieces (bullnose corner only): 0.0195 manhours each.

Allow for installing the tile in a conventional mortar bed with grout:

- For countertop installation: 0.352 manhours per square foot.

- For floor installation: 0.210 manhours per square foot.

- For wall installation: 0.270 manhours per square foot.

- Trim pieces, including bullnose, sink rail or cap, quarter round or bead, and base: 0.077 manhours per linear foot.

- Trim pieces (bullnose corner only): 0.039 manhours each.

Underlayment

Probably the most crucial step in a flooring system is the underlayment. In fact, this step alone can make or break the long-lasting performance of the finish floor. It's up to you to protect yourself by protecting that manufacturer's warranty! Always follow the manufacturer's recommended specifications for the proper underlayment to use with a specific product. It could save a lot of headaches later if something should go wrong.

At one time or another, we've all used particleboard as an underlayment, but it does have its limitations. For example, it's great under carpet or for applications where water's not involved. But if water or concrete floors factor into your next project, perhaps one of the following products may be better suited for your next flooring job.

WonderBoard

WonderBoard (Custom Building Products) is one of the most dependable ceramic tile backerboards on the market. The open fiber, mesh-wrapped portland cement product is available $1/4$ and $1/2$ inch thick. It cuts with a score and snap method, which eliminates any dust.

The $1/4$-inch board is available 36 inches wide and in lengths of 48 and 60 inches. It weighs in at 2 pounds a square foot, making a piece 36 × 60 inches weigh 30 pounds. It's primarily designed for use in bathroom and kitchen remodeling projects as a substrate for countertops and flooring.

The medium-duty $1/2$-inch backerboard is ideal for residential and commercial projects, primarily interior tub/shower enclosures, ceilings and floors, but you can also use it on exterior projects. It comes in 36 inch widths and lengths of 48, 60, and 96 inches, weighing in at 3 pounds per square foot. A 36 × 60-inch panel weighs 45 pounds. Its $1/2$-inch thickness matches up with $1/2$-inch thick wallboard for smooth transitions.

Construction Notes — Here are the steps to follow for installing WonderBoard on countertops under ceramic tile (Figure 10-15A):

1. Seal the substrate (exterior grade plywood) with a troweled waterproof membrane material. The company carries a product called *Trowel & Seal* designed for this.

2. Apply waterproof membrane.

3. Set backerboard in a level bed of flexible latex or acrylic modified thinset mortar.

4. Install $1^1/4$-inch (minimum) corrosion-resistant screws 6 to 8 inches on center throughout the field and perimeter of backerboard material.

5. Reinforce all seams with fiberglass joint tape.

For installation as underlayment for residential interior flooring under ceramic tile, brick, pavers, or stone (Figure 10-15B):

1. Use a minimum $5/8$-inch exterior grade plywood with joists 16 inches on center.

2. Set backerboard in a level bed of flexible latex or acrylic modified thinset mortar.

3. Offset backerboard joints to those of the plywood and leave $1/8$-inch gaps between the sheets.

4. Install $1^1/4$-inch (minimum) corrosion-resistant screws 6 to 8 inches on center throughout the field and perimeter of backerboard material. Fasten into joists whenever possible.

For wall application over wood or metal studs in wet areas under ceramic tile, brick, pavers, or stone (Figure 10-15C):

1. Construct the wall with wood studs or 20-gauge metal studs 16 inches on center.

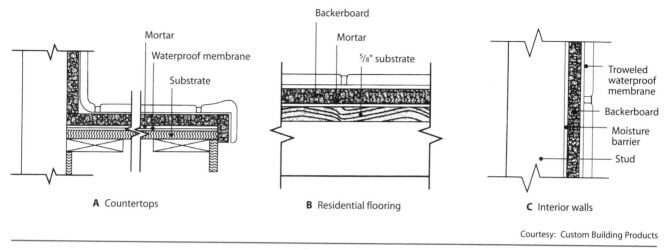

A Countertops   **B** Residential flooring   **C** Interior walls

Courtesy: Custom Building Products

Figure 10-15
Installing WonderBoard on countertops, floors and walls

2. Apply a moisture barrier using 4-mil polyethylene or 15-pound roofing felt.

3. Install backerboard with corrosion-resistant screws fastened 6 to 8 inches on center and fastened into the studs.

4. Seal the backerboard with some type of troweled waterproof membrane material.

For other types of installations, consult the company's project specification sheet.

Manhours — For scheduling purposes, estimate that a crew of two can install, tape, and apply the skim coat on $1/2$-inch backerboard in an 8-hour day. That's for 180 square feet of countertops, 525 square feet of floor, or 350 square feet of wall. That works out to the following:

▌ Countertops: 0.090 manhours per square foot.

▌ Floors: 0.030 manhours per square foot.

▌ Walls: 0.045 manhours per square foot.

Hardibacker

Hardibacker Ceramic Tile Backerboard (James Hardie Building Products, Inc.) is a fiber-reinforced backerboard that arrived on the market in the late 1980s. The material itself is composed of portland cement, ground sand, cellulose fiber, additives and water. It weighs 1.9 pounds per square foot and

comes in three sizes: 3×5, 4×4, and 4×8 feet, all $1/4$ inch thick. The surface is dotted as a guide for fastener placement.

Construction Notes — To install as an underlayment for ceramic tile:

1. Install a minimum $3/4$-inch exterior plywood fastened to joists set 16 inches on center.

2. Attach backerboard sheets in a brick pattern (never allow four corners to meet at one point), offsetting backerboard joints with those of the plywood, and leave a $1/8$-inch gap back from the walls and around cabinet bases. Allow sheet ends and edges to touch, but don't force them tightly together.

3. Attach the backerboard to the subfloor using latex or acrylic modified thinset mortar. Secure with $1^1/4$-inch (minimum) screws spaced 6 inches on center around the perimeter and, in the field, 2 inches in from each corner and between $3/8$ and $3/4$ inch from the edges.

4. Reinforce all joints with 2-inch-wide fiberglass joint tape.

To install backerboard into a wall application:

1. Construction should include minimum 2×4 wood or 20-gauge metal studs spaced a maximum of 16 inches on center.

2. In a tub/shower enclosure, ensure that the framing is adequately reinforced at the corners. Include the use of a vapor barrier.

3. Install sheets vertically or horizontally, holding $1/4$ inch above the floor, tub, or shower pan. Fasten with $1^1/4$-inch (minimum) screws spaced 6 inches on center around the perimeter and, at intermediate studs, 2 inches in from the corners and between $3/8$ and $3/4$ inch from the edges.

4. Reinforce all joints with 2-inch-wide fiberglass joint tape.

Manhours — Tub and shower enclosures are subject to high moisture levels and constant changes in humidity. For this reason, they require careful planning and special design considerations — not to mention careful workmanship! For this type of installation, consult the company's project specification sheet. The following manhours include a crew of two for installing, taping, and applying the skim coat on $1/4$-inch backerboard:

▌ Standard 5-foot shower surround: 1.5 to 2.0 manhours per unit.

▌ Floors: 0.030 manhours per square foot.

▌ Walls: 0.045 manhours per square foot.

Paneling

When hardboard paneling was first introduced, its sole purpose was to provide an alternative to more costly products on the market, like tile, marble and solid surface material. Basically, it was a 4 × 8-foot sheet of hardboard with a photograph finish. Most of us quickly learned that this product (especially when we used it as a tub/shower enclosure) just didn't hold up. I suppose it was used mostly in rentals and low-cost projects. Over the years, the concept has undergone many changes and today there's an improved product that's worth checking out.

Aqua Tile, by ABTco, Inc., is a tileboard vastly different from previous products, mainly because more care has gone into the fiber blend substrate. It's much stiffer and appears to be of higher quality. The surface of Aqua Tile features realistic-looking

Courtesy: ABTco, Inc.

Figure 10-16
Aqua Tile used in a bathroom remodel

embossed grout lines, produced with high heat and pressure. The embossed and decorative print is protected by an infrared-cured, high-gloss acrylic topcoat that provides superior moisture resistance, washability, toughness, and flexibility. Edges are finished so there's no need for molding. There's an alkyd-melamine base coat and sealer/filler between the print and the substrate. The substrate is tempered to restrict water absorption and create a highly coatable surface. Board integrity is maintained through a hot-press embossing process with engraved plates (unlike scoring techniques which weaken panel surfaces).

Figure 10-16 shows the tileboard in a Pewter Rose Mosaic pattern that makes a handsome bathroom or kitchen wall covering. If your customer has ceramic or marble tile tastes, but not the budget, then Aqua Tile tileboard may be the answer.

Construction Notes — Store Aqua Tile indoors, with all sides fully exposed to the air (24 hours above grade and 48 hours below grade) before installation to allow it to adjust to room temperature. Skim-coat the walls using a trowel-type waterproof wall adhesive, and trowel the back of the panel with adhesive using a $^3/16$-inch notched trowel. You can get an appropriate adhesive from ABTco. When the adhesive gets tacky, position and press the panels into place. Of course, you already know not to apply any adhesive until *after* a trial fitting!

When combined with color-matching tub shelf corners, Aqua Tile makes a good candidate for a budget bathroom remodel. The molded plastic shelf takes up 7 inches on either side on a 5-foot wall, so you only need to use one 4-foot-wide panel on the back wall. To guarantee an outside finished edge, always cut the panel so the cut points to the inside corner and toward the top of the tub. Once caulked, you'd never know the sheet was cut. Apply caulk to all edges and joints *after* the adhesive has cured. As always, follow all manufacturer's recommendations.

Manhours — These manhours are based on a two-man crew, but there may be times when the job may be too small to support two, or you think you can handle it on your own. If that's the case, be sure to adjust the figures accordingly.

- To install $^1/8$-inch $\times$ 4 $\times$ 8-foot sheets: 0.034 manhours per square foot.

- Add for panel adhesive: 0.008 manhours per square foot.

- Labor for a standard 5-foot tub enclosure: 0.043 manhours per square foot.

- Installing moldings: 0.034 manhours per linear foot.

- Caulking: 0.004 manhours per linear foot.

Fixtures

While we adapt to new products and new methods of installation, we also need to keep an open mind about existing products that are being improved for everyday use. Some may look the same on the outside, but have new internal components, while still others might have received a facelift. The following products fit this category. They just might be the clincher to secure a contract for your next job.

Flushmate

Have you given serious thought about which toilet to purchase? Personally, I always left it up to the plumber after I described what the customer was looking for. He sure installed a lot of *white* toilets on my jobs! Today, purchasing a toilet isn't as simple as it used to be. Customers want fixtures that meet their individual tastes. They look for a unique style and color — something different from the plain, boring white toilet that we all grew up with — and that I always thought was just fine!

The crunch to save energy and conserve water hit sometime around the mid '80s. Conventional gravity-type (siphon bowl) toilets were consuming 3.5 to 5 gallons per flush (gpf). Then a new breed of toilets, low-consumption and Ultra-Low Flushing (ULF), came on the market. As of January 1994, the Federal Energy Act requires manufacturers to make toilets that use only 1.6 gpf — about half the water that toilets were originally designed to use. Unfortunately, using less water didn't provide 100 percent effectiveness, so some toilets had to be double-flushed. Customers were unhappy with these new water-saving toilets, so something had to change.

Sloan Valve Company designed Flushmate to eliminate problems associated with gravity-type toilets: double flushing, clogging, overflowing, and soiled bowls. However, the system couldn't be used in a gravity-type toilet, so they designed a new toilet — the Blowout Bowl. Figure 10-17 shows the difference between a pressure-assisted and gravity

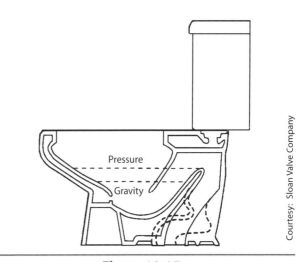

Figure 10-17
Pressure-assisted vs. gravity toilet design

	Gravity	Pressure
Water used	1.6 gpf	1.6 gpf
Water spot*	4" x 5"	10" x 12"
Trap seal depth	2"	$2^5/_8$"
Trapway size	$1^5/_8$"	$2^1/_8$"
Hydraulic bowl design	Siphon jet	Blowout
Rate of flow (gpm)	25	70
Tank sweats	Yes	No
Double flushing	Yes	No
Leakage	Yes	No

*The surface area of water in the bowl as one
looks down at it.

Figure 10-18
Pressure-assisted vs. gravity performance

toilets. The Flushmate uses *pressure* to flush the toilet and boosts 1.6 gallons to beyond the performance level of 3-gallon toilets. The manufacturer claims it forces waste at 70 gallons per minute (gpm), a rate three times that of gravity models. That's strong enough to clean the bowl with every flush and completely eliminate double flushing. Figure 10-18 compares gravity and pressure toilet performance.

But how does the Flushmate work? When the tank is ready to be flushed, the air and water pressures in the tank are equal. When the flush button is pushed, water in the valve cartridge discharges. The main

Courtesy: Mister Miser Urinal

Figure 10-19
The Mister Miser fits between studs 16 inches on center

valve opens and flush water surges into the bowl, causing it to empty within 4 seconds. The waste is carried through the drainline at the crest of the "turbo-charged" torrent of water. Finally, the flush valve closes and the tank begins to refill. If you have any questions, contact Flushmate at the address given in the Appendix.

Manhours — The Flushmate is sold under different marketing names from the leading fixture manufacturers. There's no need to worry about manhours because Flushmate is already installed in the tank when the toilet is purchased. Installation is the same as for a conventional toilet. A crew of two should be able to install it, including water supply, set, trim, and seat, in about 2.65 manhours per unit.

Mister Miser

Well, I thought I'd seen just about everything on the market, but I was surprised when information on the Mister Miser urinal landed on my desk. It's a wall urinal made from recyclable ABS-structural plastic that weighs only 6.5 pounds. It's a great product for any shop, but I think it should be included in residential bathrooms as well. It only needs 10 ounces of water to flush and it comes in 12 colors to fit any decor. The unit is designed to prevent splashback, a problem with conventional toilets. And there won't be any more arguments about who left the toilet seat up because the system won't flush unless the door is closed. But check it out with your local building officials before you install it to make sure it's approved for your area.

What's unique about this product is that it fits between studs framed at a standard 16 inches on center (Figure 10-19). The added cleats are 2 × 2s held back $1^{15}/_{16}$ inches from the face of the wall studs. That allows for $^1/_2$-inch wallboard. If the same stud space has to hold a vent as well as the urinal, you can cut off the right ears and use the second set of ears for installation. Also, cutting $^1/_4$ inch off the ears on the left side will allow the urinal to slide to the left, leaving more room on the right for the vent pipe.

Manhours — When calculating manhours, remember that the code in force in your area may require a licensed plumber to install the waste, vent, supply and do the final hookup. Of course, this may also apply to other fixtures. The figures given assume a one-man crew and unit installation only, no finishing of wall surface or rough-in:

▌ New construction: 1.0 manhour per unit.

▌ Existing structures: 3.0 to 5.0 manhours per unit.

Soft Bathtub

There's nothing like a good long soak in a hot tub, and this is a tub designed just for that pleasure. The Soft Bathtub (International Cushioned Products Inc.) uniquely and *comfortably* conforms to you once the hot water is added (Figure 10-20). A high-density foam core insulates the tub so it retains hot water temperatures longer. According to the manufacturer, tests show only 1 degree F of heat loss every 10 minutes. It's been estimated that a cast iron or cultured marble bathtub could be three times less efficient.

The elastomeric polyurethane coating resists corrosion and dulling longer than other coatings. But it's probably harder to clean than a porcelain tub, right? Wrong! The manufacturer claims its nonporous surface is "eight times easier to keep clean than porcelain." After all, cushioned tubs and whirlpools are used in hospitals, where hygienic conditions must be strictly maintained and accidents avoided. The nonslip and soft surface of this tub could minimize the risk of a fall and the chance of injury if a fall should occur. Don't overlook these features when considering a new bathtub in the residential market, especially for the accessible market.

The standard size Soft Bathtub weighs in at just over 100 pounds. It's easy to install in a standard 5-foot bathtub opening. Its cushioned perimeter allows it to butt up against an uneven surface and form a seal, or you can put it into any pedestal installation. The manufacturer offers a ten-year warranty on the product when installed in residential and five years on commercial and institutional projects.

Manhours —

▌ For new construction: 2.0 manhours for a crew of two for supply fittings but no rough-in.

▌ For remodeling of existing space: 6.67 manhours for a crew of two for tear-out, fixture, and supply fittings. This doesn't include finishing the wall surface.

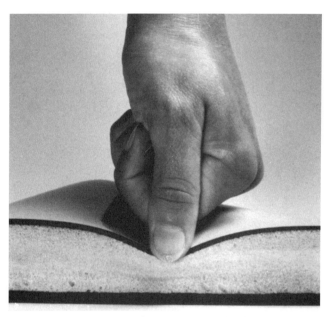

Courtesy: International Cushioned Products Inc.

Figure 10-20
The foam core makes this bathtub soft
and it also insulates

Tile Shower Base

The time will come (maybe you've already been there) when a customer will want a shower base all in tile. If you haven't had this experience yet, don't be too quick to make that bid. There's a lot to this installation. If the customer insists, I suggest you find a subcontractor who specializes in tile shower bases. But there are other options available that make an attractive shower base without the difficult installation. One that comes to mind is a solid-cast polymer shower floor with inlaid 2-inch ceramic tile and sanded epoxy grout, from the Laurel Company (Figure 10-21). They claim to be the first manufacturer to produce a shower base that incorporates these features. The tile shower base is available in a variety of popular colors to coordinate with wall and plumbing fixtures — and the quality is impressive.

The shower base is nonporous, stain resistant and waterproof, and the tile provides a safe slip-resistant surface. This product installs like other preformed shower bases. You can get rectangle, square, and neo-angle shower bases, plus a unit that will replace the standard width of a 5-foot tub (30 × 60 inches with a left- or right-handed drain). The product carries a five-year limited warranty.

Courtesy: Laurel Company

Figure 10-21
Polymer shower floor with inlaid ceramic tile

Manhours — Figure 5.0 manhours with a crew of two for removing an existing tub or shower base and installing the new base. This doesn't include rough-in, hookup, or finishing the wall surface.

Composite Kitchen Sink

Consumers have been requesting an attractive and durable, yet affordable, kitchen sink. Sterling Plumbing Group has answered that request with a composite double-bowl kitchen sink. The 33 × 22-inch sink is made of tough, nonscratch VIKRELL, a composite material that resists chipping, cracking, and peeling. It features two 8$^1/_4$-inch-deep offset bowls, including one oversized bowl. It has a light-luster matte finish and solid color molded throughout, available in almond, white, almond granite, and gray granite. The lightweight design makes for easy installation, and multiple faucet knockouts are available to accommodate a variety of faucets. The composite sink installs just like any other self-rimming sink.

Sterling warrants to the original purchaser/user that the sink will be free of manufacturing defects for the life of the product, or three years from the date of the sale if the sinks are used in commercial buildings (hotels, motels or rental property).

Manhours — For a complete installation, including faucet with sprayer, two strainer drain fittings and valve supplies, use 3.06 manhours per unit for a crew of two.

Helpful Accessory Items

From time to time you come across a job where you have to make do with what's available. Unfortunately, that's time-consuming and almost always creates a lot of extra work. Haven't you wished for products that could speed the process and just make the overall installation easier? As I worked on this book, a couple of products came my way that could have made those jobs a whole lot simpler.

Colored Caulk

As I worked with laminates, solid surface materials, and self-rimming sinks, I always looked for a sealant company that could supply caulk to match the material — too often without success. A few years ago, a dentist hired to me to move some cabinets I had installed in his old office into a new office. The dentist wanted everything to be just right, including the caulk on the backsplash, where it met both the wall and the countertop. He wanted caulk the same color as the countertop. At the time, caulk in matching colors wasn't available. I ended up mixing two latex caulks together, in trial-and-error proportions, until I got the color I needed to complete the job. The job turned out looking great, but what a hassle it was! I remember wishing that I could just go buy a caulk that matched.

Colored caulking products have reached the market since then, but it was just recently that I ran across a company (Gloucester Co., Inc.) that offers 21 coordinated colors. Colortones (part of the Phenoseal product line) are bath and kitchen sealants specially designed to coordinate or match with most designer colors and surfaces in bathrooms or kitchens. They're available in a 6-fluid-ounce squeeze tube. The sealant dries in 12 to 48 hours to a glossy surface that's both scrubbable and flexible. And it cleans up easily after installation with just soap and water — if you do it before the sealant sets. The manufacturer also claims it's highly mildew- and rust-resistant (a mildewcide has been added).

Manhours — For labor costs, figure the following:

■ $^1/_8$-inch bead: 0.018 manhours per linear foot.

■ $^1/_4$-inch bead: 0.022 manhours per linear foot.

■ $^3/_8$-inch bead: 0.027 manhours per linear foot.

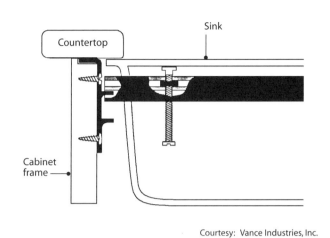

Courtesy: Vance Industries, Inc.

Figure 10-22
Cross-section of the Sink Undermounter

Courtesy: Vance Industries, Inc.

Figure 10-23
Bird's eye view of an Undermounter installation

Sink Undermounter

What would you say if I told you I found a product that may simplify your next undercounter sink installation by 50 percent? With a little practice, you may even be able to beat that time.

The Undermounter is an aluminum bar that's adjustable to two different heights located on the support brackets. These brackets hang from the face frame and back of the cabinet. If you plan to install a solid surface countertop directly to the cabinet without using filler pieces, you'd have to recess the bracket flush to the cabinet's top edge (Figure 10-22).

The bars are available in lengths of 25, 36, and 50 inches, and you can trim them to fit unusual sizes or custom-made cabinets. Inside the bars are leveling bolts that carry the sink and hold it in place up against the underside of the countertop (Figure 10-23). Using this system should save you both time and money.

After cutting the sink hole, you can install the Undermounter and sink (only) in an hour (after you've done it a few times). Installing the Undermounter itself takes about 0.25 manhours per pair.

Heat and Ventilation

Heat and ventilation are crucial to your customer's comfort, but how much thought do you give to the mechanical devices involved? When I did bathroom and kitchen remodels, heat and ventilation were primary concerns. The most common (and frustrating!) problem I experienced concerned the location of the heat register. It always seemed to be in the wrong spot, especially when it was time to install new cabinets. This was especially true of older homes, where I'd find heat registers underneath pedestal-style and wall-hung wash basins.

Alternative Heat Sources

The first decision I had to make was what to do with the heat. If the home had a forced-air system, redoing or rerouting ductwork wasn't worth the time, energy or expense. It was easier to close off the vent altogether and install an electrical heat source directly in the toe-kick of the cabinet. But before doing this, I recommend consulting a heating, ventilating, and air-conditioning (HVAC) contractor. You want to make sure that the system maintains efficiency if the vent is blocked off.

Two good choices to consider here are electric and hydronic systems. (Hydronic systems work with heat transfer by hot water.) Either system can be the primary source of heat in the room, or just a secondary source, to "take the chill off." These systems can eliminate conventional heaters, recessed or baseboard, that take up valuable wall space.

The recommendation that recirculating water be used is one drawback to a hydronic system. Most states will allow a connection to a loop off a domestic hot water heater, but you'll need to install a pump to effectively use the water for heating. Unless the house is already set up for this, the expense probably isn't worth it.

Heat Recovery Systems

Excess moisture, pollutants, and dangerous contaminants are found more and more frequently in tightly-built energy-efficient homes. They simply don't have a natural exchange of fresh air. These homes require a mechanical ventilation system to regularly replace stale indoor air, like a high-performance heat recovery ventilator (HRV) that works in conjunction with the furnace. The HRV provides balanced airflow by exhausting stale air to the outside while drawing in an equal amount of fresh, filtered, and tempered air. During cold weather, an HRV recovers heat from the exhaust air to heat the fresh air. Some types also exchange humidity as well as heat, a particular advantage in hot humid climates because it reduces the load on the air conditioning.

As you research these systems for your next project, ask yourself:

▌ Is this the only system to consider?

▌ Is air quality the only issue here?

▌ What about removing moisture from bathrooms, hot tub areas, kitchens, and laundry rooms?

▌ Are there alternatives?

The next couple of products I'm going to introduce might answer these questions. If you consider using one of them, contact the manufacturer's technical support to make sure it's the right unit for your particular project.

MPV

Multi-Port Ventilators could eventually replace conventional electric bathroom fans. This system by American Aldes Ventilation Corporation is apparently so quiet that you might not be able to tell if the unit is actually running. They recommend installing a delay switch (timer) or a switch with a pilot light, so the homeowner will know the unit is on.

The MPV 300 can handle a house up to 6,000 square feet, including units in up to six bathrooms. Designed for both residential and light commercial markets, the fan can easily ventilate bathrooms. But if the fan runs continuously, or for several hours each day, it can also provide whole-house ventilation for improved indoor air quality.

You can install the unit in the attic, basement, or crawl space, but insulated ductwork is highly recommended if you put the unit in the attic or other unheated area. The ductwork comes with an accessory kit, including a sleeve with a backdraft damper to block air flow during the off-cycle to reduce condensation within the pipe. Condensation is created when warm moist air (from the room) penetrates slowly up into a pipe without a damper. MPV carries a three-year warranty, though it's limited to the original user.

Notice in Figure 10-24 that the master bath uses a 6-inch duct — the room is larger than 7 × 10 feet. A 50 cfm intermittent exhaust rate (4-inch duct) would be standard under most model code organizations for this size room. However, the *Uniform Building Code* requires five air changes an hour, so 6-inch duct is used to meet code and handle the larger volume. While we're on the subject of codes, first check to see if flexible duct meets compliance in your area before you decide to install an MPV.

Manhours — Installing this system may require a licensed electrician for the final hookup. The manhours I'll suggest are based on the third time around, no rough-in electrical or hookup. Of course, the labor estimate for an existing home can vary, depending on the complexity of the structure, whether it's a one- or two-story home, or whether the walls are open. These figures are for the MPV 300 system with four to six multi-ports (up to 6,000 square feet) for continuous operation:

▌ New construction: 5.0 manhours for a crew of one.

▌ Existing structure: 8.0 manhours for a crew of two.

VMP-K

The VMP-K, also by American Aldes, is designed for use as a continuous exhaust ventilation system. The central unit itself resembles your car's distributor cap. Its center port has a 6-inch opening, so when the system runs at normal speed (low), the port

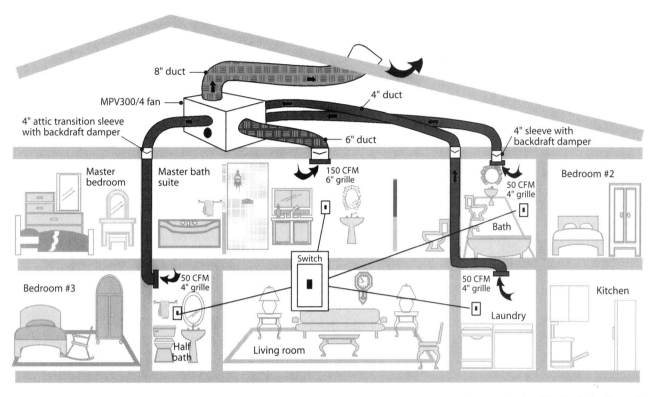

Courtesy: American Aldes Ventilation Corporation

Figure 10-24
Multi-Port Ventilator

delivers a constant 30 cfm. When the unit runs on high, the cfm increases to 100. The remaining ports, located on the outside, are 3 inches in diameter and rated 20 and 10 cfm. The K-3 unit has four ports, and the K-5 has six.

The K-5 unit will handle a house up to 2,400 square feet, and you can "Y" the center 6-inch port to provide higher flow for two bathrooms. There's another advantage to this system: You can hook it up to a solid and perforated pipe in the crawl space or under a concrete floor. It will suck up any radon and exhaust it out through a 6-inch exhaust duct up on the roof.

In the attic installation shown in Figure 10-25, notice that the unit hangs in place with a tri-hook. Installing the exhaust duct in the soffit area is an option.

However, using a system like this requires careful planning. A central exhaust system is often called an exhaust-only system, meaning that only the exhaust

is fan-powered. This creates a slight negative pressure in the home, which will draw in outside air. You can provide fresh air inlets in the exterior walls of main living areas and bedrooms, or a fresh air duct to the return of a forced air heating system. If you don't plan inlets, you'll get unintentional leakage from any vulnerable areas of the structure.

This negative pressure can also cause backdraft of flue gases on atmospherically-vented furnaces, water heaters, and fireplaces. You've probably noticed how hard it is to start a fireplace when cold air is flowing down the chimney, or likely you've experienced smoke spilling into a room. This isn't just an annoyance, it can be unsafe. And the same thing happens with other combustion appliances, whether you can see it or not. The problem is aggravated when the home is very tightly built and the exhaust rates are high.

If you use newer direct-vented combustion appliances, exhaust-only systems are often practical alternatives to more costly systems, like heat recovery ventilators (HRVs). But in a house with traditional

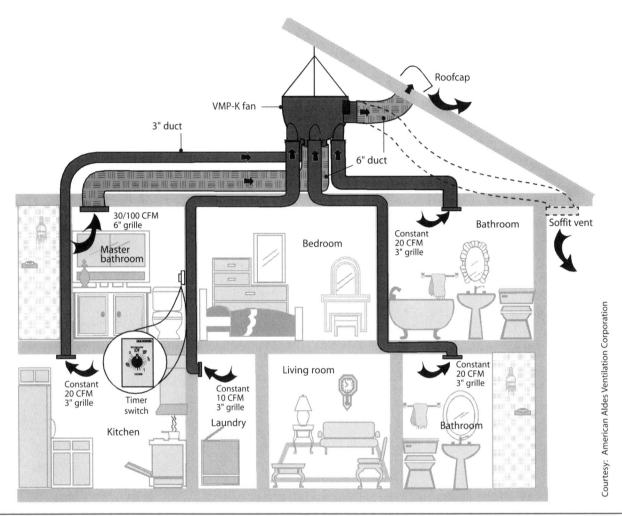

Figure 10-25
VMP-K ventilator hanging from a tri-hook

atmospherically-vented combustion appliances, you need a balanced system of ventilation, like an HRV, or a supply-only system.

How do you figure all of this out? It's simple. Work with the manufacturer and use an HVAC sub with the right kind of experience.

Manhours — Again, installation time in an existing home will vary depending on the conditions you face. Here are the estimated manhours for a six-port system for the K-5 (up to 2,400 square feet):

▌ New construction: 6.0 manhours for a crew of one.

▌ Existing structure: 10.0 manhours for a crew of two.

Let There Be Light!

I'm talking about light that comes in from the outside, compliments of Mother Nature herself — not light fixtures. Specifically, my interest is windows, for a couple of reasons. Especially in bathrooms, windows serve two purposes. First, I've found that a metal-framed opening window (3 feet × 1 foot), installed close to the ceiling in the tub/shower area, helps to ventilate a lot of moisture and bring in fresh air. Second, of course, is the natural light it brings in. Even if the window is restricted by the overhang, it still brings in light, sunshine, and warmth. It just makes it more pleasant and comfortable for the customer to walk into this area.

Anything you can do to bring in the outdoors makes any bathroom or kitchen feel more open. But the downside of that is that windows can have a tendency to make customers feel their privacy will be compromised. So how do you bring in light and air while at the same time making the customer feel comfortable and secure?

Glass block, I believe, is just the ticket. They've been available for centuries, of course, primarily in commercial and industrial buildings. But now they're "fashionable" again for the residential market. The good news about glass blocks is that they transmit light while retaining privacy, for both exterior and interior applications. And your choices are more varied than ever. You'll find different sizes, and specialized blocks you install with unique mortarless systems that use metal frames with colors to match most decorating schemes. Other systems have a completed unit that will fit into a framed opening. Consider these products if light and privacy are considerations in your next project.

IBP Glass Block Grid System

IBP Glass Block Grid System comes ready to install just like any window. It's designed to eliminate the need for a traditional mortar installation of glass block. There are some situations where a mortar installation is just what you want. But the ease of installation for this new system makes it practical in many applications where mortar installation isn't. The frame comes preassembled and ready to install, including the grids. Once you've installed the frame, either vertically in a wall or horizontally in a roof or floor, or as a partition or shower enclosure, then you put in the glass blocks. First, wrap each block with insulated foam tape (provided) and insert it into the grid. Then fill grout lines with 100 percent silicone sealant.

You may want to tape off the sides of the grout lines to prevent any excess silicone from getting on the glass blocks. With so many intersections, it may be easier to caulk all horizontal lines at once and then caulk the vertical lines the next day (Figure 10-26).

The grid system comes in a wide variety of stock sizes and shapes that you can curve to any desired radius. The frames are available in bronze, chrome,

Courtesy: Acme Brick Company

Figure 10-26
Apply the caulk carefully for a professional-looking job

brass, red, black, and white. Custom sizes and colors are available, as well as design assistance. In Figure 10-27, you can see how the light through the glass block installed in the backsplash area makes the countertop feel wide open. There are no framing members in this area, so a header needs to be installed. Be sure to consult a structural engineer or your local building department.

Just like any other glass product, these blocks will break. To repair, simply cut away the sealant, remove the damaged block, insert a new block, and reseal. From a security standpoint, even if a block is damaged or removed, the remaining grid still makes it difficult to enter a room.

Courtesy: Acme Brick Company

Figure 10-27
Glass block floods this countertop with light

Manhours — The following manhours are based on a 4 × 4-foot window completely installed, including setting blocks in a frame, but no finish work, for a crew of one:

▌ New construction: 1.0 to 2.0 manhours.

▌ Existing opening of the same size (including removal of existing window in wood siding, but not brick, aluminum or vinyl): 4.0 to 6.0 manhours.

HY-LITE Block Windows

Just when you think you've seen it all, someone comes along with a product that makes you do a double-take. HY-LITE Block Windows has been manufacturing lightweight acrylic blocks since 1987. They've created some unique systems that can make your job a whole lot easier.

One clear advantage they have over glass blocks is that acrylic blocks are 75 percent lighter. A unit can come from the factory fully assembled, with blocks and synthetic rubber grout, so all you have to do is install it as you would any other type of window. Incorporating a triangular block lets you create octagons, pyramids, and triangles that aren't possible with glass blocks.

Another interesting possibility involves using accessory items such as low-voltage lighting or fiber optics to create dramatic effects. Just imagine how that would look in peach, rose, or blue colored blocks! The aluminum frames are available in painted tones of white, tan, and bronze, as well as clear anodized. Extrusion accessories simply attach to the main window frame to create free-standing partitions, light columns, or any other design you can think of.

Now they've introduced a line of awning and casement windows, part of their Prestige collection. The frame (available in white) is made of vinyl and is glazed with acrylic blocks. This is the first window that I'm aware of that uses blocks. Your customer gets two advantages: the look and feel of glass blocks and the opportunity to ventilate the room.

Courtesy: HY-LITE Block Windows

Figure 10-28
The acrylic block awning window unit

Figure 10-28 shows an awning unit with dual locking handles for added security. Like the casement, it's backed by the manufacturer with a 10-year limited warranty. Install these as you would any other vinyl window. The manhour estimates are based on a crew of one, but I'd consider a two-man crew.

Manhours — To set awning or casement windows only, with no rough-opening or finish work:

▌ Up to 8 square feet: 1.00 manhour per unit.

▌ From 8 square feet to 16 square feet: 1.50 manhours per unit.

▌ Over 16 square feet: 2.00 manhours per unit.

As you can see by the materials and options in this chapter, you have all kinds of opportunities and choices when designing or remodeling a bathroom or kitchen. But don't limit these products to just these areas. Be creative, and see where these products might fit into other areas of a home. Let your customers know they do have options: there's always an alternative around the corner!

Now let's move on to the next chapter for a look at new ideas in decking. It's packed full of alternative products that will help in the design and layout of your next deck project.

11

Decking

Perhaps the peak season has already arrived in your area and you're frantically answering the phone, bidding jobs, digging out your tools, and checking out new materials. Calls for new decks are a sure sign of spring. If it's not the deck season now, get ready for it. It's coming!

Figure 11-1 shows a few of the newest decking products available today. This chapter will introduce you to some of the latest materials on the market, but remember that this is by no means the final word on the subject. By the time you read this book, I'm confident that other innovative products will be out there just waiting to be installed. Always be on the lookout for new products that make your jobs — or your sales — easier.

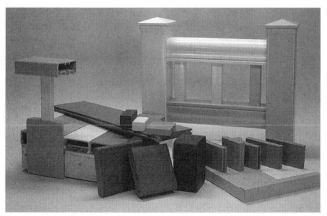

Courtesy: C.R.S., Inc.

Figure 11-1

Alternative decking products come in many shapes, sizes and colors

Alternative Products

Today's customers are open to new ideas and creative approaches to suit their particular needs or lifestyles. As a professional, you're in a position to educate them, so do your homework and request information and samples of the new products. Be sure to check the track records and written company warranties of these products. Verify that the products have undergone safety testing, structural integrity evaluations, long-term field performance tests, and that quality assurance standards are established. Don't install any product until you're satisfied that it will continue to do the job once you walk away.

As you work with products made from recycled materials, you shouldn't find any metal or other foreign material in them. But, depending on the quality of the sorting operation where the recycled materials are ground, stray pieces might sneak in from to time to time. These very small pieces are generally no cause for concern, but they can be a danger if you're using power tools. Always wear eye protection, just in case there *is* some stray metal in the material.

For working with these alternative materials, there's one tool, besides an impact driver, that stands out above the rest: the miter saw. If you want to add to your tool collection, I recommend you consider a slide dual compound miter saw. Figure 11-2 shows a twin-rails system by Makita (LS1013). I tested it while working on this book, and from the standpoint of convenience, efficient dust collection, and ergonomics, this tool can handle all those interesting cuts required by custom deck projects.

Figure 11-2
Twin-rails miter saw

Manhours

If this is your first experience working with alternative decking materials, you'll soon discover they handle and install a lot like wood products. And you may be able to install them faster than the traditional materials — after you've had the opportunity to install them a few times. It'll probably take a few jobs to understand any new product's character. Then you may be able to cut your labor time, but there are no guarantees. True savings come in the overall finished project: some from labor, others from less maintenance, while still others come from fewer callbacks.

To help you estimate your labor time, I've included some basic manhours to consider when figuring costs for the products described in this chapter. The figures apply to most of the materials I'll describe. For the materials where they don't apply, look for appropriate manhours at the end of that product's section. All of the manhours are based on a crew of two who've had some experience with the material — say the third time around.

These costs assume 5/4 or 2 × 6 decking material supported by 2 × 8 pressure-treated joists spaced 12 or 16 inches on center over 4 × 8 pressure-treated beams. Beams are supported by 4 × 4 × 2-foot pressure-treated posts set 6 feet on center in concrete. Fasteners are screws, galvanized steel joist hangers,

and post anchors. Decking is spaced at $^1/_8$ or $^3/_{16}$ inch (depending on the product) to allow it to drain. Costs are based on a 15 × 10 foot (150 square foot) deck with one 15-foot side attached to an existing structure.

- Concrete footing 12 inches wide by code requirement for depth for post supports, includes digging, pouring, and bracket: 0.50 manhours per linear foot.

- Installing decking material: 0.180 manhours per square foot.

- Add for diagonal pattern: 0.010 manhours per square foot.

- Add for stain and sealer finish: 0.006 manhours per square foot.

- 42-inch handrails which consist of 2 × 2 balusters $5^1/_2$ inches on center, 2 × 4 vertical supports lag bolted to rim joist and side of joists, two horizontal 2 × 4 top and bottom rails (both turned on edge), and 2 × 4 or 2 × 6 cap: 0.50 manhours per linear foot.

- Concrete pad to support stairs 36 × 36 inches (roughly) by code requirements for depth; includes digging, pouring, finishing, and bracket: 0.530 manhours per 12 linear inches.

- Deck stairs, assuming one or two steps, 12-inch tread, either box or stringers (2 × 12):

  1. Open risers: 0.270 manhours per linear foot.

  2. Closed risers: 0.350 manhours per linear foot.

- Deck stairs, three or more treads high by 36 inches wide. Stringers cut from 2 × 12 and using 2 × 4s on both sides of the middle stringer and on the inside of both the left- and right-hand stringer:

  1. Open risers: 0.75 manhours per riser.

  2. Closed riser: 0.90 manhours per riser.

Now that you have a rough idea of labor costs, let's get down to business by reviewing some of the new products on the market. Remember to check with your local building department when it comes to structure design. And keep in mind that most alternative materials aren't designed for structural use.

TREX

TREX is an innovative product that was first introduced to the market in the 1980s, and has been produced under a couple of different names. Now it's a product of TREX Company, LLC.

TREX is made from approximately 50 percent recycled and/or reclaimed plastic (grocery bags and industrial stretch film) and 50 percent waste wood (wood fiber from sawdust and used pallets made of hardwoods). It contains no virgin wood or preservatives. Because it contains both wood and plastic, it has the strengths of both materials. The wood gives it the advantages of low thermal expansion/contraction and natural UV stability. It's also slip-resistant and paintable. Because of its plastic content, TREX resists moisture (including saltwater) as well as insects and solvents, so it won't rot or deteriorate.

If you're a nonbeliever, try this experiment. Soak a sample for ten minutes in hot water and watch how it appears to absorb water. Then take it out of the water. Within minutes, the material's surface will dry back to its original color. TREX isn't as rigid as wood, so it's not intended for use as columns, beams, joists, stringers, or other primary load-bearing members.

Changes in the die design have produced a much improved decking board, both in overall dimension and appearance. The top surface has a slight crown, so be certain to install that side up. TREX has also been evaluated and listed with NES (National Evaluation Service, Inc.), which encompasses the three widely-accepted model building codes: BOCA, ICBO, and SBCCI.

TREX comes in three colors:

- Natural, a light brown that fades to a driftwood gray after 6-12 weeks of exposure to sunlight and water.

- Winchester Gray, a dark brown that fades to a deep rich gray after 6 to 12 weeks.

- Woodland Brown, a colorfast brown that stays dark.

The manufacturer advises that you can accelerate the weathering process by periodically spraying water on the deck.

You can purchase 8-, 12-, and 16-foot lengths in the normal wood dimensions: 1 × 6, 2 × 2, 2 × 4, 2 × 6, 2 × 8, and 5/4 × 6 (1$\frac{1}{8}$ × 5$\frac{1}{2}$ inches). There are some other less-commonly-used sizes available, as well as custom profiles. The product is very dense and contains no toxic preservatives, grains, or knots (Figure 11-3). The 5/4 board features radius edges.

I've installed TREX, and I've never seen it check, crack or splinter. You'll soon discover it's not necessary to predrill, even when placing deck screws close to the edge. (But the manufacturer and I recommend you predrill anyway, especially in cold weather.) If you countersink a screw, the material lifts up around the head. Tap the area with a hammer to flatten the raised (mushroom) portion. If you barely cut the surface with the countersunk bit, the screw will pull itself and the decking material down flush to the surface, leaving a clean, countersunk look. I prefer the countersunk look to the flattened-down look.

You can nail it either manually or with an air gun. When nailing manually, be sure to set the nail at least $\frac{1}{2}$ to $\frac{3}{4}$ inch deep before letting go of it, or the nail could fly out full force.

TREX sands, routs, and cuts just about like natural wood. But because of its density, don't expect to cut through it as easily as you cut redwood or cedar. While a 40-tooth carbide saw blade gives a clean cut, the manufacturer recommends an 18- or 24-tooth carbide blade to keep both the blade and the material cooler.

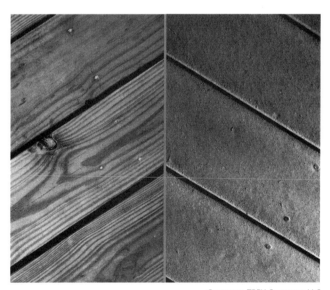

Courtesy: TREX Company, LLC

Figure 11-3
One-year-old wood vs. two-year old TREX

The company offers a ten-year fully-transferable limited warranty against checking, splitting, splintering, rot, insect or fungal decay. Nobody knows yet how long it will actually last, but ongoing accelerated laboratory testing has surpassed 20 years and it's still going strong.

It's easy to take care of a TREX deck. It just needs periodic cleaning. TREX readily accepts paint and stain, according to the manufacturer, and is splinter- and knot-free. No sealants are required.

Gaps Required

Like wood, TREX expands and contracts with temperature changes. It's a good idea to have the material delivered to the job site at least a week before installation. Stack it on 2 × 4 sleepers laid every 2 feet and as level as possible to keep it up off the ground. Place 1 × 2 strips (stickers) at 2-foot intervals between the layers of material to create air spaces. The stickers prevent each layer of boards from sagging, and heavy blocks on the top of the pile help stabilize the upper pieces. Using both sleepers and stickers allows air to circulate between the boards and the material to adjust to the temperature.

When you install it, leave gaps between the boards in both width and length. For width, a good rule of thumb is to leave at least ⅛-inch space between boards. Because pressure-treated lumber arrives wet, most contractors leave no gap between boards. However, with TREX, gapping width-to-width is required.

For boards up to 16 feet long, the manufacturer recommends that you allow a ¹/₁₆-inch end gap for every 20 degrees F difference between installation temperature and the highest temperature expected during the year. Along the width of the board, leave a ⅛-inch minimum gap. Base your spacing on the manufacturer's formulas.

TREX Thermal/Moisture Expansion and Contraction coefficients are published for both width and length. While both are important, we'll work out an example based on gapping along a board's length. To calculate the thermal expansion and contraction of TREX, use the following formula:

coefficient of thermal expansion × temperature change × length of original material in inches

For length, the TREX coefficient of thermal expansion/contraction is $15 × 10^{-6}$ °F, or 0.000015. The temperature change is the difference between the temperature at installation and the hottest temperature expected during the year. Keep in mind when using this formula that the material must reach the outdoor temperature before you install it.

Let's take, for example, a 2 × 6 board 16 feet long. The outside temperature is 60 degrees F and the temperature will reach 100 degrees F during the hottest part of the summer, a 40 degrees F temperature increase. The increase in length would be:

$$(15 × 10^{-6} °F) × 40°F × (16' × 12'')$$

$$0.000015 × 40°F × 192'' = 0.1152''$$

The resulting number, 0.1152, is between ⁷/₆₄ and ⅛ inch, so at 60 degrees F you would install the boards with a gap just under ⅛ inch.

Here's a rule of thumb you can use to meet the manufacturer's recommendation: Allow ¹/₁₆ inch for every 20 degree F difference. There are two 20 degree F differences in the 40 degree F temperature difference in our example. So you'd allow 2 × ¹/₁₆ inch, or ⅛ inch. That's a lot easier than that long calculation.

Construction Notes — For their 5/4 material, the building code requires joists 16 inches on center for a 100 psf load. But I have a couple of suggestions to help stabilize the decking material between joists to prevent deflection. First, place the joists on 12-inch centers (especially on the hottest side of a house). Second, bridge each joist with blocks. The extra joist helps keep the decking boards in tight, especially if you use screws, not nails.

Because TREX is more flexible than wood, you'll soon discover that handling a 16-foot piece of this product requires a second pair of hands, especially in hot weather. After you gain experience and feel you understand the characteristics of TREX, you may be able to install it in half the time of traditional decking.

If you're able to start from the house with full-width boards, there's no problem with installation. But if the layout requires you to trim the board to fit (less than 3 inches), string a line to make sure this first piece is installed straight. Otherwise, it will con-

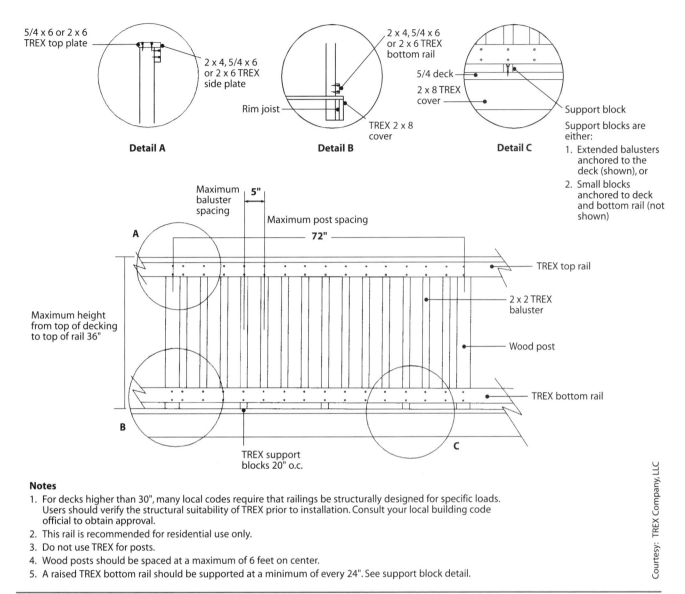

Detail A

5/4 x 6 or 2 x 6 TREX top plate

2 x 4, 5/4 x 6 or 2 x 6 TREX side plate

Detail B

2 x 4, 5/4 x 6 or 2 x 6 TREX bottom rail

Rim joist

TREX 2 x 8 cover

Detail C

2 x 4, 5/4 x 6 or 2 x 6 TREX bottom rail

5/4 deck

2 x 8 TREX cover

Support block

Support blocks are either:
1. Extended balusters anchored to the deck (shown), or
2. Small blocks anchored to deck and bottom rail (not shown)

Maximum baluster spacing

5"

Maximum post spacing

72"

A

TREX top rail

2 x 2 TREX baluster

Maximum height from top of decking to top of rail 36"

Wood post

TREX bottom rail

B

TREX support blocks 20" o.c.

C

Courtesy: TREX Company, LLC

Notes
1. For decks higher than 30", many local codes require that railings be structurally designed for specific loads. Users should verify the structural suitability of TREX prior to installation. Consult your local building code official to obtain approval.
2. This rail is recommended for residential use only.
3. Do not use TREX for posts.
4. Wood posts should be spaced at a maximum of 6 feet on center.
5. A raised TREX bottom rail should be supported at a minimum of every 24". See support block detail.

Figure 11-4
2 x 2 raised TREX rail

form to irregularities on the house and every board will follow suit. Follow this practice with any decking boards less than 3 inches wide.

While the company offers balusters, caps and railings, TREX, combined with top and bottom railings of natural wood like cedar or redwood, produces a great-looking deck. If you choose to use all of their products, there are some added measures you need to take. Since this product isn't intended for use as a structural member, use wood for the railing posts and place support blocks on 18-inch centers under the bottom rails. Figure 11-4 shows a recommended railing

system. You can find other systems and useful information in the TREX Contractor's Handbook.

TimberTech

Crane Plastics' *TimberTech* is made up of 70 percent wood and 30 percent plastic, a composite which lightens to the color of driftwood with exposure to sunlight and water. It expands and contracts as temperatures change, and resists damage from termites, carpenter ants and marine borers. It's new to the market and backed by a 10-year limited warranty.

Courtesy: TimberTech/Crane Plastics Company

Figure 11-5
End view of TimberTech composite decking

What makes TimberTech different is the unusual design. As you can see from Figure 11-5, it's open at the bottom. That makes it much lighter to handle than other decking materials on the market, while maintaining structural integrity. Four "feet" run the length of the material. It's 6 inches wide across the finish top, available in lengths of 12 and 16 feet (plus 20 feet in some markets).

The $1^{1}/_{2}$-inch thickness can span 24 inches on center for residential use, but 16 inches is required for commercial purposes. (Personally I'd use 16 inches on center or less for all applications.) It's not designed for use in any load-bearing situation, like joists, stringers or beams.

Its extra width and the tongue-and-groove design help reduce your labor costs and eliminate fastening through the top surface. And because of its tongue-and-groove design, gapping isn't required. I noticed that this product has drain holes cut into the tongue. This allows any top water to drain to the bottom groove and out. But the manufacturer recommends that you slope the deck's frame away from the house at 1 inch per 10-foot run to allow for proper drainage.

The top surface features the design and feel of a woodgrain pattern. For aesthetics, use the manufacturer's trim accessories or consider installing trim pieces around the deck's perimeter and flush with the

top of the decking material. To eliminate fastening through the surface in the starting piece, install their starter strip (U-channel) along the leading edge of the deck.

You don't need paint or stain to maintain the deck's longevity, but if the customer wants it to blend in, the manufacturer recommends using a quality oil-based latex paint or solid color stain. Let the completed deck weather for 8 to 10 weeks before applying the paint or stain, to ensure a good bond. Of course, once you paint or stain it, it'll have to be maintained to keep up its appearance.

Re-Source Lumber

Would your customers appreciate a deck that carries a 20-year limited warranty not to rot, split, crack or splinter? Re-Source is made from 100 percent recycled waste, including plastic milk containers. This synthetic ridged-plastic material consists of HDPE (high-density polyethylene) resins, UV-inhibiting pigment systems, and selected additives. With no painting or sealing required, this is about as maintenance-free as you can get.

Re-Source is manufactured in many dimensional lumber sizes, shapes and colors, and can be installed with standard woodworking tools. A 72-tooth carbide blade cuts through it easily, and routing is a snap. In fact, a newly-routed edge blends in perfectly with the rest of the material. But if you use a router bit with a roller bearing, a plastic film will soon wrap around the bearing. This film comes off easily, but you'll need to check your bit regularly. Using a countersunk bit before installing screws gives a cleaner-looking job, or you can use plugs supplied by the manufacturer.

Re-Source is available in tongue-and-groove boards for a quicker installation (Figure 11-6). Some force is required to make sure one T&G board fits tightly into the next. You may need a specialized tool to help you achieve that snug fit. The BoWrench by Cepco Tool Company (Figure 11-7) is the easy and quick solution to pulling any kind of decking board into place (as well as T&G plywood). The BoWrench binds itself to any 2-by joist, freeing your hands to install the decking material. The tool in the photo has the optional T&G cam. You can also get an adjustable joist gripper for joists, studs, or rafters up to $1^{1}/_{2}$ to $6^{1}/_{4}$ inches wide.

Attach the boards through the tongue and into the joist with 2-inch galvanized finishing nails at a 45-degree angle. For best results, use a nailset to properly sink the nail below the surface level. When fastening to steel joists, use #6 self-tapping screws through the groove at a 30-degree angle.

You can use decking clips for the installation, so you don't have to fasten through the surface. This system also allows for improved expansion and contraction tolerance of the decking surface, because an 8-foot piece may expand or contract $1/4$ inch over a 50 degree F temperature change. If the deck material is installed at 50 degrees F and the temperature increases to 100 degrees F, the material will expand $1/4$ inch. Because of ReSource's minimal shrinkage, you can build a deck without the usual gaps between the boards.

The T&G boards aren't 100 percent watertight, so water will either run through or run off. Any water left on the surface is really beneficial since the product actually provides better traction when it's wet than when it's coated with dry dust. If safety is a concern, check out their anti-skid surface, applied at the factory or in the field.

Re-Source products aren't designed for use as supporting members or joists. If you use them in a railing system, place blocks under the bottom rails and between the posts (not more than 3 feet apart) to prevent sagging during hot weather. Install the $3/4 \times 5^1/2$-inch decking material on a joist system that's 12 inches on center. Bridge the joists with blocking material to tightly secure the entire framing system.

The company has introduced a 20-gauge galvanized joist system for their decking. There are two main profiles that make up the system — the "U" that measures $2^1/2 \times 6$ inches and the "C" that measures $2 \times 6^1/16$ inches. The ledger board and rim joist use the "C" channel, while the joist uses the "U" channel. I've never used a "U" channel for a joist, but I see the potential for standing water inside the channel. I've been told that this isn't generally a problem. If standing water does become an issue, you can drill small weep holes into the "U" channel on the underside. The manufacturer recommends spacing joists no more than 12 inches on center and a distance of not more than 8 feet between beams. Use a friction blade to cut the galvanized steel profile, then apply galvanized steel primer to all cuts.

Courtesy: Re-Source Building Products, Ltd.

Figure 11-6
Installing Re-Source plastic decking

You can construct beams using both profiles. Fit the "U" channel on the inside of the "C" profile and fasten them together through the top and bottom face of the "C" and into the two flanges of the "U" with #8 self-tapping screws.

This product may attractively coordinate with a mobile home or a home with vinyl siding. It comes in weathered redwood, light oak, cedar, gray, and white, allowing your customers to create color accents that enhance their home. When fastening through the surface, use their color-coordinated screws.

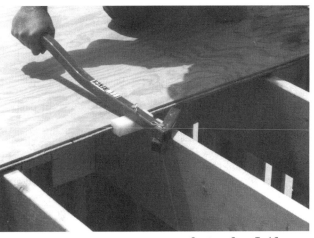

Courtesy: Cepco Tool Company

Figure 11-7
Using the BoWrench to pull decking board into place

The manufacturer recommends white for the railing system (with matching screws) but not for the decking surface. White will show dirt very quickly, requiring regular washing with soap and water. White decks also have a greater tendency to reflect light than darker decking material. Of course, this depends on the location of the deck with relation to the sun. Don't rule white out entirely. In some cases, it just may be what the surrounding area needs, and white is much cooler to walk on barefoot. The vital questions are: "What's important to your customer?" and "What's right for the situation?"

E-Z Deck

First introduced in the Canadian market by ZCL Composites Inc. in 1992, E-Z Deck arrived in the U.S. market in 1995. It's a fiberglass-reinforced composite, which means that it's made from new materials rather than recycled. E-Z Deck is available in lengths from 8 to 40 feet in 2-foot (even) intervals, with rails from 8 to 40 feet in 4-foot (even) intervals. Two board widths are available: 2 × 4 (actually $1^1/_2$ × $3^7/_8$ inches) and 2 × 6 (actually $1^1/_2$ × $5^7/_8$ inches) in Classic Gray, Sandalwood, and Arctic White.

You snap them into place over retaining clips (available in 8-foot lengths) fastened to the top of the joist and available for both 45-degree and 90-degree applications. Butt joints require a second support (double joist) for the retaining clips.

E-Z Deck structural boards are designed and engineered to be used as a direct replacement for dimensional lumber, laid flat for flat deck areas. When you use it like dimensional lumber (side-by-side flat placement over supports which are 16 or 24 inches on center), it's designed to provide stiffness and strength that's equal or greater than select-quality cedar decking. They recommend joist spacing at 16 inches on center for the 45-degree application. You may want to consider blocking the joists for a stiffer framing system. When installed according to the manufacturer's instructions, E-Z Deck products may be used for construction of balconies meeting the National Building Code standards in the same manner as dimensional lumber.

The manufacturer doesn't expect normal exposure to sunlight to have any significant effect on the structural strength of E-Z Deck within a normal 20- to 30-year life span. Over time, the effects of weathering may produce an even mellowing of the surface, but because the product is still fairly new, they're not sure how long it takes.

For your customer's peace of mind, E-Z Deck carries a lifetime limited warranty for the original owner for residential application only. The fine print states "ZCL warrants E-Z Deck to be structurally unaffected by ultraviolet degradation and natural weathering for a period of fifteen years. In addition, E-Z Deck will maintain its appearance for a period of ten years from the date of installation and will not check, flake, peel, or blister."

Two years ago they introduced their premium line, the same product as E-Z Deck with a traction coating applied for a superior surface. This added feature (Tex-Plus) makes sense: not only does it enhance the product, but it's now more practical in wet areas near hot tubs, docks, wheelchair ramps and swimming pools.

What I found unique about this product is its complete hand/guard rail system made of the same material. You place the balusters (43 and 50 inches in length) in precut slots on the bottom rail, which serves as a fascia board as well. Then fit the top rail over the balusters and fasten it with self-drilling screws (available from the manufacturer) at a 45-degree angle upward through the underside of the deck rail through the groove.

I haven't used their railing system, but after studying the installation guide and photos, it appears there could be a better way to tie the top rails together at the corners. They use two flat L-brackets and self-drilling screws to hold the system together (Figure 11-8). Since the handrail is hollow (Figure 11-9), it's possible that a molded 90-degree piece could be fabricated to fit the inside to help lock the corner together as a solid unit. Maybe the manufacturer will pick up on this. But overall, it's a good-looking handrail system.

Phoenix Recycled Plastics

Phoenix Recycled Plastics, Inc., entered the market in the early '90s with a permanently-colored plastic lumber made from 100 percent recycled plastics. They also distribute plastic lumber from other

Figure 11-8
The underside of the E-Z Deck top handrail

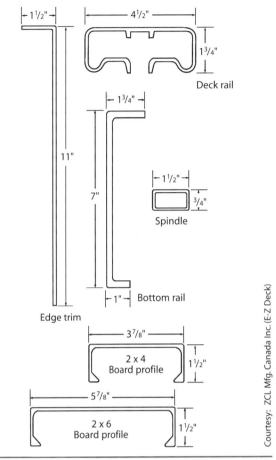

Figure 11-9
Various E-Z Deck profiles and dimensions

companies as well. They offer three grades of plastic to meet a wide range of customer needs:

- Premium Grade — 1 × 6 decking, 1 × 6 T&G, 2 × 2 balusters (nominal dimensional sizes), and $\frac{1}{4}$ inch × 16 foot trim board.

- B-Grade — 1 × 6 , 5/4 × 6, (nominal dimensional sizes).

- Structural Grade — for deck joists, beams, and posts; includes nominal dimensional sizes clear up to 10 × 10.

Each grade, of course, features special characteristics for particular applications. Architectural and engineering services are available to assist you with your project. Contact Phoenix for samples.

Phoenix's products feature a built-in woodgrain texture with consistent color all the way through. Standard colors such as white, black, gray, teak, weathered teak and redwood are available, but they can also provide custom colors for a special project. Their ^{3}/4-inch premium grade material is flexible enough for use in a curved application. In Figure 11-10, notice the curved white material that bands the edge of the T&G decking boards. This structure features a framing system that's 12 inches on center.

Most nominal sizes are available for decking boards, T&G, round and square posts, and structural fiberglass-reinforced lumber in lengths up to 48 feet. If you're considering this product for structural

Figure 11-10
Curving deck blends into environment

applications, be sure to consult a customer service representative before you begin, and check with your local building department.

Because the manufacturer offers such a wide selection of colors, sizes, and products, opportunities to use this product are practically unlimited. Some areas to consider besides decks include:

▌ Boardwalks

▌ Docks

▌ Handrails

▌ Benches

▌ Tables

▌ Furniture

▌ Trim

▌ Landscaping

▌ Fencing

Phoenix Plastic Lumber is more flexible than wood. When using 1 × 6 or 1 × 6 T&G decking material, place the boards on joists spaced 12 inches on center. You can use the 2 × 6 decking on joists at 16 inches on center. Consider the linear expansion and contraction when installing this material. An 8-foot section may expand or contract $^1/4$ inch over a 50 percent temperature change. If you install the decking material in temperatures of 50 degrees F and the temperature increases to 100 degrees F, it'll expand 1/4 inch. To calculate expansion and contraction in length dimension, use the following formula:

0.00007 × length of board in inches ×
°F of temperature change =
inches of expansion or contraction

Ecoboard

American Ecoboard Inc. is doing its part to clean up the environment by recycling polyethylene materials to manufacture their products. They entered the market in the fall of 1993 with a product made from a blend of only HDPE and LDPE (low-density/bags, film or wrapping). They claim that with this homogenous blend of materials, there's no chance that Ecoboard will split, warp, peel or flake. A foaming agent added during the manufacturing process creates a honeycomb effect in the center of each board, simultaneously adding strength and reducing weight.

Available in four standard colors (slate gray, sandstone brown, mission brown and redwood), Ecoboard has a wood-like appearance. Optional colors include driftwood white and desert sand. Nominal dimensional sizes include 2 × 2 up to 6 × 6, including 5/4 × 6 with lengths from 8 to 18 feet. Tongue & groove products, available in 2 × 10 and 3 × 10, are great for bulkheading and docking projects.

Cutting an Ecoboard 2 × 4 with a 96-carbide tooth blade yielded a clean, smooth surface. The manufacturer recommends building the framework 16 inches on center for both 2-by and 5/4 material. For heavier loads, consider spacing the framing members closer and adding extra support. The product will expand up to $^1/4$ inch over an 8-foot span, so plan for this movement in your design. An ideal installation temperature is between 40 degrees F and 75 degrees F. When installing in temperatures over 75 degrees F, butt ends as closely as possible and space boards $^1/8$ to $^3/16$ inch.

Ecoboard isn't designed for structural application. Support the bottom plate for railings every 2 feet to prevent sagging. To give you an idea of the weight of this product, a 12-foot length of 5/4 × 6 weighs 30 pounds, or 108 pounds for a 6 × 6, which may be comparable to the weight of green or treated material. In other words, it's heavy.

Carefree Decking System

Carefree Decking System by U.S. Plastic Lumber, Ltd. is similar to other HDPE products mentioned in this chapter. It's also made from 100 percent recycled post-consumer plastic. The system offers a 50-year limited warranty and comes in six standard colors: cedar, weathered wood, gray, sand, teak and white. I bring Carefree Decking System to your attention because of its unusual design.

There are many ways to attach the decking board to the joist. Figure 11-11 shows their EC Deck Clip that was designed specifically to work with their decking material. It's an innovative stainless steel fastening system that allows for thermal expansion and contraction in most severe climates. This system takes about 5 percent more labor time than nails or screws, but it eliminates exposed fastener heads on the deck surface. That's something your customer will appreciate.

The manufacturer claims the new design ($1^1/8$ × $5^1/8$ inches × 12 feet) makes deck boards lighter and stronger, though it isn't recommended for structural applications. They have some good-looking custom profiles: rounded edges for the rail cap, stair tread with bullnose and knurled surface, and balusters with cove detail edges. And they offer nominal dimensional sizes, such as 4 × 4 and 6 × 6 posts. The knurled surface on their decking boards is great for around pools, spas, floating docks and wheelchair ramps.

The manufacturer says that joist spacing either 16 or 12 inches on center is acceptable (depending on the application), but consider 12 inches on center. Also keep in mind the expansion and contraction of recycled plastic lumber. At a temperature of 70 degrees F, the decking board requires a $1/4$-inch gap between ends every 12 feet. The manufacturer recommends additional space for installation in climates where there's a large temperature swing, or in conditions where the deck is built between two walls.

One last thought: while you can use nails to fasten components, screws are a better choice as an overall fastener. And while it's not necessary to predrill, it's still a good habit to get into with all products. It helps prevent the screw from snapping off, extends the life of the tool, and overall, you just get a better finish job.

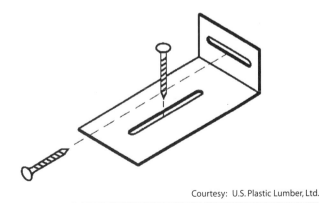

Courtesy: U.S. Plastic Lumber, Ltd.

Figure 11-11
Use 6D stainless steel or
galvanized nails or $1^1/4$-inch deck screws

Brock Deck

CTB Inc. has been manufacturing vinyl products since 1985. (We'll discuss their Triple Crown fence system in the next chapter.) In about 1986, they introduced Brock Dock, a vinyl decking material used in marinas and resorts. To carry over into the deck market in the early '90s, the product known as Brock Dock became Brock *Deck*. They added a vinyl residential decking railing system using galvanized steel post mounts that are installed before the decking. Both Triple Crown and Brock Deck are now being manufactured by Royal Crown Ltd.

Their low-maintenance, slip-resistant, color-retaining Hi-Polymer grade PVC planks carry a 20-year warranty against abnormal weathering and discoloration. They're available in three basic colors (tan, white, and gray) specially formulated for outdoor exposure where color and holding its shape are important. It has ultraviolet inhibitors to protect it from the sun's harmful rays. Impact modifiers provide added strength and impact resistance in all weather conditions while preventing dents, splits, cracks, or breaks.

Brock Deck's slip-resistant pattern exceeds UL standards and the space between planks meets the ADA surface code. You can buy it in lengths from 2 to 20 feet in even 1-foot increments. The six channels (per plank) help with drainage and drying of surface water. The planks measure $1^1/2$ × $5^7/8$ inches and the overall design allows two to come together to an even 12-inch width. Also available are vinyl rim covers for 2 × 8 or 2 × 10 framing members to resist the effects of weather and moisture. Depending on the building code in your area, the decking material can span joists spaced between 12 and 24 inches on center.

What I found interesting is that the decking material isn't secured to the joist with screws through the surface. It's held in place by aluminum clip-strips fastened to the tops of joists (Figure 11-12). The strips are available in 3-, 4-, 4.5-, 5-, 5.5-, and 6-foot lengths. A start clip is used up against the structure and a finish clip is used at the end of the deck. Aluminum snap locks and a spacing template are available to prevent placing the snap locks too close together. Once you've secured the clip-strips with stainless steel screws, the plank is ready to install into the snap lock system.

Courtesy: Royal Crown Ltd.

Figure 11-12
Brock Deck's snap-lock decking system

To attach Brock Deck, set one side of the plank's leg into the clip-strip and strike with a DBH (dead blow hammer) to set the leg. Then rotate the leg into the clip and strike the surface again with the DBH. With the plank securely in place, set the center leg by striking the center with your rubber mallet. Once you've done a few, you'll get the hang of it. When butting two blanks together, double the joist and install a 3-inch snap lock specially designed for this type of application. After all the decking material is

Courtesy: Royal Crown Ltd.

Figure 11-13
A four-sided, 4-foot square vinyl picnic table

in place, a C-channel fits over the end, held in place with $1/4$-inch plastic push lock pins. Or you can install end caps with one lock pin.

One way to rise above the competition in deck construction is to provide some little extras, like building a simple privacy screen, trellis, perhaps a flower box or two. How about a picnic table to match the decking surface? Brock Deck is one step ahead of you. They offer a variety of tables in kits that would make a great deck-warming gift (Figure 11-13).

Wood Products

New wood products are coming to market that you should definitely consider for deck construction. Some of these products are remilled from existing stock materials, others are imported, while still others add newly-formulated preservatives to treated lumber to make them more environmentally safe. What this does mean for you? It gives both you and your customer more choices.

Supreme Decking

The manufacturer of Supreme Decking remills select cedar, pine, and redwood to eliminate most of the defects, creating a smoother, cleaner decking material that's a pleasure to install. Each remilled decking board has a $1^{1}/4$-inch finished edge, while the middle is $1^{3}/8$ inches high. This $1/8$-inch crown allows water to drain, discouraging surface puddling and cupping, and extending the life of the boards. It also has a great massage-type feeling as you walk barefoot over it!

The decking boards feature lengthwise relief cuts on their undersides (similar to hardwood flooring) to help keep them flat. These cuts also allow the wood to breathe and adjust to temperature and humidity changes to minimize warping, cupping, and twisting (Figure 11-14). A prime cut (top) from the original wood gives a consistently higher-quality product, and the smoothest finish available on any stock decking material. The manufacturer says the material isn't weakened by the undercuts, since it's still 1 inch thick at its thinnest point.

Supreme Decking is a high-quality product that will complement any customer's home and surroundings. But keep in mind that a quality product doesn't come cheap. Figure on about 10 percent more than cedar or redwood decking material. It installs just like any other wood decking material.

A new product in Supreme Decking's line is Deck Strap fasteners. These are "L" brackets you install to the underside of the marked decking board. After installing two brackets per joist (based on roughly 6-inch-wide material), lift the decking board into place and fasten it to the joist to create a deck surface that's virtually free of nails or screws. The drawback to this system is that it requires more labor time — possibly up to 50 percent. But it may be just the thing for the customer who's willing to pay for a premium deck.

Iron Woods

Customers fret over having the best woods when it comes to hardwood floors. Why don't these same concerns carry over into the deck market? How about offering your customers hardwoods as an alternative to the standard wooden decking materials currently on the market? There's a new breed of woods, the "hardwoods" of decking material. These woods, formerly used mainly in the commercial industry, are now crossing over into the residential market. Timber Holdings Ltd. has imported these materials under the Iron Woods trademark for over 35 years.

One famous project constructed of these materials is the Boardwalk in Atlantic City. A few years back my family and I had a chance to walk the Boardwalk. I remember it vividly because I couldn't figure out what type of wood was used for the surface. It appeared to be some type of hardwood, but I couldn't put my finger on it. I later discovered it was Ipe and Cumaru, both woods offered under the Iron Woods name. Other species include Ekki (used in engineered structures, including bridges), Jarrah (great for interior hardwood floors), Macaranduba, Dinizia, and Bangkarai. All of these woods are selectively harvested or farmed in Brazil, Australia, Indonesia and Africa.

These hardwoods feature a 25-year-plus durability rating (Figure 11-15). Along with their incredible strength, hardness and resistance to splintering, they have a natural resistance to chemicals, rot, termites and marine borers. The wood is very dense and heavy. Depending on the species, it can weigh as

Courtesy: Supreme Decking Inc.

Figure 11-14
Remilled decking board with an ⅛-inch crown

A CCA-treated pine after 5 years on the Boardwalk

B Iron Woods decking after 24 years on the Boardwalk

Courtesy: Timber Holdings Ltd.

Figure 11-15
Traditional CCA-treated pine vs. Iron Woods

much as 69 pounds per cubic foot for green material (air dried). To compare that with domestic woods, redwood weighs 28 pounds and Southern pine is 35 pounds. They carry the NFPA (National Fire Protection Association) and *UBC* (*Uniform Building Code*) Class A-1 rating (the same as steel or concrete).

Interestingly enough, these woods aren't kiln dried. In fact, they're milled after air drying. The wholesaler can expect to receive the product with a moisture content around 18 to 20 percent. I've been told this moisture content helps to minimize expansion. However, the wood will experience some shrinkage — about $1/16$ inch or less in width for a 4-inch wide board, with no measurable shrinkage in length. Getting a 16-foot stick isn't uncommon (20 feet or longer by special order), and neither is clear all-heart. Iron Woods are available in four grades: FEQ (First Export Quality) All Heart, FEQ First One Face, #2 Second Grade, and FAS (First And Second) Mill Run.

Ipe and Cumaru both have a rich golden-red color which can be maintained with a minimum of oiling. However, if left natural, they turn to a soft silver over a period of months depending on its exposure to the sun. These woods, along with the rest of the Iron Woods family, resist splintering, twisting, cupping, and checking.

Courtesy: Timber Holdings Ltd.

Figure 11-16
Boardwalk project using Iron Woods

Timber Holdings Ltd. is committed to the sustainable management of these unique timbers. They only use sources that are registered with IBAMA (Brazilian Institute for the Environment and Renewable Natural Resources) for harvesting from well-managed, sustained-yield forest lands. They comply with the sustainable forest management guidelines of the ITTO (International Tropical Timber Organization) and Brazilian Forest Code Law 4.771. Using these unique hardwoods gives the standing timber added value. This helps the local economy and provides incentives to properly manage these renewable resources rather than burning them to clear fields for subsistence farming, commercial agriculture, or fuel wood.

Construction Notes — Because you're dealing with an extremely dense wood, it's important to follow the manufacturers' suggestions for proper installation of their products. Keep in mind that you're still dealing with wood, so be sure to leave $1/8$ or $3/16$ inch between decking materials. Here are some other things to consider:

▌ Use carbide-tipped saw blades and high-quality drill bits. Predrill ends when using any type of fasteners to avoid splitting. Screws, including self-tapping, need pilot holes and countersinking.

▌ Immediately after cutting, seal all ends with a clear aqueous wax end sealer such as Anchorseal to help reduce end checking.

▌ Stainless steel fasteners eliminate the potential for chemical reactions or staining around the fastener head. You can use coated or galvanized steel fasteners, but they generally have shorter service lives than stainless steel fasteners. Some installers use stainless steel #7 trim head screws with a $9/64$-inch pilot hole to eliminate predrilling of countersink holes.

▌ The nailing pattern for 1 × 4s, 5/4 × 6s, 1 × 6s, and 2×4s is two per board (either nails or screws).

▌ At 100 psf and nominal dimensional sizes, 1 × 4 or 1 × 6 can span 24 inches, 5/4 × 6 can span 32 inches, and 2 × 4 can span 48 inches (Figure 11-16). This boardwalk project is at Avon By The Sea, New Jersey. Even with the large timbers used for the joist, blocking is used throughout the framework.

A Joists are lifted off the concrete to allow circulation

B The same project near completion

Courtesy: Greenheart-Durawoods, Inc.

Figure 11-17
Japanese boardwalk project of Ipe 2 x 4s has a 40-year life expectancy

It's best to seal Iron Woods products after installation to assist in adjustment to the environment and to help reduce the potential of surface checking. Their natural density and alkaline content can cause reactions with certain finishes and affect their drying and adhesion. Many oil- and water-based coatings have caused a color change in the wood, so the manufacturer recommends using penetration oil-based finishes. Do a sample check with several boards to determine the compatibility. Before applying sealer, make certain the wood is dry, then brush and wash the surface to remove dirt, dust and sticker residue.

Manhours — Based on installing 200 square feet of 1 × 4s, 1 × 6 or 5/4 (only) on joists 24 inches on center with a crew of two, you could use 0.014 as your manhour figure per square foot. If you're predrilling, bump this figure up at least 25 percent, to 0.0175 manhour per square foot. Timber Holdings Ltd. wants to point out that because the product is clear and straight, you spend less time sorting and straightening, plus there are no callbacks and no yard returns.

Pau Lope

In the last ten years, Greenheart-Durawoods Inc. has also been distributing Ipe under the trademarked name of Pau Lope. This hardwood decking material delivers a smooth texture, highlights, richness, and appearance similar to teakwood. The boardwalk under construction in Figure 11-17 is in a waterfront park in Yokohama, Japan. In part A, you can see how the joists have been lifted up off the concrete to allow for required circulation. Part B shows the same project near completion. If you look closely, you'll see that the Ipe 2 × 4s are fastened with only one screw per joist. The life expectancy is 40+ years.

Greenheart-Durawoods' products are harvested from Brazil's Amazon rain forest using the highest acceptable standards for forest management. Among the characteristics of Pau Lope are:

- Decay and insect resistance — Natural insect and decay resistance throughout all fibers of the wood, including termite resistance.

- Durability of wear surface — Extremely hard and dense; resists all types of punishment.

▌ Appearance and weathering — Natural variations of rich wood colors with little or no knots. Clear lengths up to 16 feet long. Maintains rich color with minimal oiling. If left natural, turns a soft silver over time. Resistant to shrinking, splintering, twisting, cupping, and checking.

▌ Fastening — Fasten with marine grade stainless steel screws or nails recessed into the surface. Will not pull loose or pop up and there is no tannic acid reaction with stainless steel fasteners.

▌ Other — Extremely strong. Natural fire resistance rating: Class A-1 NFPA and UBC.

The distributor recommends the following:

▌ Keep the product out of direct sunlight until it's ready to install. Apply a wood stabilizing agent to the deck exposed surfaces prior to direct sun exposure.

▌ Sticker the product to allow it to adjust to the environment.

▌ Proper air circulation beneath the deck is critical to dimensional stability.

▌ Because of the high density of this product, slower speeds are recommended when using power tools. To give an idea of its density, a 20-inch piece of 2 × 4 weighs in at around 4.5 pounds Because the wood is so stable, one screw in the center of a 2 × 4, for example, in each joist is sufficient.

▌ Predrilling is a must! Drill pilot holes close to the actual size of the fastener, especially at the ends to prevent splits.

▌ Exposed ends on a crosscut should be sealed with a clear paraffin wax-based end sealer.

Use the same manhours listed under Iron Woods above.

ACQ Preserve

Some customers are leery of products on the market (such as pressure-treated lumber) that use preservatives known as CCA (chromated copper arsenate) or ACZA (ammoniacal copper zinc arsenate). Even though lumber treated with these preservatives in a vacuum-pressure process is very stable, several decades of field testing have shown some leaching of the components. Although the amounts were insignificant, in today's market, many consumers are concerned about the impact of these products on the environment, and the possible risk to their health. Who wants their one-year-old crawling around on, and possible sucking on, arsenic-impregnated wood? Alternative products are being developed every day, and hold a lot of appeal to people who want to stay clear of any kind of poison or preservatives. It's your job to stay abreast of what's available, and to be able to offer them.

One preservative that's been on the market for about five years is ACQ (ammoniacal copper quaternary), marketed as ACQ Preserve by Chemical Specialties, Inc. They claim this preservative contains no arsenic, chromium, or other EPA-listed hazardous compounds. Extensive independent testing has shown that ACQ Preserve offers the same durability, strength and protection as chemicals used in other pressure-treated products. Lumber products treated with ACQ Preserve are environmentally safe for use in above- or below-ground construction. They weather from green to a natural brown tone and carry a lifetime limited warranty. Expect to pay about 10 percent more than traditional decking material.

Manhours — This product installs just like any other decking material, so use 0.014 manhour per square foot for a crew of two. This is based on 2 × 4s (only) on joists set 16 inches on center.

Hardware

Not only does the right hardware expedite the project (in some cases), but it helps you produce the professional-looking jobs your customers expect and deserve. Some hardware products are specially designed to help prolong the life of certain materials. You'll want to stay on top by checking out new products coming to the market. Contact the manufacturer or your local distributor for samples, then try them out to see how they can improve your next project.

Deckmaster

One problem inherent with wooden decking material is that its installation requires screws piercing its surface. That allows moisture to penetrate the material. Also, depending on the material being used, surface screws can just spoil the look of the deck. Now there's another way to install that deck that solves both of those problems.

Check out the Deckmaster (distributed by GRAB-BER Construction Products), an impressive 22-inch-long bracket, available in galvanized or stainless steel. It rests on top of the joist, and is fastened to the side. Installing the brackets alternately to the opposite side of each joist from the previous course creates a very solid system. Then fasten the decking boards from the underside with two screws per board, set on an angle. In Figure 11-18, you can see how clean and attractive the deck looks with no fasteners installed through the surface. It has the elegant look of a hardwood floor, and no moisture penetration problems.

You can get a surface almost 100 percent free of deck screws, depending on the design of the deck's frame, the size of the joists, and the distance between the bottom of the joist and the ground. The exception is areas directly over the length of a beam when the joist is a 2 × 6 or less and 12 inches on center. That area will have screws through the top surface because there's just not enough room to use the angle drill required to install these brackets. You'll soon discover the wider (higher) the joist is, the more room you'll have to maneuver the tool. Also, if you're using blocking between your framework, cut the top corners on an angle to allow the bracket to slide by.

If you spray the tops of the brackets with flat black paint, it prevents the metal from reflecting between the deck boards. If the deck is high enough to be accessible from the underside, spray the brackets before you install them. But if the deck is low to the ground (no one will see from the underside), then spray the tops once the brackets are in place, because any paint overspray on the joists won't be visible.

Manhours — The material costs for the hardware will run about 50 percent higher than using screw fasteners. It takes about 20 brackets per 60 square feet. After installing them a few times, I found labor costs to be about the same as installing screws. For the first time around, use 1.5 to 2.0 manhours per 60 square feet with a crew of one, including painting the

Courtesy: Deckmaster

Figure 11-18
To avoid damage to the joist, use the BoWrench
(Figure 11-7) instead of a chisel

brackets and installing joists and decking. After the third time around you may be able to do it in 1.0 manhour per 60 square feet.

EB-TY Fasteners

Because it mounts from the topside, this fastener may be just the ticket for your next project — especially if it's a deck with very little clearance under it. I'm told that EB-TY (Blue Heron Enterprises L.L.C.) was created out of frustration after attempts to make similar products work. This is a complete new system that still leaves the deck surface free of fasteners. The concept is nothing new to the woodworking industry, but it's exciting to see it applied to the construction industry.

The EB-TY is a plastic (UV-resistant polypropylene) biscuit with two tabs that help maintain a $3/32$-inch gap between decking boards. You use a biscuit joiner to cut slots in the decking centered over the joist. Then install the biscuit, using a screw to secure it to the joist through a predrilled hole in the biscuit.

Look at Figure 11-19. The installer in section A is using a drill, but an impact driver will work best, especially in hardwoods. The manufacturer recommends drilling a pilot hole at a 45-degree angle through the biscuit and into the edge of the decking material. Then install a #7 $2^1/4$-inch stainless steel screw.

A Install stainless steel screws at a 45-degree angle

B Biscuit installed with tabs down

Courtesy: Blue Heron Enterprises L.L.C.

Figure 11-19
EB-TY plastic biscuit spaces and installs the decking material

The tabs on the biscuit need to point downward (toward the joist). They're colored black so they won't show in the shadow of the gap. The contractor in section B of Figure 11-19 is using construction adhesive. The recommended product is PLPremium. While you don't have to use exterior construction adhesive with EB-TY, the manufacturer says that using it gives you these advantages:

▌ Added holding power

▌ Eliminating any squeaks

▌ Filling any voids

Manhours — Manhours are difficult to figure because it's a time-consuming product that takes a while to learn, but based on a 350 square foot deck (2 × 6s spaced 24 inches on center), figure 8.5 manhours for a crew of two the first time around. This includes cutting the slot, inserting the biscuit, drilling the pilot hole, and installing the fastener.

Dec-Klip

There are other deck clips on the market, but the Dec-Klip (Ben Manufacturing, Inc.) is the original that was first developed in 1984 (Figure 11-20, part A). The original concept wasn't to speed production, but rather to eliminate the use of fasteners through the face of the decking material. They're made of $1/8$-inch hot-rolled steel, treated with zinc plating to help repel rust. The company offers two sizes to cover both 5/4 and 2-by material. Built-in pads rest on top of the framing member to keep the decking material off the joist so air can circulate between the bottom of the decking material and the top of the joist.

Construction Notes — For installing the first decking board against the structure, use a hacksaw to modify a Dec-Klip by cutting off the trailing leg. Then you can nail it against the structure and on top of the joist as shown in part B of Figure 11-20. Install the deck board into the points using either a beater board or the BoWrench (Figure 11-7). Using this "starter" Dec-Klip eliminates nails in the surface. Now you can slip a row of regular Dec-Klips beneath the leading edge of the first decking board. Make sure the Dec-Klip is snug up against the edge before driving a #10 galvanized box nail through the slot, toenailing it at a 50-degree angle into the decking board edge and the top of the joist. Then nail the Klip to the joist using a #8 box nail through the anchor hole in the exposed leg.

A The installed Dec-Klip

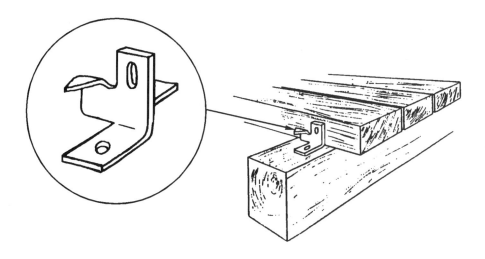

B Cross section of a Dec-Klip installation

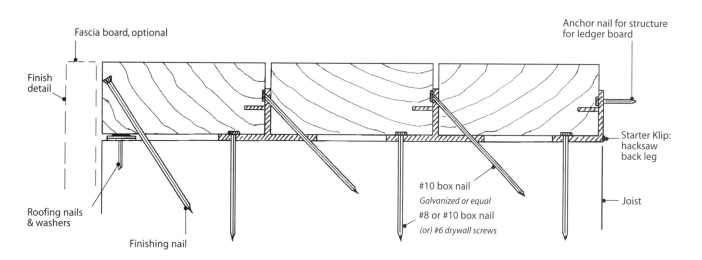

Fascia board, optional

Anchor nail for structure for ledger board

Finish detail

Roofing nails & washers

Finishing nail

#10 box nail
Galvanized or equal
#8 or #10 box nail
(or) #6 drywall screws

Starter Klip: hacksaw back leg

Joist

Klip requirements: (Approx. only)

Board Dimensions		Joist Spacing	No. Klips per sq. foot
2 x 6		2 ft. centers	1.0
2 x 6	5/4 x 6	16 inch centers	1.5
2 x 4		2 ft. centers	1.8
2 x 4	5/4 x 6	16 inch centers	2.3

Courtesy: Ben Manufacturing, Inc.

Figure 11-20
Dec-Klips install decking with no surface fasteners

When you reach the last decking board, place a couple of galvanized washers (equal to $1/8$ inch) and nail them to the top surface of the joist close to the end. Finally, toenail the edge to complete the fastener-free system.

The installed Dec-Klip leaves a $1/8$-inch gap between the decking boards. Make sure your wood is seasoned before you install it, or this gap will widen. If you choose to bypass this step, squeeze the boards tightly together to allow for the shrinkage.

Manhours — This is another product that takes a while to learn, and whether you use nails or screws will affect your productivity and labor costs. Another variable that influences labor cost is whether you use a beater board or the BoWrench. To give you a general idea, the following figures are based on a crew of one building a 10 × 10-foot deck using 2 × 6s, 24 inches on center, and after the third installation:

▌ Using a beater board with nails: 1.5 manhours per 100 square feet.

▌ Using a BoWrench with nails: 1.0 manhour per 100 square feet.

▌ Using a BoWrench with screws and no pilot holes: 0.75 manhour per 100 square feet.

▌ Using a beater board with screws and pilot holes: 2.0 to 3.0 manhours per 100 square feet.

▌ Using a BoWrench with screws and pilot holes: 1.75 to 2.75 manhours per 100 square feet.

Counter Snap

For those who insist on fastening from the top, here's a new product that's worth looking at. The Counter Snap headless deck screw (O'Berry Enterprises, Inc.) is a deck screw with an interesting twist (pun intended!). After driving the specially-scored screw into place, you snap the head off, leaving the screw $1/8$ inch below the surface with a hole only $1/8$ inch in diameter. It's almost like having a finish nail with threads.

To use Counter Snap screws, you need a special fixture with a countersunk hole in the center (Figure 11-21). First place the fixture on the deck surface and slide a screw into the hole. Then install the screw using a #2 recessed square driver bit. When the head seats into the countersunk hole, it will snap off under pressure. It doesn't take much pressure to snap the screw. When you're finished, there are 2 inches of screw holding the wood. That's fine for 5/4 material, but not quite enough for a 2 × 4.

There's one disadvantage to having to use the fixture to install Counter Snaps. In softwoods that have a slight crown, the bottom of the fixture leaves an impression about 1 inch in diameter. If the wood is perfectly flat it's not as noticeable and eventually the wood will return to its original state. The company is currently working to redesign the base of the fixture to make it wider. In the meantime, you can avoid this situation by drilling a $5/16$-inch hole in the center of a $3/32$-inch steel plate that's at least $4^1/2 \times 4^1/2$ inches.

A Special fixture for installing the Counter Snap

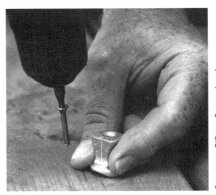

B The head will snap off after installation

Courtesy: O'Berry Enterprises, Inc.

Figure 11-21
Installing the Counter Snap headless deck screw

Then countersink the existing hole in the fixture by $3/32$ inch (the thickness of the steel), using a countersink bit. Now place the fixture on the steel base and you're ready to fasten deck boards. You'll lose some labor time using this plate, but it will absorb and distribute the pressure, eliminating the marks.

In softwoods you don't have to drill a pilot hole except at the ends. In hardwood, however, you need a pilot hole for installation of all fasteners. You can go back and fill the holes, but they'll begin to shrink within 30 minutes in a softwood that wasn't predrilled. I tried a little hot water on the hole and within a few minutes the hole just about disappeared. If you prefer to install from the top, these headless screws will leave a better-looking finish surface than a screw with a head.

What happens if a decking board needs replacing? Not a problem! But before you tackle the project, you'll need a plug cutter (long enough to cut through the thickness of the decking material) and a precision drill stand that attaches to an electric drill. Use the drill stand with the plug cutter to cut around each of the screws, lift the board, and remove the standing screws fastened to the joists. Then you're ready to insert the new decking board.

Manhours — These manhour figures are based on a 10×10-foot deck, eight joists 16 inches on center, and 20 2×6s. That's 160 intersections between decking material and joists. If each intersection has two screws, you'll have to install 320 screws. Figure 1.0 manhour for a crew of two if you don't have to drill any pilot holes, and add 0.50 manhours if you do.

Stratton Bracket

During normal deck construction, you have to cut the house siding to allow installation of flashing and the ledger board. If you've done this, you'll know how easy it is to skin your knuckles or smack your fingers when attaching joist hangers. One way to save your hands is by using the Stratton Bracket system by P. A. Stratton & Co., Inc. (Figure 11-22). It gives you the option of building the entire deck framework on the ground and then, with extra helping hands, lifting the frame up and into the brackets.

A good way to try the system is to use a starter kit that contains both brackets and lag bolts. This process eliminates attaching a ledger board to the

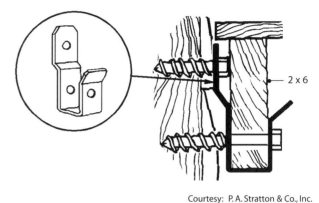

Courtesy: P. A. Stratton & Co., Inc.

Figure 11-22
The Stratton Bracket installed with lag screws

house and having to cut the siding before installing the brackets. It works extremely well when installing directly to concrete walls. All you need to do is drill for a lead anchor to accept a $1/2$-inch lag bolt. Whether you're attaching to concrete or the exterior siding, remember to caulk behind the bracket on the top hole.

The manufacturer recommends spacing the brackets 6 to 12 inches in from the ends of the ledger board. Intermediate brackets should be 48 inches on center for 2×6 construction, 24 inches on center for 2×8s, and 16 inches on center for 2×10s or 2×12s. Consult your local building department for their required bracket spacing, even though these brackets were designed to meet or exceed the building code requirements.

One thing you'll like about the system is that it allows you to shift the framework within the bracket to center and square it to the house. Once the framework is in position, secure it to each bracket with a $1/2$-inch lag or bolt/nut (depending on if the bracket is fastened to wood or concrete) through the bottom hole. It also lets air circulate between the house, the decking and the ledger board. This prevents water buildup, which could cause rot or decay or permit bug infestation. And it allows the ledger board to expand and contract with the seasons without any damage.

The Stratton Bracket is so versatile that you'll find many uses for it. If you can't think of any right off, consider these: stair and flower box installation, mounting of solid shutters over windows, creating a collapsible work bench, or to provide door security for a 2 × 4 cross door. If you need a temporary installation, it's perfect!

Manhours — Using a one-man crew to drill the hole and secure the bracket, figure 0.25 manhours per bracket for wood application and 0.50 manhours per bracket for concrete application.

Deck Bracket

The ledger board, where a deck attaches directly to the side of a house, is always a cause for concern. If you don't do this step correctly, water can get trapped there. Then if there's no way for that water to drain out or dry, the ledger board will begin to deteriorate. Of course, the likelihood of damage depends on many factors, including whether there's a roof covering the deck, whether the siding is embedded into caulk that's directly on the deck platform — and whether the customer gets carried away with the garden hose.

Crawford manufactures an aluminum deck bracket that provides air space between the back of the ledger board and the face of the siding or sheathing. It helps prevent sill rot and insect damage in this area. The bracket (Figure 11-23) measures 8 inches wide on the back (the side facing the house) and 4 inches on the front (facing the ledger). It's 5 inches high and 4 inches deep, designed to be attached to the rim joist with lag bolts. The siding fits around the neck of the bracket, then the area is caulked.

Courtesy: C.R.S., Inc.

Figure 11-23
Aluminum deck bracket
provides air space under ledger board

When using these brackets, make sure you attach them to a surface that's structurally sound and at least 6 inches thick. The span from the dwelling to the first support determines the number of deck brackets you'll use. If the distance is less than 8 feet, space the brackets at 8-foot intervals. From 8 to 12 feet, space the brackets every 6 feet. From 12 to 16 feet, space the brackets every 4 feet. Never space them more than 8 feet apart. Observing these precautions when using the brackets can save time, money, and most importantly, prevent callbacks.

Manhours — These figures are based on a one-man-crew installing only the brackets and don't include siding over the bracket:

▐ New construction: 1.0 manhour per bracket.

▐ Existing structure: 1.5 to 2.5 manhours per bracket (includes removing siding).

▐ Concrete application: 2.0 manhours per bracket.

Dekmate

The Dekmate Bench Bracket (Canadian Dekbrands) is a quick and simple way to incorporate a bench and railing all in one system (Figure 11-24). Made from polyethylene, a structural foam-injected plastic, it's maintenance-free; there's no worry about rust or rot. Its design also makes it great for use on docks, concrete patios, and existing decks. Once the bracket is in place (it attaches from the top as well as from the side), it only takes up $4^{1/2}$ inches of the deck's surface area. The tail (support) of the bracket can easily be mounted into concrete. Would your customer like a standalone bench in the garden? The angled back makes the bench more comfortable.

The company guarantees the product to the original purchaser for ten years from the date of purchase. If it proves defective during that time period, they'll replace the brackets only at no cost.

Manhours — To install the bracket on a flush deck (no overhang), figure that a one-man crew can install each bracket in about 20 minutes (0.333 manhours).

Protection

Perhaps one day you'll be called on to build a deck up on a roof platform or to install some type of protective material over an existing deck. Products are

Figure 11-24
Dekmate bench brackets

available to accomplish just that — after you prepare the project area with a solid substrate like exterior plywood or concrete so it's ready to accept the protective material. I've located a couple of products that may be just the ticket. The first is a two-component batch-mix urethane coating. The second is a single-ply PVC waterproof membrane.

GacoFlex

Gaco Western, Inc. has several products that may be just right for some of your jobs. The unique thing about their systems is the use of ground walnut shell as the granule. When a deck system uses sand or other hard granule, there's a tendency for the granule to scuff out. Within months, a deck can be absolutely smooth. Since walnut shell's hardness is identical to the coating, it wears with the coating instead of scuffing out. The texture lasts as long as the deck, which is frequently 10 years.

Another unique feature is that they don't use a fabric sheet as part of the system. Their coatings, when applied and cured correctly, are strong enough that they don't need the fabric to give them internal strength.

Gaco Western is in the final developmental stages of the next generation of urethane coatings for decking. Watch for products that are solvent-free with no volatile organic compounds, and plural-component urethanes (not water-based) that are UV stable and can be applied by brush or roller.

One current product that's worth considering is GacoFlex UB-64 base coat, a two-component (1:1 ratio) urethane. When properly combined and applied, it cures to a tough, puncture-resistant synthetic rubber membrane with high tensile strength. The coating will expand and contract with the substrate as the temperature changes. The system has to be installed by factory-approved installers. Whether you're using it on wood or concrete, make sure all seams and cracks are properly reinforced before they apply the base coat.

If you're going to use GacoFlex in a traffic flow area, it takes two coats of UB-64, then adding the granules in the first of two topcoats of GacoFlex U-66. The U-66, available in colors, provides a skid-resistant and UV-resistant coating when combined with organic granules.

Allow each coat to dry until tack-free and sufficiently cured for foot traffic before you apply any additional coatings. The curing will take anywhere from less than two hours to overnight, depending on drying conditions, temperature, and the product applied (Figure 11-25).

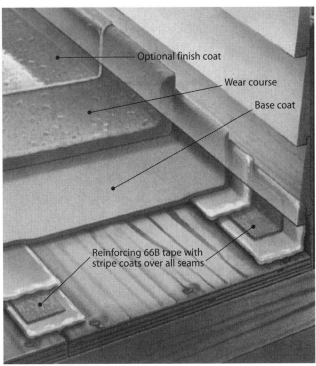

Optional finish coat

Wear course

Base coat

Reinforcing 66B tape with stripe coats over all seams

Courtesy: Gaco Western, Inc.

Figure 11-25
Gaco deck protection system

The second Gaco product I want to bring to your attention is an acrylic decking system that uses GacoFlex A-56. This is a semi-pro decking system for small contractors in situations where base coatings are required. It's a water-based single-component elastomer coating. It's water resistant, extremely hard, very strong, and has good adhesion qualities. Actually, this product can stand on its own as both a base coat and a topcoat. It's important that you don't apply it to a surface that is, or will be, cooler than 50 degrees F within 12 hours. It requires complete evaporation of water to achieve a cure. GacoFlex A-56 has essentially a dull or flat finish, available in several colors.

New to the market is GacoFlex A-58. It's a tough, durable, and glossy topcoat that can be applied over A-56. It has many of the physical characteristics of a urethane in a water-based formula.

Manhours — You can apply any of these systems by spray, roller or brush. Remember, the UB-64 has to be installed by a factory-approved installer, while you can install the A-56 and A-58. To get an idea of the labor costs, here are manhours based on 100 square feet:

- UB-64: In an 8-hour day, a one-man crew can cover 2,500 square feet; however, because it's a two-component product, any coverage above 500 square feet requires a crew of three.

- A-56 or A-58: In an 8-hour day, a one-man crew can cover 3,000 square feet.

VersaDek

Most membranes aren't intended for use in areas with foot traffic, and they usually require experienced installers. If you opt to apply a membrane, choose a product that can handle foot traffic, and that has seams that you don't need to overlap.

Versadek Industries produces a high-quality line of interior and exterior PVC vinyls with a polyester backing that are both decorative and waterproof. The company has a 20-year history and their Ultra (commercial weight, open texture vinyl that resembles a level looped carpet) carries a lifetime warranty for residential and five years limited for commercial applications. It exceeds the standard for slip-resistance set by the ADA, which means you can install it on any walking surface, including ramps.

Ultra is 72 inches wide and suitable for installation over almost any solid substrate including previously-coated surfaces. It's tough and durable, yet flexible enough to expand and contract with the substrate without cracking or splitting. Install it just like you'd lay vinyl flooring in a kitchen. You trowel on the adhesive, which achieves an incredible grip on wood, concrete, and most other building materials. It's actually both an adhesive and a sealant, which helps the vinyl wear layer to waterproof the substrate. Don't install it directly over asphaltic-based products (tar, roof sealant, driveways, etc.) because it will discolor.

Construction Notes — When installing the adhesive, begin with an 1/8-inch V-notch trowel and apply enough adhesive to completely saturate the backing. If you have seams, keep the adhesive back 6 inches on either side of the seam (centerline) until the seam has been double-cut. Carefully unroll the vinyl into the adhesive to prevent wrinkles and bubbles. You can't just flop it down! Take care to smooth out any bubbles with your hands as you unroll. The manufacturer recommends rolling the vinyl with a lightweight roller (75 pounds or less).

You have two choices when it comes to seaming: the standard overlap head seam or a cold weld butt seam. The latter is a perfect choice for a completely flat look. To accomplish it, make a double-cut through the center of the overlap using a straightedge and utility knife. Make sure you're using a sharp razor blade. Fold the material back on itself and apply adhesive in the seam area (6 inches on both sides of the centerline), then roll.

Immediately clean any adhesive that oozes out with a medium bristle brush (fingernail brush) and water. Use paper towels to soak up the residue and let the surface dry. Now you're ready to apply the cold weld seam sealer.

This is a two-component/two-step process. First use Type A, a thin-viscosity primer that softens the vinyl, allowing the edges to be fused together. Then use Type C, a thick coating applied to the seaming area over Type A after it has dried about 10 minutes. Type C softens the vinyl slightly and fills any gaps or holes to make the seam both strong and weatherproof. It dries in one to two hours.

You need to spread both components at least $\frac{1}{8}$ inch thick and $\frac{1}{4}$ inch wide over the centerline of the seam. The seaming system works best in temperatures between 50 degrees F and 85 degrees F. Never put sealant on seams in direct sunlight with surface temperatures over 90 degrees F. The seam isn't totally invisible, but it wasn't intended to disappear. The manufacturer designed this cold weld system to be strong (which it is), durable, waterproof, and relatively easy to accomplish. Of course, like anything you do in this field, practice makes perfect! The more you work with this seaming kit, the closer you'll get to making the seam almost disappear.

The manufacturer has gone to great lengths to create an installation manual that covers such topics as:

■ Surface preparation

■ Perimeter installation

■ Types of adhesive and application

■ Installation of the vinyl

■ Seaming

■ Caulking

■ Drains

■ Ventilation

Figure 11-26 shows how to do the inside edges. On new construction, if there are gaps where the substrate meets a vertical wall, use a cant strip (a piece of 2 × 2 lumber cut at a 45-degree angle). If there's no gap, you can use a bead of caulk smoothed with your finger to make the transition. Allow the vinyl membrane to turn up the wall at least 3 inches and to lap under the building paper and exterior siding.

Versadek vinyls are ideal for covering not only decks, but balconies, patios, swimming pool and hot tub areas, and docks. They come in four colors: mist, gray, cinnamon and sandbar.

Manhours — These figures are based on a deck size of 10 × 30 feet, a crew of two, and include the seaming. Note that these figures are based on an experienced crew — the third time around:

■ New deck without handrails: 3.0 manhours per 300 square feet.

■ Existing deck with handrails: 4.0 to 5.5 manhours (depending on railing design) per 300 square feet.

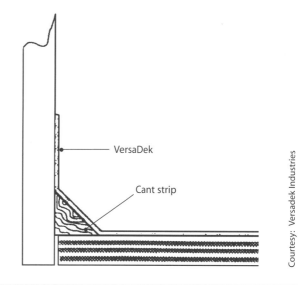

VersaDek

Cant strip

Courtesy: Versadek Industries

Figure 11-26
Installing VersaDek vinyl in a corner

▌ Coving: There's no way to pinpoint a manhour cost because of the variation in siding materials. However, if the structure has vinyl or aluminum siding, it could take upwards of 1.0 manhour for 20 linear feet for a two-man crew.

Railing

It's best if you and your customer have a clear picture of the type of railing system you'll install before you put down the decking material. If not, Feeney Wire Rope's *Cable•Rail* is a product that your customer may find appealing. It's a definite alternative to standard railing systems on the market today.

The ⅛-inch stainless steel cables used in this system can improve the looks of any home or deck while preserving the view. The photo in Figure 11-27 shows the cable spaced at 6 inches on center. Many cities and counties are changing from the 6-inch intermediate rail spacing requirements to the new 4-inch requirements, so be sure to check your building department's local requirements.

The Cable•Rail system combines two special fittings to attach and tension each cable. The threaded terminal is fitted at the factory to one end of each cable. The cable comes in 5- to 80-foot lengths, and custom lengths are available. The installer inserts the terminal through holes in the end post and secures it with a washer and hex nuts. Cable tension is adjusted by tightening or loosening those nuts.

The other end of the cable uses a quick-connect fitting, a special one-way wedge design that allows the cable to slide easily through the fitting in one direction. Once the cable has been released, the wedge grabs and locks onto the cable. Then you just cut off any excess cable.

A single corner post poses some design problems, especially if you want a continuous pass of the cable. The cables will have to be offset at least ½ inch to allow clearance for the fittings. Using two posts allows the cable to pass continuously around the corner, which provides a more attractive finish look (Figure 11-28). Also, if the end of the post that has the threaded terminal is exposed, I'd recess the nuts and cap in the exposed end with a 1-by finish material.

Courtesy: Feeney Wire Rope & Rigging, Inc.

Figure 11-27
Cable•Rail deck railings don't obstruct the view

Courtesy: Feeney Wire Rope & Rigging, Inc.

Figure 11-28
Use two posts to make a smooth corner transition

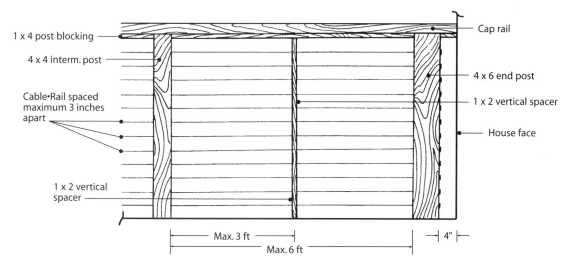

A Standard 42" high, 12 strand Cable•Rail frame detail

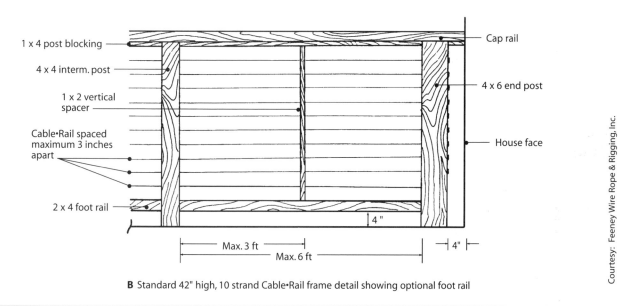

B Standard 42" high, 10 strand Cable•Rail frame detail showing optional foot rail

Courtesy: Feeney Wire Rope & Rigging, Inc.

Figure 11-29
Installing Cable•Rail wire railings

Figure 11-29 gives an overall view of two kinds of installations. Notice that to meet the 4-inch code requirements, the Cable•Rail is spaced 3 inches on center. The company also carries optional hardware to help when unique and unusual decks present design challenges.

Manhours — You'll soon discover that lacing the cables doesn't take much time. Most of your labor time will be spent drilling the holes in the supports. To achieve uniform holes, I suggest that you drill the posts in a shop with a drill press.

These estimated manhours are based on a 12 × 12-foot deck with railings on three sides and no steps. That's 36 feet of railing. It takes 12 cables for a 42-inch-high post (spaced 3 inches apart) with no bottom rail. The posts are spaced no more than 3 feet apart, with two posts on each corner.

▮ To lace and secure the cables with a one-man crew: 2.0 manhours.

▮ To drill the 15 posts for a crew of one: 4.5 to 6.0 manhours.

Landscaping

You know that a deck alone doesn't make a masterpiece. It takes a whole package: design, type of wood, and the surrounding area. All these factors combine to create a great deck project. With that in mind, it's just possible that you need to sell an entire package to the customer — one that includes landscaping. Whether you do the work yourself or hire it out, there are a few products on the market you may want to recommend to your customers. They just could be the ticket to help that new deck blend into and complement its surroundings.

Vinylattice

Many of us have used standard- or architectural-grade wooden lattice for camouflaging the area underneath a deck or for other decorative purposes. Have you ever considered using vinyl lattice? Cross Industries, Inc., produces a durable PVC product, called Cross Vinylattice. It comes complete with matching vinyl accessories, including frame channel, panel channel, cap molding, angle, and stainless screws with plastic caps. The lattice has the appearance of a fine wood product with a flat enamel finish when installed (Figure 11-30).

Courtesy: Cross Industries, Inc.

Figure 11-30
Using Vinylattice to screen a hot tub

For exterior applications in vertical planes, you can hang the lattice from the top by about four fasteners in the middle portion of the width. Leave the sides and bottom edges free to expand and contract between frame channels or panel joiners. Add braces for maximum 4-foot unsupported lengths.

For porch rail panels 3 feet high by lengths up to 5 feet, fasten at every other perimeter joint or strip. Longer horizontal lengths up to 8 feet should be fastened top and bottom in the middle 4 feet of length. Leave the ends free to expand and contract within the frame channels or panel joiners.

Here are some of the advantages of Vinylattice over traditional wood:

▌ Sturdy construction; it stays together as you work with it.

▌ No splintering or splitting.

▌ It cleans easily.

▌ The joints are sealed.

▌ There's no need to paint it.

▌ It comes in standard colors and you can order custom colors.

▌ It's durable and resistant to both mildew and termites.

▌ It's available in different panel lengths, widths and thicknesses.

▌ It's available in many patterns, or you can custom order a pattern to fit your design.

▌ It lasts longer.

Manhours — When figuring your labor costs, keep in mind that vinyl lattice is no different to install than cedar or redwood lattice. Use 0.030 manhours per linear foot for a crew of one. This includes handling and securing but not any molding you may use to trim the lattice.

DesignWood

DesignWood Interlocking Timbers by Thompson Industries, Inc., are landscaping timbers with male and female ends that interlock with a joint pin. They're pressure-treated southern yellow pine available in 4 × 4 and 6 × 6 with lengths of 2, 4, 6, and

Courtesy: Thompson Industries, Inc.

Figure 11-31
DesignWood interlocking landscaping timbers

Courtesy: Thompson Industries, Inc.

Figure 11-32
Flexible Decking Panels work around obstacles

8 feet, allowing you to create decorative landscape bordering, planters and retaining walls. Other projects could include a planter/bench top combination using their prebuilt 2 × 2-foot and 2 × 4-foot bench tops that work with the 4 × 4 timbers, a raised-bed garden, or a sandbox/bench top. This list can be as long as your imagination can carry you. In Figure 11-31, notice how well the DesignWood blends in with the environment and the deck, both as a border for the landscaping and to hold the pavers for the steps.

Manhours — Use these manhours for a landscaping project, using a crew of one and not including any excavating:

▌ 4 × 4: 0.045 manhours per linear foot.

▌ 6 × 6: 0.060 manhours per linear foot.

Decking Panels

Thompson Industries, Inc., also markets Flexible Decking Panels made from treated southern yellow pine and stained to resemble redwood. The slats are a true 1 × 1½-inch dimension in a 2 × 2-foot panel. The slats are locked together with a flexible PVC tubing which makes the panel great for contours, walkways, or pathways through the flower garden. The system doesn't require any fasteners to hold the panels in place — the weight alone does the job. These are great for ground level installation and resurface applications (Figure 11-32).

Manhours — For a one-man crew, figure 0.0833 manhours per panel, which doesn't include any framework or border material.

Have an open mind toward the alternative products now entering the market, and keep up with new developments. Your customers may already be aware of these products and eager to put them to use on their projects. Don't lose a sale because you're not aware of the latest in alternative products. You already know the deck market is very competitive, but being adaptable and flexible will help keep you on top of the competition. Now, let's look at products that can help to tie the surrounding environment to the structure.

Outdoor Products

This chapter is about venturing outdoors with some ideas to help tie together an entire project. We'll take a look at some interesting products that can have multiple uses. That's where your creativity comes in. Take, for example, C-LOC by Crane Plastics. It's an engineered vinyl sheet piling originally designed for use in creating bulkheads for water retention and erosion control. But when I saw the product, I wondered why you couldn't use it as a finish foundation material. Some products, such as vinyl fencing, can add a dynamite look to a ho-hum fence. Best of all, they're maintenance-free and extremely user-friendly. The invisible pavement products provide ground support for an area so vehicles can drive over a grassy area without damaging the grass.

There are a lot of interesting items throughout this chapter, so you should have no trouble finding one or two to help complete your next project. Don't limit yourself and your customers. Be creative and mix products to design an award-winning project. You'll satisfy your current customers, and bring in new ones. It's all up to you and your willingness to use alternative materials.

Railing and Fencing Systems

Who says you have to stay with traditional products when it comes to railings or fencing? Open up your mind to other types of materials instead of wood. How about vinyl or urethane? I've had the opportunity to look over these interesting products, and have reviewed their installation. The urethane products look great when mixed with traditional materials, but most important — they're virtually maintenance-free: resistant to weather damage, insects, and decay. Check out these products. You just never know which one could work in your next project.

Balustrade System

We've already talked about Style-Mark, the manufacturer of the Balustrade System, with reference to their high-density urethane products. Now they've created balusters, rails, newels, and porch posts that are all specially designed and reinforced to meet structural requirements. Balusters are reinforced with fiberglass, rails and newels with PVC, and porch posts with steel. The system is available in a variety of styles and sizes ($4^1/2$ and 12 inches) that offer flexibility for use indoors or out. You can use it to put the finishing touch on porches, patios, balconies, and stairways (Figure 12-1).

Style-Mark's *Registered CAD/dxf Library Files* include over 1,300 urethane architectural millwork items. These files were developed to make it easier to work with their products, whether you're designing architecture, creating elevations or specifying products. You can import these drawings into your layout, then show it to your customer. It helps them get a clear vision of the project.

Construction Notes — The $4^1/2$-inch balustrade system installs about like traditional wood materials when it comes to building a railing system. However, because these aren't traditional materials, it's important that you check with your local building department to make sure the recommended procedure

Courtesy: Style-Mark Inc.

Figure 12-1
The Baulstrade System brings elegance to a porch project

Courtesy: Style-Mark Inc.

Figure 12-2
Drilling holes in the bottom rail to insert the balusters

meets code. It's possible that simple alterations may be required to meet the codes in force in your area. Note that the $4^1/2$-inch balustrade system has been tested and meets the following BOCA National Building Code/1993 Criteria: 1615.8.2 Guard Design and Construction 1615.8.2.1 In-Fill Areas. But be aware the manufacturer warrants only its products, not the installation.

Courtesy: Style-Mark Inc.

Figure 12-3
Installing the top and bottom rails to the wall

Use the instructions supplied by the manufacturer as your guidelines for successful installation. High-density urethane material makes drilling holes for the insertion of balusters in the $4^1/2$-inch bottom rail a simple task (Figure 12-2). Using a drill press in the shop will help you make accurate and precise holes.

Here's a quick overview to give you an idea of how to install this system:

1. Place rail support blocks under the bottom rail at every span of 48 inches or less.

2. You can attach the top and bottom rails to the wall or post from the underside of the rails using the manufacturer's specially-designed angle brackets. But for aesthetic reasons, I recommend that you install the railing the same as wood railings, with one exception. Instead of toenailing with finish nails, use galvanized screws. Predrill and countersink them before installing the screws (Figure 12-3). Then fill the holds with automotive body filler.

3. If you use any angle brackets, purchase the optional decorative trim collar for the bottom and top newel or porch posts. This trim will hide the brackets and give the project a polished look.

4. Install newel posts on the deck with a steel base mounting plate with $3/8$-inch threaded rods.

5. Use a top-mounting plate with hardware to secure the new post cap with urethane-based construction adhesive.

6. When constructing the rail sections, remember to use adhesive at the ends of each baluster. Square the system with a framing square before clamping and then use strap-type clamps. Allow it to stand at least 12 hours before handling. Consider painting each section prior to installation.

Manhours — If you've installed a railing system before, you know that a successful installation requires a crew of two. This one isn't any exception: figure on a two-man crew. The complexity of the project could affect the manhours listed below, but these figures should provide you with a starting point to bid the project:

▌ Newel post, $5^1/2$ inches wide, using the manufacturer's hardware and gluing cap: 1.22 manhours per post.

▌ Porch post, $5^1/2$ inches wide, using angle brackets and trim collars: 1.50 manhours per post.

▌ Balusters (24, 28, or 32 inches), including drilling, applying adhesive and clamping: 0.556 manhours each.

▌ Railings, installing section: 0.267 manhours per linear foot.

▌ Bottom rail support block: 0.333 manhours each.

PVC Railings

Re-Source Building Products, Ltd. makes PVC Railings cleverly designed for mounting on the deck surface. The $3^1/2 \times 3^1/2$-inch white PVC octagonal newel post is hollow, so it slips over a galvanized steel support bracket. Be sure to check with your local building department before you install a system like this. You may need to add extra mounting supports in the areas where you're installing the fasteners to secure the steel insert in place. The $^1/8$-inch steel plate welded to the bottom of the insert is predrilled for easy installation to any deck surface, including concrete. For a more rigid feel to the overall system, you can add concrete to the post. A trim collar slides into place, completely covering the steel plate. No one would have a clue how this system was installed!

Courtesy: Re-Source Building Products, Ltd.

Figure 12-4
The PVC Railing slips over the galvanized steel support bracket

The railing shown in Figure 12-4 is attached to the Re-Source decking product and has Metropolitan interlocking slats. You can get the rails in lengths of 6 and 8 feet. The top and bottom rails are prepunched to accept the $^7/8 \times 1^1/2$-inch or 3-inch balusters, which are available in heights of 36, 42, and 48 inches to the top of handrail when installed. There's a steel insert for the top rail that provides added support. You can also use it as the bottom rail if you feel it's needed. The newel post is 2 inches higher, and the newel post bracket comes in lengths of 28 or 38 inches. These railings install cleanly, and the system is virtually maintenance-free.

Construction Notes — From an installation standpoint, this has to be one of the easiest systems to put in. First secure the top and bottom rails to the structure with wall brackets, eliminating a newel post. Drill $^1/4$-inch holes, spaced 24 inches apart, under the bottom rail for water drainage. Because the rail inserts go slightly less than halfway through the post (but not less than $1^1/4$ inches), you can build each railing section and then slide it into the posts as a unit. Finally, the posts are prepunched to accept the rails, so be sure to maintain a level height off the deck. Make sure the height meets the code in force in your area.

Remember to space the baluster and post the same distance on both the left-hand and the right-hand side (with the rail holes centered between the posts). Secure the caps with PVC cement.

One additional thought: If the decking surface is on a slant, you'll need to adjust the steel support bracket. This system is prewelded for a plumb application, so you might have to place steel shims under the front or sides of the steel plate. Consider using stainless steel washers for the shims. To make it plumb, you'll have to cut the bottom of the post on an angle.

Manhours — In general, this is a simple railing system to install. You're likely to spend more time properly mounting supports and setting the posts in cement. Although it can easily be installed by one person, you can cut the time required in half using a two-man crew. The following figures from the manufacturer are based on a crew of one, and don't include planning time or time for cutting. They also assume you've installed the product at least three times and everything is precut. I suggest you consider adding at least 50 percent to your manhour figure the first time around:

I Installing steel support bracket including PVC newel post, gluing cap, and installing trim collar: 0.08 manhours per post.

I Installing balusters in top and bottom railings and installing in post: 0.20 manhours each per 8-foot section.

I Installing wall brackets: 0.08 manhours for a set of two.

Triple Crown Fence

As I drive through the countryside, I always enjoy the view of white fences spread throughout the farmlands. However, I definitely don't want to be the person who has to paint those miles of fences! It's estimated that it costs about $.70 per foot per year to maintain a typical wood fence. That expense can be eliminated completely if you install maintenance-free fencing. Triple Crown fence by Royal Crown, Ltd. is vinyl, which makes it virtually indestructible. Their special design simplifies installation (no screws, nails, or brackets required to attach railings to posts). Under extreme stress, the rail lock will release, eliminating any danger to horses and riders, for instance.

The hi-polymer (polyvinyl chloride) engineered vinyl material is backed by a limited lifetime warranty against abnormal weathering and discoloration. The fencing is available in various sizes, in white, gray, and tan. The plank rails are a single-piece extrusion, with internal support ribs for added strength and impact-resistance. Their design includes prepunched posts that permit the rails to snap into place with flexible rail locks (Figure 12-5). You can get both square and round posts in several sizes, and you can use the plank rail with both kinds of posts. They also offer hub rail fencing for the race track as well as fencing for dressage arenas.

Installing a Triple Crown fencing system will initially cost more than most. However, any fencing your customer chooses represents a sizable invest-

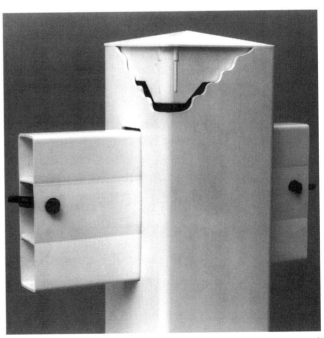

Courtesy: Royal Crown Ltd.

Figure 12-5
Cross section of the Triple Crown Fence

ment. Because it's maintenance-free, it can return the customer's investment in seven years or less. And this is precisely the approach that you should take when selling a product like this.

Construction Notes — A three- or four-rail fence is recommended for most applications for confining livestock. Use a five-rail style when separating stallions or bulls from other livestock. To learn more about the five-rail system, contact the manufacturer. The basic outline here is for the two-, three- or four-rail system:

1. Each post hole should be roughly 8 to 12 inches in diameter. For a two-rail post, the hole should be 30 inches deep; for three- and/or four-rail systems, 36 inches deep.

2. Space post holes 8 feet on center.

3. When filling the hole with crushed stone, concrete mix and crushed stone, pea gravel, or ready-mix concrete, ensure the post is plumb from all directions.

4. Rather than set all the posts first, set four or five at a time and then install the rails. This will help keep the posts properly spaced and aligned.

5. Plank rails come in 16 feet lengths, so be sure to stagger rails so no two joints are on top of each other. When shortening rails, cut them 2 inches shorter than the center line distance between the posts.

6. Rails come predrilled and ready for rail lock installation. You'll need to drill two $1/2$-inch holes on either side of the plank rail for the rail lock. Measure $3/4$ inches in from the end and 3 inches up from the bottom of the rail.

7. When installing the post caps, use a spot of PVC cement on the outside flange of the cap on either side.

8. Gate posts require a special steel insert to carry the weight of the gate.

Manhours — Overall, Triple Crown is an easy system to install with a crew of two (and a third pair of hands wouldn't hurt). You'll probably spend the most time digging post holes and building the gates.

You can save time by using a gas-powered auger. Manhours listed are for a crew of two unless otherwise listed:

▌ Light to medium soil, a laborer working with a hand auger: 4.0 manhours per hole.

▌ Medium to heavy soil, a laborer working with a breaking bar and hand auger: 2.0 manhours per hole.

▌ Medium to heavy soil, using a gas-powered auger: 0.066 manhours per hole (15 holes per hour).

▌ Setting post with concrete (ready-mix): 0.50 manhours per post.

▌ Installing railings: 0.333 manhours per 16 feet.

▌ Setting post caps with PVC cement: 0.15 manhours per cap for a one-man crew.

▌ Setting gate post with concrete (ready mixed): 1.0 manhour per post.

▌ Assemble gate: 0.50 to 0.75 manhours per unit for a one-man crew.

▌ Installing gate: 0.50 to 0.75 manhours per unit.

No-Dig Fencing System

This fencing system by Genova Products is interesting, to say the least. From a remodeler's standpoint, it could provide a quick solution to solving your customer's fencing blues. They've gone to great lengths to make their system easy to install. The fencing components, made of pure virgin vinyl, actually slide over existing 4 × 4 posts or steel T-stakes. Of course, you can install new ones, but what a clever way to take advantage of existing fencing posts and turn them into a work of art. Like magic, you'll erect a complete maintenance-free fence within a very short time. Actually, you'll spend more time removing the old fencing. Just think of the side business you could create by giving facelifts to existing fences!

Genova Products have created some interesting cap designs for both pickets and posts (Figure 12-6). There's nothing to prevent you from mixing and matching your caps to come up with an interesting pattern that'll set it apart from other fences in the neighborhood. The company offers everything you need to construct an entire system, including accessories like electric fence wire holders for livestock

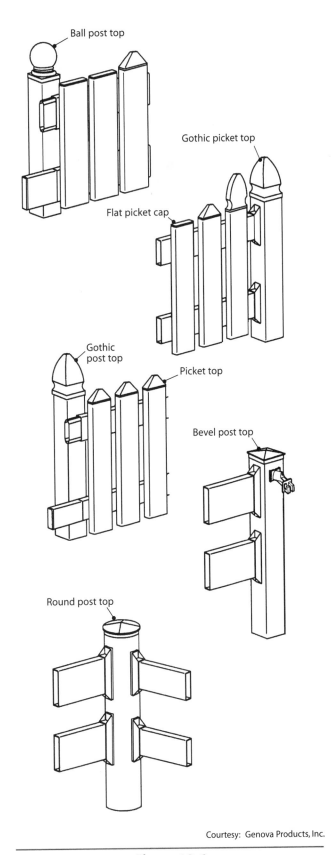

Ball post top

Gothic picket top

Flat picket cap

Gothic post top

Picket top

Bevel post top

Round post top

Courtesy: Genova Products, Inc.

Figure 12-6
Cap designs for Genova Vinyl Fencing

applications, and screw covers to hide screw heads. They even have their own line of screws for different phases of the project, gate hardware, and white vinyl adhesive for permanent attachment of fence fittings to posts and rails.

Construction Notes — If you're installing over existing 4 × 4 posts or steel T-stakes, it's important that they be sound and properly installed in the ground. If not, correct any problems before you go any further. It's important that the steel T-stakes all face the same direction. The tail of the T has to face toward the inside of the yard being fenced. Be sure to check that they're also plumb — it's required for a successful project. The screws supplied are self-drilling and self-tapping, so the tool to use is an impact driver with a #2 Phillips bit. Here's an overview of how to install this system:

1. When installing over steel T-stakes, tap two stake post mount fittings over the stake, one toward the bottom and the other at the top. To get the first stake post mount fittings down toward the bottom of the stake, use a $2^{1}/_{2}$-inch PVC pipe the length of the stake, a beater board, and a 4-pound hammer. The stakes all face the same direction; notice the positioning of the fittings — the post fits snugly over the fittings. Then slide the vinyl fence post over these fittings (Figure 12-7).

2. If you need to set new posts, figure a hole 10 to 12 inches in diameter at a depth of 18 inches for a 3-foot post, 24 inches for a 4-foot post, and 30 inches for a 6-foot post.

3. After the post is in place, you're ready to install the rails. Figure 12-8 shows a couple of options for installing these rails. A side mount requires two rail post mounts on either side of the post. The mounts are held in place with two #10 × $^{3}/_{4}$-inch screws in the predrilled holes in the upper and lower part of the mount (Figure 12-8, section A). Space posts at 8 feet $4^{1}/_{2}$ inches on center.

4. Front mounts require two #8 × $2^{1}/_{8}$-inch screws through the face of the rail and into the post. You'll need a rail coupler when a seam is involved (Figure 12-8, section B). Space posts at 8 feet $^{1}/_{2}$ inch on center.

5. Secure the caps with Genova's vinyl adhesive.

A It's important when putting in the steel T-stakes that they all face the same direction. Face the tail of the "T" toward the inside of the yard being fenced.

B To get the first Stake Post Mount fittings down toward the bottom of the stake, use a 2^1/$_2$-inch PVC pipe the length of the stake, a beater board, and a 4-pound hammer.

C The stakes all face the same direction; notice the positioning of the fittings — the post fits snugly over the fittings.

Courtesy: Genova Products, Inc.

Figure 12-7
Installing the Genova Vinyl Fence over steel T-stakes

Manhours — Installing this system should be simple — easily handled by a one-man crew. However, your labor charge will depend on the design: privacy, picket, or post and rail spacing. The best spacing for the posts is 8 feet 4^1/$_2$ inches on center. If they're less, all you have to do is cut down the 1^1/$_2$ × 5^1/$_2$-inch rails. Manhours are based on a crew of one installing the picket design (^{3}/$_4$ × 4 inches), pickets spaced 3^5/$_{16}$ inches apart, rails mounted to sides of posts, and working with existing posts. They also include cementing in flat tops and caps for each picket and caps for posts.

▌ 3 feet high, two rails spaced at 18 inches: 0.110 manhours per linear foot.

▌ 4 feet high, two rails spaced at 26 inches: 0.120 manhours per linear foot.

Column Concepts

Looking for a creative way to dress up a chain link or wrought iron fence? *Quick Column*, a high-density polymer with a cementitious finish, is designed in

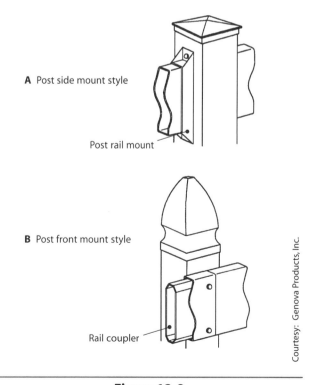

A Post side mount style

Post rail mount

B Post front mount style

Rail coupler

Courtesy: Genova Products, Inc.

Figure 12-8
Two options for installing the Genova rails

Figure 12-9
Dressing up fence posts with Quick Column

Figure 12-10
With matching paint and stone veneer,
you can't tell this is a prefab unit

two halves that make it easy to install around a fence post (Figure 12-9). Just line up the predrilled holes and fasten with the self-aligning pins. You could incorporate this split column into just about any fencing system.

The cementitious finish comes in an off-white color. To match the existing structure, use a good latex primer and finish coat. The manufacturer offers three different cap styles to blend in with the existing fence and surrounding environment. Other products they offer include columns (hole center for rebar and cement), base and trim, chimney caps, light modules, barbecue islands, planters, fencing sections, gazebos, mailboxes (Figure 12-10), and specialty items. With brick, stone, stucco, and glass blocks available, you can mix and match component pieces to create a refreshing and unique design.

Construction Notes — All the products offered by Column Concepts are easy to install. But depending on the size of the unit and the type of material used, you could have to deal with considerable weight. Some pieces weigh upwards of 300 pounds! You'll probably need to use lifting equipment to set some units in place.

When installing the mailbox, for instance, you'll need a footing and pier. This process normally entails removing a section of topsoil 5 inches deep and 3 inches larger than the base of the column. Then you dig a hole 36 inches deep and 8 inches in diameter in the center. Fill it with concrete, and while it's still wet, set and level the base.

The split columns will require some handwork (cutting) in order to fit around rails. For example, on a chain link fence you may have to trim both sides of the column where it meets the chain link and the upper and lower rails. All the pieces have alignment pins. Just apply construction adhesive in the predrilled holes, plus a $3/8$-inch-wide bead 1 inch in from the outer edge. For the split column (with the exception of the cap), you don't need the bead of adhesive. Just put adhesive in the predrilled holes.

Manhours — All of these products require two-man crews. You'll have handwork for digging footings, although for a pier there's the option of using a gas auger to save time. On the split columns, have the manufacturer do the detail work from the factory. Otherwise you'll need the proper tools to cut through the cementitious surface. The manhour estimates given here are based on the second or third post when installing a split column. Figure at least 45 minutes to an hour for the first post, and thereafter:

▌ Split column — fitting and applying adhesive for pin holes and cap: 0.333 manhours per unit.

▌ Prefabricated mailboxes — screw four anchor bolts into base and wet-set into concrete footing: 0.416 manhours per unit.

Gate Hardware

We've been talking about maintenance-free products for fences, but haven't mentioned hardware except for the manufacturer's gate kits. But I want to bring to your attention some items produced by D&D Technologies. First, from a safety standpoint, they'll *help keep children safe* by keeping them from areas where they shouldn't be, such as swimming pools (Figure 12-11). Second, they're injection-molded glass-fiber reinforced polymers, not metal like most hinges and latches. This makes then 100 percent rust free for life. It's not enough to build a fence around the pool to discourage little ones from entering. You have to have the proper hardware for controlling the gate.

Tru-Close

A tension-adjustable, self-closing hinge, Tru-Close offers over 30 models to choose from. They're available in two categories, Regular and Heavy Duty, to handle gates made from vinyl, wood, aluminum, and heavier metal that weigh up to 132 pounds. They can be adjusted in a matter of minutes. The spring is made of stainless steel to ensure powerful, reliable closure.

Why is tension adjustment important and required? Gates differ in size and weight and will experience different wind loads. Lighter gates should close softly while heavier gates (including those in windy areas) require greater closing tension. Pool inspectors, as the manufacturer points out, require a gate to close from any position, including at rest on the latching mechanism. If it doesn't meet the test, the inspector will require hinge replacement. Fixed tension hinges aren't recommended for swimming pool or other safety gates. As you can see in Figure 12-12, it's a simple matter of removing the cap by taking out the screw, then adjusting the tension with a screwdriver.

Courtesy: D&D Technologies

Figure 12-11
The right gate hardware helps protect children around swimming pools

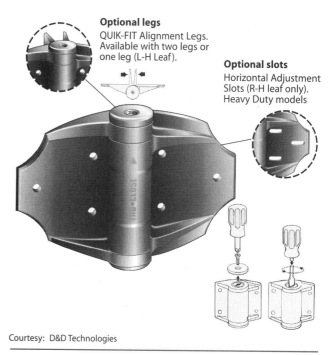

Optional legs
QUIK-FIT Alignment Legs. Available with two legs or one leg (L-H Leaf).

Optional slots
Horizontal Adjustment Slots (R-H leaf only). Heavy Duty models

Courtesy: D&D Technologies

Figure 12-12
Adjust tension in both hinges

Tru-Close hinges carry a limited lifetime warranty and have been tested to meet BOCA, CSPC (Consumer Products Safety Commission), AFA (American Fence Association), and NSPI (National Spa & Pool Institute) barrier codes. The "quick-fit alignment legs" give added strength, and allow for quick and easy hinge alignment on both the post and the gate. They were designed to set a gap which is large enough not to jam children's fingers. Regular models are available with two legs (or none) and fixed screw holes or vertical slots. The heavy-duty range includes two legs, one leg (or none), and fixed screw holes or horizontal slots for sag adjustment.

Magna-Latch

Magna-Latch is a two-part latch system, designed so when the gate swings shut, a powerful "permanent" magnet draws a latch bolt from one housing into the other, latching it securely. I was amazed to see and feel the magnet work. However, I should point out that no product, including this one, can guarantee to keep children out of an area if they really want to get in. This latch isn't child-proof, but it is *child resistant*. For use on swimming pool gates, consult your local building department.

The Top Pull is the ideal latch for swimming pool gates (refer back to Figure 12-11). The latch release knob is designed to sit above the fence, out of reach of small children, and it's key-lockable for added security. The only way to open the gate is to pull up on the knob at the top of the latch. Easy for you, but extremely difficult for a young child. The manufacturer also offers the Vertical Pull, a shorter version of the Top Pull (Figure 12-13). The small Side Pull latch is less obvious, and it's suitable for most household gates. (On swimming pool gates, to keep toddlers from entering, you can use both the Vertical Pull and Side Pull models in conjunction with appropriate gate latch shields, or according to building code regulations.)

The Magna-Latch carries a limited lifetime warranty and meets the same codes as Tru-Close hinges. Both products guarantee a positive lock — every time. The molded components are UV-stabilized and the aluminum tubing is powder coated. The overall product is rust-free. If you're in the pool business, you need to bring these two products to your customers' attention.

Construction Notes — For the Tru-Close, pan- or hex-head noncorrosive stainless steel bolts or screws are recommended. If you're not going to use a latch, install a gate stop to prevent the gate from closing beyond the 180-degree angle, which would damage the hinge. Don't countersink the predrilled holes. Use through-bolts and nuts for metal or vinyl gates. Vinyl posts and gates may require support for the hardware and fasteners, depending on what vinyl product you're using. Remember, each hinge must experience equal tension at all times to ensure smooth, reliable gate closure.

The Magna-Latch system also has an optional post-fixing bracket. With it, you can install the unit on existing posts that don't reach the required height. They also offer special adapter kits for chain link fences, and horizontal adjustment plates for wider gaps (Figure 12-13). Be sure to check with your local building department for proper installation requirements. Specifications could require the latch release knob be at least 4 feet 11 inches above the finished ground or fixing surface and the gate must open outward, away from the pool. Of course, the latch must be fitted to the outside of the gate. Remember to verify the code requirement for the proper height and location before installing. Once the main latch body is pressed down fully onto the mounting bracket, it's difficult to remove.

Manhours — There's nothing hard about installing either the hinges or latch system. Hinges will require an extra pair of hands, especially when hanging the gate. And be sure to allow extra time for tension adjustments. Consider whether or not backing material will need to be installed in vinyl products. That'll increase your manhours. Based on the first time around, follow these suggestions:

- Hinges, including hanging of gate with a two-man crew for wood application: 1.0 manhour per pair.

- Hinges, including hanging of gate with a two-man crew for steel application: 1.5 manhours per pair.

- Latch with a one-man crew for wood application: 0.75 manhours per unit.

- Latch with a one-man crew for steel application: 1.25 manhours per unit.

Outdoor Products **275**

Courtesy: D&D Technologies

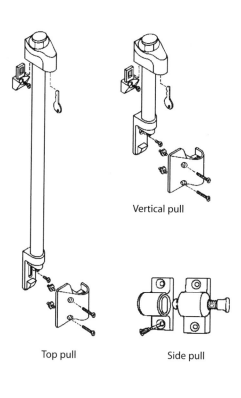

Vertical pull

Top pull

Side pull

Optional
Post-fixing bracket — Latch can be mounted to fences without extra-high posts.

Optional
Round post adaptors — Three adaptors allow latch to be mounted to fences and gates with round posts (see top adaptor too).

Optional
Adjustment plate — Caters for the common offset between gate and fence and allows horizontal adjustment of striker body (before and after installation).

Gate   Fence

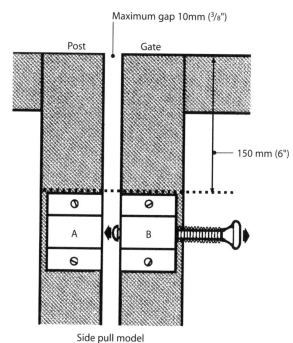

Maximum gap 10mm ($^3/_8$")

Post   Gate

150 mm (6")

A

B

Side pull model

A Three types of Magna-Latch

B Optional fittings for the Top Pull model

Figure 12-13
The Magna-Latch system for swimming pool gates

Bulkheads/Retaining Walls

Bulkheads and retaining walls sometimes play a dual role as a protective barrier — to prevent the advance (by retention and erosion control) of earth or water. Traditionally, they were made of mass timber that just didn't have any eye appeal. Today, with new products entering the market, we have choices to help us create projects that blend in with the environment.

I opened this chapter with a hint about C-LOC by Crane Plastics, an engineered vinyl sheet piling designed for use in creating bulkheads. It's an interesting product that can be very pleasing to the eye. Then there's Keystone, which brings to mind the pyramids. It's a mortarless interlocking system that provides both function and design flexibility. In my hometown this product is used in both residential and commercial projects. They look great no matter where they're used.

C-LOC

An unusual interlocking system, C-LOC is produced by Crane Plastics Company, the same company that manufactures TimberTech (decking material). They've designed engineered vinyl panels that interlock too. Because C-LOC panels interlock along the vertical slope of each panel's angled sides, they create a surface with no breaks along the face (Figure 12-14). When back pressure or force is applied to the face, the joints compress, increasing the structural integrity of the wall.

These panels are made of natural clay-colored heavy-duty ($7/32$ inch thick) exterior grade vinyl (80 percent of which is post-industrial regrind), which allows you to create inside or outside curves that follow natural contours. The overall design really lends itself for use in shoreline bulkheads, wave breakers, retaining walls, or anyplace where there's a need to manage water or retain soil. Accessory 90-degree corners and wall intersects are available.

C-LOC is a good alternative to replace timber, concrete, or even stone. It requires no painting, it's virtually maintenance-free, and it carries a 50-year limited warranty. In addition, it won't rot, rust, corrode, crack, or peel, and it's impervious to sunlight, saltwater, and marine borers. Golf courses use it to control soil erosion along fairways and greens. In Perth Amboy, New Jersey, city officials chose it to replace a destroyed bulkhead. And it has been used

in marinas and along waterfront properties to protect against erosion because it didn't disrupt the landscape. To learn more about this product, contact the company for physical property data information and test results, and be sure to ask for a product sample as well. The address is in the Appendix.

Construction Notes — Depending on the complexity of the project, the equipment required to drive the panels may range from simple to complex. Here's some of the pile-driving equipment you may need:

1. A vibratory hammer mounted on a backhoe.

2. A portable, air compressor-driven jackhammer with a sheet shoe.

3. Conventional land-based or barge-mounted drop hammer.

4. A waterjet driven by a high-output pump, either manually held or suspended from a crane.

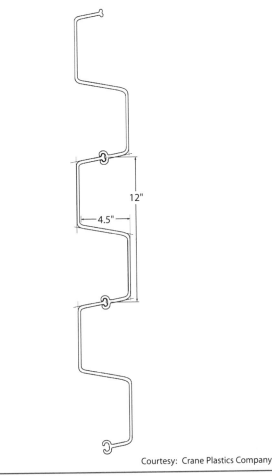

Courtesy: Crane Plastics Company

Figure 12-14
The C-LOC panels interlock invisibly

Hardware required for most installations includes galvanized tie rods, bolts, spacers, lag screws, and washers. You can get optional hardware (including helical tieback anchors and expanding earth anchors) at most marine construction suppliers.

If your customer wants you to install a bulkhead, it's important that you consult an engineer who specializes in this field. There are so many different types of bulkheads that you need to know the correct one for your particular application. Unfortunately, because each bulkhead system is different and complex, it's difficult to go into any great detail on the installation of each system. But you can't overlook the importance of estimating.

As with any construction project, accurately estimating the amount of C-LOC sheet piling material required for your specific location and conditions is vital. It saves both time and money to purchase the right quantities in the right dimensions. These general guidelines may be of assistance:

1. Measure the linear footage of the area you intend to protect using C-LOC.

2. Using a plumb line, measure the distance from the mud line to the height of the finished seawall to identify wall height. Measure every 6 feet to make sure you catch any variations in depth.

3. When determining the necessary length of sheet piling required for the project, a rough rule of thumb is that 50 percent of the length of each panel should be located beneath the mud line.

4. To determine the number of panels required, divide the linear footage to be covered by 12 inches if you're using CL-4500 panels, or 24 inches for the CL-9000 panels.

5. When you're calculating the number of panels needed, don't forget to order the accessory corner pieces when they're required.

Manhours — To help give you a better understanding of manhours, the manufacturer offers the following information to start you off. Consider renting any equipment you need. Personally, I'd find a subcontractor who specializes in this trade who already has the proper tools.

Manhours vary with the size of the wall being constructed, the number of walls required, and soil conditions. On average, however, you should be able

Figure 12-15
A commercial project using Keystone standard units

Courtesy: Keystone Retaining Wall Systems, Inc.

to install 100 linear feet of C-LOC in 8 manhours with a crew of three. This figure assumes you've installed the product at least three times.

Keystone

Keystone mortarless retaining walls are eye-pleasing, blend in with the environment, and are simple in their structural design. Essentially, Keystone is a specially-designed concrete block that offers an affordable solution to your customers' retaining wall and decorative landscaping needs.

Figure 12-15 shows a commercial project using Keystone standard units. Look closely at the fiberglass pins to see the geogrid reinforcement hooked over the pins. Geogrids are made from high-density polyethylene or polyester. Depending on wall height, soil types, and surcharge loads (a load imposed on the soil directly behind the wall which exerts an additional loading on the structure), you'll need to place geogrid reinforcements horizontally in the soil behind the wall units at designated height placement.

Compact the soil over the geogrid reinforcement to reinforce the soil mass. This interlocks the entire system together. Essentially, this composite forms a larger gravity wall structure, a principle which has been around since the time of the pyramids.

I watched my neighbor (who is in the landscaping business) install Keystone units in both his front and back yards in a very short time. One reason they're so quick is that they're mortarless. Another reason is that they don't require any special tools. In most cases (depending on whether or not you're installing a structural retaining wall), you simply interlock the units with fiberglass pins to the required height and top off the wall with caps.

The product is offered in a variety of colors and heavy face textures, in units 4 and 8 inches high. Angled sides allow for graceful curves, concave and convex shapes, and serpentines. The overall design provides so much flexibility that you'll have to restrain yourself to stay within your customer's guidelines. Don't overlook the deep shadows you can create by careful block placement. If you're looking for places to use these blocks, consider the following:

▌ Residential

▌ Apartment complexes

▌ Hotels

▌ Restaurants

▌ Shopping centers

▌ Commercial and industrial developments

▌ Parks

▌ Golf courses

▌ Marinas

▌ Ponds and waterways

▌ Roads and highways

A mortarless system like this provides built-in drainage for hydrostatic water pressures that can destroy traditional walls. The system allows water to pass through, so there's no buildup behind the wall that could possibly blow it out. In addition, Keystone's high-strength, low-absorption concrete withstands Mother Nature's freeze-thaw effects.

The two most common retaining wall structures built today are simple gravity walls and larger composite reinforced-soil walls. For both structural types, Keystone offers products and specific designs that conform to local building codes and national standards.

Construction Notes — In the design of any earth-retaining structure, there are two primary areas of concern:

1. Resisting lateral earth pressures

2. Providing adequate foundation-bearing capacity

Lateral earth pressure and soil instability exist whenever soil is placed at an angle greater than its natural slope. The magnitude of earth pressure is a function of the soil, wall geometry, and surcharge loadings involved. Bearing capacity is the foundation soil's ability to support the wall system without failure or excessive settlements.

One of the most basic types of retaining walls, the gravity wall, relies on its mass and cross-sectional geometry to resist the earth pressure that attempts to move the structure laterally. Keystone Retaining Wall Systems resist lateral pressure with their hefty weight and deep embedding shape. For low, noncritical applications, these products make highly cost-effective gravity wall structures. Maximum wall height for noncritical walls depends on the wall's block position for degree of slope, soil loads affecting the walls, and site conditions, including drainage.

You can combine Keystone units with soil reinforcement options (like geogrids, geotextiles, anchors or galvanized steel grid reinforcing) to create larger composite structures. With a properly-designed combination, the reinforced soil mass can support greater earth pressure and surcharge loads for retaining walls over 50 feet high. If you find yourself involved in construction of a large retaining wall, consult with a Keystone representative or a civil engineer.

Keystone offers an excellent Design and Construction Manual that shows step-by-step installation for their Standard and Compac units. It also gives instructions for installing guardrails, fences with metal and wood posts, and lighting applications, to name a few. Contact Keystone or your local distributor to get a copy for your library.

I'll just cover the basics of Keystone's installation:

1. Prepare the base leveling pad. Excavate a shallow trench according to the designed length and width of your Keystone wall. Leave enough space behind the units for a granular backfill drainage zone (which will vary with site, soil, or engineering requirements). The prepared base should be level, with 6 inches of well-compacted granular fill (sand, gravel, or $1/2$ to $3/4$ inch crushed stone) at 95 percent Standard Proctor compaction or greater. Keystone recommends additional trench depth for below-grade placement on a ratio of 1 inch below grade for each 8 inches of wall height above grade.

2. Install the base course. Set the first course of units side by side (with sides touching) on the prepared base, with the kidney-shaped void facing down and pin holes facing up. Make certain each unit is level both side to side and front to back (Figure 12-16). The first course is critical for accurate results. For straight walls, use the pins or the straight back edge of the unit for alignment. For curved walls, align the front face plane to the radius line. It takes 12 blocks to form a 360-degree circle using either Standard or Compac blocks.

3. Install the interlocking pins. Place a reinforced fiberglass pin into the paired holes in each unit. Two pins are required per unit (pins of adjoining units should be 12 inches on center). Once in place, the pins create an automatic setback for the additional courses. According to wall requirements and design, place pins in the front holes for a near vertical setback and the rear holes for $1^{1}/4$ inch setback.

4. Install and compact backfill. Fill in all voids with $1/2$ to $3/4$ inch crushed stone and compact the fill to eliminate settling. Pea gravel isn't recommended because it doesn't compact well. Use existing soils for backfill behind the gravel drainage zone. (Avoid heavy clays or organic soils because of their water-retention properties.) Compact to a minimum of 95 percent Standard Proctor compaction, placing fill in 8-inch lifts on a course-by-course basis. (Use only walk-behind mechanical compaction equipment within 3 feet behind the units.) Sweep off any pebbles or debris so the units rest evenly upon each other.

Figure 12-16
A bird's-eye view of the course layout

Courtesy: Keystone Retaining Wall Systems, Inc.

5. Install additional courses. Place the next course of units over the fiberglass pins, fitting the pins into the kidney-shaped recesses. Center each unit over the two underlying units. Visually sight down into the kidney-shaped recess for accurate pin positioning. Pull the Keystone module toward the face of the wall until it makes full contact with both pins.

6. Follow the same procedures for each course until you achieve the required height.

7. To complete the wall, install caps specially designed to finish the wall. In areas of high public traffic, secure the cap units in place with KapSeal Adhesive on the top surface of the last course before applying cap units. Place the cap over the pins on the underlying unit. Pull the cap forward to the automatic setback position. Backfill and compact to finish grade.

Manhours — This is an easy product to work with and install, so the manhours required will depend on the job's complexity and what needs to be accomplished. For reinforced soil walls, I strongly recommend you hire a subcontractor who specializes in this type of work. You should be able to handle noncritical gravity walls yourself. Once the area is prepped and ready for block installation (not including backfill), a one-man crew can install approximately 100 square feet of Standard or Compac units in 8 hours. These manhours, which include a crew of two, may be a little on the conservative side, but better safe than sorry:

■ Standard units, $8 \times 18 \times 21$ inches: 0.111 manhours per square foot.

■ Compac units, $4 \times 18 \times 12$ inches: 0.093 manhours per square foot.

Patios

While most patios are made of concrete, you can use pavers in unique patterns to create a more interesting project. What about quarry tile or bricks placed on end laced with concrete? Certainly these products make wonderful patios, but they've all been used many times over. Are you game for something new? How about a product you can install either indoors or out, a product that can transform existing floors, walkways or patios with an attractive, comfortable, spike-resistant surface? What about imprinted concrete for patios or walkways? Here's more about these interesting products.

Carlisle

Carlisle Tire & Wheel Company, a division of Carlisle Corporation, has been around since 1917, originally manufacturing automobile and bicycle inner tubes and industrial tires. Over the years, the corporation has grown to include roofing systems, high-performance wire and cable, specialized trailers, rubber and plastic automotive components, and many more products. They're one of the first U.S. manufacturers to produce resilient tile from reclaimed rubber. They offer two tile products, SoftPave and PlayGuard, both fabricated from recycled tire rubber and other rubber products. They're made from 50 percent post-consumer and 80 percent post-industrial rubber.

SoftPave

SoftPave, a $1 \times 24 \times 24$-inch tile, is made from recycled rubber and a tough urethane binder. And they're not all black! It's available in five colors: charcoal, red flecked, green flecked, rust, and black. The charcoal and rust tiles will show some color changes with time, so they're not recommended for high-traffic areas.

The tiles are uniformly manufactured in standard, grid, and brick patterns out of small granules bonded together. Even though the standard surface is flat, there's no need to worry about slippage because these uniform rubber granules create pockets that provide surface traction. The underside has feet $7/8$ inch in diameter, $1/4$ inch high and spaced $1/4$ inch apart. That makes the tile more comfortable underfoot, as well as allowing air circulation and drainage. Carlisle also makes accessory items like diagonal tiles, step tiles (without the feet), stair nosing, and ramps.

Besides a residential patio, these tiles would be great for factory floors, aerobics centers, recreational decks, docks, play areas, pro shops, utility rooms, anywhere golfers will be wearing their spikes, and (my favorite) weight and fitness rooms. These are just a few areas to consider — you can probably come up with others. In fact, consider them for any area that requires an attractive, comfortable, resilient surface.

SoftPave does have some limitations. For example, it can't be used in areas of concentrated petroleum exposure or in areas exposed to continual moisture and high temperatures, such as pools and spas. I suggest you consult with Carlisle if you're considering installation in any questionable areas.

PlayGuard

This shock-absorbing resilient tile provides a permanent, maintenance-free aid to playground safety and accessibility requirements that comply with guidelines established by the U.S. Consumer Product Safety Commission (CPSC). The difference between this tile and SoftPave is the thickness. PlayGuard is available in thicknesses of $2^1/4$ and $3^3/4$ inches. The $2^1/4$-inch thickness meets CPSC guidelines for 6-foot drop heights, while the $3^3/4$-inch width provides the resiliency you need for 8-foot drop heights. That's good news for playgrounds throughout the country.

PlayGuard is available in red, gray and black, plus a SuperTop version with a colored, granulated PVC wear surface. Looking for molded hopscotch tiles to incorporate into the standard tiles? Yes, they have them. And their specialty ramp accessories can help make playgrounds accessible for children with special needs. PlayGuard complies with the ADA requirements by providing a firm, accessible surface for individuals using wheelchairs or other mobility aids. Now, physically-impaired children who have faced obstacles in the past can enter playgrounds surfaced with PlayGuard and its accessories with ease and just have fun!

And there's one more use we should talk about — PlayGuard makes an excellent surface for rooftops. It's completely compatible with all roof surfaces and its rubber conical base won't damage the underlying roof. PlayGuard's exclusive mechanical fastening system holds tile tightly in place without requiring attachment to the roof deck. If required, a UL Class A or Class B fire-rated version is available.

Construction Notes — SoftPave and PlayGuard are virtually the same to install, either inside or out. Site preparation may vary and rooftop areas and large outdoor playgrounds have extra requirements to make the installation successful. Generally speaking, you can install SoftPave tile over existing asphalt or concrete surfaces or over a new asphalt, concrete, or compacted gravel base. SoftPave isn't recommended for rooftop applications. Contact Carlisle for their installation guides that give preparation guidelines for a new base or stairs. For this book, I'll zero in on SoftPave for an outdoor installation:

1. Determine a starting point for the first course of tile that best suits the site area. Consider this just like you would if you were installing floor or wall tile. Depending on the project, you could be starting from the center, in a corner, or along one edge. Your first row will dictate how succeeding rows will follow. Make sure that reference lines are visible and true.

2. The only method recommended for outdoor use is to fully adhere SoftPave. Carlisle offers a two-part polyurethane Versaseal tile adhesive that you can use over concrete and asphalt.

3. Tiles and substrate must be dry prior to and during installation with adhesive. A low temperature formula adhesive is available for cold weather applications under 50 degrees F.

4. Once the adhesive has been blended, pour from the container onto the substrate in a straight line, forming an adhesive "ribbon" approximately 3 inches wide by 8 to 10 feet long and in the general direction of the first course of tile to be laid.

5. Using a $1/4$-inch V-notched or $1/8$-inch square-notched trowel, spread the adhesive out slightly wider than the tile being placed. Place tiles into the adhesive bed following the course lines you've established in advance.

6. Avoid leaving the cut edge of a tile exposed to the naked eye. If the factory radius edge is removed, back this edge using a SoftPave transition ramp or masonry, unless you're placing the cut edge against a wall or other vertical member.

7. Make accurate cuts using a heavy-duty utility knife and straightedge. (Make sure you have enough sharp blades at the job site.) For radius or free-form cuts, use a jigsaw that carries at least a 4.5 amp rating. The saw also needs to have a 1-inch stroke, variable speed, and an orbital setting. Use a wood blade with about 7 to 10 teeth per inch.

8. Ramps installed at corners should be mitered.

Figure 12-17 shows a typical installation. That adhesive is sticky! You may want to wear rubber gloves to spread the adhesive, but don't handle the tiles wearing gloves. Let the adhesive cure 24 hours before allowing traffic onto the tiles.

Manhours — Once you've prepped the surface for tile installation, the manufacturer advises that a three-man crew can install 100 square feet in an hour. This assumes it's a straightforward job with a crew that has installed these tiles at least three times. You may have to increase your manhour estimate if the project is complex. Consider how many cuts need to be made and whether they're straight, radius, or free-form. If there are stairs involved, or if you're

Courtesy: Carlisle Tire & Wheel Company

Figure 12-17
A typical SoftPave installation

Courtesy: Bomanite Corporation

Figure 12-18
Imprinted concrete with a Canyon Stone pattern

working around any obstacles, it's going to take even more time. When you're bidding a job, take all these factors into account before you give a price.

Bomanite

In 1955 Brad Bowman developed the Bomanite process of imprinting concrete. Bomanite is cast-in-place, colored, imprinted and textured concrete. The finished product is unbelievable — I've seen it used in some areas in my hometown, and the finished result is very realistic-looking. Bomanite Corporation offers your customer a choice of more than 90 standard patterns and 25 standard colors in a wide variety of textures. These include used brick, slate, granite, limestone, cobblestone and sandstone (Figure 12-18).

The manufacturer tells me that one big advantage their product has over natural materials or other paving products is its lower cost. Another is the stability of a monolithic pour, which ends the problems with loose materials that you can get with individual masonry products. Bomanite has been tested time and time again under the most extreme environmental conditions, both in high-traffic areas and harsh climates around the world. It's durable and suitable for indoors or out, residential or commercial applications.

Here are some places you could consider installing Bomanite:

▌ Interior floors

▌ Walkways

▌ Patios

▌ Entries

▌ Pool decks

▌ Showroom floors

▌ Lobbies

▌ Plazas

Construction Notes — All Bomanite products are installed by factory-trained, licensed contractors. I haven't seen an installation in progress, so Bomanite shared this information. The system begins with a concrete mix poured into forms to create a monolithic slab just the right thickness and mixture for your

A Tapping the face of the imprinting tool

B The finished result looks natural

Courtesy: Bomanite Corporation

Figure 12-19
Installing Bomanite imprinted concrete

geographical area. Depending on the size of the slab, you may need physical expansion joints or simple crack control joints.

After the initial floating, a dry-shake color hardener is hand-cast evenly across the plastic surface of the wet concrete and then uniformly floated into the surface. A second shake of the color is applied and then the surface is refloated. After this has been completed, the surface is troweled smooth. The color hardener is a blend of three principal ingredients: mineral oxide pigments, cement, and graded silica aggregates. There are two grades of Bomanite Color Hardener available: Regular and Heavy Duty. Regular is recommended for all applications except areas subject to heavy traffic. Heavy Duty includes a specially-graded emery (aluminum oxide) to increase wear resistance, so it's recommended in heavy traffic areas.

The next step is the imprinting. Bomanite offers two types of imprinting tools to its licensees: (1) Bomanite tools with a smooth surface (such as tile or brick patterns) and (2) Bomacron tools that reproduce the natural texture of a specific surface (such as granite, slate or wood patterns). For the textured tools, a release agent is applied across the surface of the colored concrete to help the tools detach from the concrete and provide color variation to the textured surface.

After the imprinting is complete and the concrete is well cured, a sealer is applied to protect the Bomanite. It will deepen and beautify the color as well as prevent stains from penetrating. If you want to go one step further for a realistic look after the initial curing period, you could grout the imprinted joints using a sand/cement/water mixture.

In Figure 12-19, section A, the contractor uses a tamper to tap the face of the imprinting tools, insuring a good impression in the concrete. In section B, he's cleaning the grout line around the edge of the imprinting tool to protect the edges of the finished impressions while the imprinting tool is being removed. It's hard to tell the difference between this and the real thing, isn't it?

Manhours — I think you'll agree that it would be wise to just hire the licensed contractor who specializes in this field. Figure that a five-man crew can install 1,000 square feet in 8 manhours after the area has been prepped and ready for concrete.

Invisible Pavements

There are few things more enjoyable than walking through a park and letting the grass run up between your toes. And we all know how great it looks in the spring with its new rich green color. As an added benefit, grass supplies oxygen for our environment, and creates drainage. It seems to me that more grassy areas are being replaced with asphalt or concrete. Some of the downtown areas I've visited across the country are pretty much concrete from building to building. A little grass in those areas would sure be an improvement.

Wouldn't it be nice to replace some of the parking lots with grass? The drawback, of course, is that the first time a vehicle drives on the grass, it leaves tire tracks in the ground. That's not a pretty sight. As I hinted at the beginning of this chapter, there are products on the market that can provide ground support and yet allow grass to grow. This is especially helpful in a residential area where a customer needs a driveway to the backyard to store a boat but would prefer not to have a visible driveway across the grass. Systems like those discussed below allow the grass to completely grow within the product. After its first cutting, the customer (and his neighbors) would never know there was a driveway in this area!

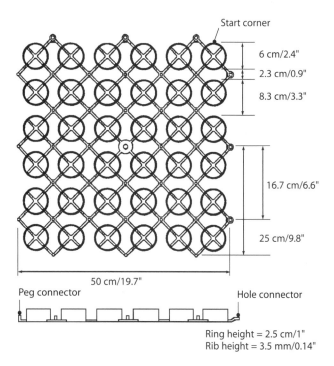

Courtesy: Invisible Structures, Inc.

Figure 12-20
Two views of Grasspave^2

Grasspave^2

Invisible Structures has over 40 years of combined experience in landscape architecture and porous paving system design, development and application. Their products can meet project specifications for a variety of structural and environmental applications, including porous paving, erosion control, and drainage systems. They also custom-design products to fit the unique requirements of individual projects. One product I'd like to introduce is Grasspave^2. This "invisible" porous paving technology combines recycled raw materials, sound structural engineering, efficient production techniques, and proven horticultural methods to give designers a living turf alternative to asphalt for traffic-bearing projects.

The manufacturer points out that this product enhances the environment in three ways:

1. Grasspave^2 is made from 100 percent recycled HDPE plastic. That keeps many common consumer and industrial products out of landfills.

2. Grass paving directly improves the environment by recharging water tables on site (reducing flooding hazards downstream), reducing sources of oils and solvents from asphalt, absorbing carbon dioxide, and creating oxygen.

3. It enhances the beauty and quality of the built environment, replacing hot asphalt-paved areas with cool, green lawn-like spaces.

What is it? Basically, Grasspave^2 is composed of thin-walled independent plastic rings connected by an interlocking geogrid structure. Installed below the lawn's surface, it's completely invisible. The rings are rigid, but the grid itself is flexible, which makes it easy to install on uneven grades and reduces the usual cut and fill requirements (Figure 12-20). The system works because the rings transfer loads from the surface to the grid structure and then to an engineered base of coarse material below. That prevents the upper root zone of the grass from compacting. Small loads, like pedestrians, are supported by a single ring. Tires and large loads are supported by several rings. Also the rings act to contain the root zone medium (usually sand) and prevent lateral migration away from tires, feet, or other loads. This protects the grass root system, enabling roots to grow deep into the porous base course. The result is healthy, green turf at the surface — and there's nothing like lush green grass.

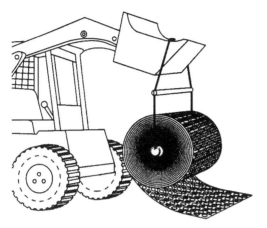

Courtesy: Invisible Structures, Inc.

Figure 12-21
Using a tractor to install Grasspave[2]

Courtesy: Invisible Structures, Inc.

Figure 12-22
Spreading sand over the Grasspave[2]

Construction Notes — The product is only available in rolls, in widths from 3.3 to 8.2 feet and lengths from 32.8 to 164 feet. The rolls cover between 108 and 1,346 square feet. With the exception of three heaviest rolls, a two-man crew can install rolls manually. On smaller rolls, the weight starts at 41 pounds, but larger rolls can weigh as much as 513 pounds. Those three heavier rolls require a lifting machine, like a backhoe (Figure 12-21). Use a chain and two rods capable of holding the weight of the roll. It's best to start at the far end so the tractor will reverse over the base course as the roll is laid down.

To help you understand basic installation, the manufacturer supplied the following:

1. Depending on the project, the excavation depth might have to be determined by a Geotechnical Engineer, but generally, it will be the same depth as required for asphalt.

2. Place and compact a sandy gravel base course. To ensure the base is porous, run a hose to check that the water flows into the base and drains away. Add subsurface drainage as necessary to those areas that don't drain properly. If you're interested, Invisible Structures can provide a chart that shows the difference in runoff between asphalt and Grasspave[2] over various soils.

3. Spread the company's supplied hydrogrow mix (fertilizer and soil polymer mix) over the base by hand or use a small fertilizer spreader.

4. Place Grasspave[2] units, rings up, directly over the base course. Use pegs and holes provided to interlock units. Work away from the corner of the first paver. You can easily shape the units with pruning shears or a utility knife to cut the grid between the rings. Units placed on curves and slopes can be anchored to the base course, using 16d common nails with fender washers, as required, to secure them in place.

5. Install sand in the rings as they're laid in sections by "backdumping" directly from a dump truck, or from buckets mounted on tractors. Then the truck or tractor can exit the site by driving over rings already filled with sand. Spread the sand laterally from the pile using a flat-bottomed shovel or a wide "asphalt rake" to fill the rings. Use a stiff-bristled broom for final finishing of the sand. Compact the sand using water from a hose or irrigation heads (or rainfall, if you're lucky) with the finish grade at least at the top of rings and no more than $1/8$ inch above top of rings (Figure 12-22).

6. Install grass seed and mulch over the sand-filled rings with commercial hydroseeding equipment. Coverage must be uniform and complete. Keep the seeded areas fertilized and moist during development of the turf plants. After installation, protect the grass from traffic until its root system is well-established. After that, only simple maintenance is required — mowing, irrigation, and fertilizing.

Manhours — This looks like an easy system to install, but professionally speaking, I'd want a landscaping or paving subcontractor to handle any large projects. I might try installations in the residential market, knowing I could rely on subcontractors for certain phases of the project where specialized equipment was required. You could bid the project as an overall package or let the customer be responsible for seeding. The manufacturer indicates this product can be installed and finished on most applications in 1.0 manhour per 325 square feet. This assumes a crew of three to four, and doesn't include excavating.

Grasscrete

Grasscrete, a grass/concrete "invisible" porous pavement grid system, is manufactured by Bomanite Corporation. It's similar to Grasspave[2], but in concrete instead of plastic. It was designed for use anywhere impervious paving products (asphalt and/or concrete) would be appropriate, especially for the commercial and municipal market. Those areas would include access routes, drainage ditches, and driveways, to name a few. It's also a recognized product for use in erosion control, reduction of storm water runoff, and to meet landscaping zoning requirements. Grasscrete can also solve problems created by natural storm water runoff patterns.

In water management applications, including erosion control, Grasscrete is a viable and cost-effective solution — and it looks a whole lot better than concrete. The overall design helps to release hydrostatic pressure while water infiltrates the system, allowing the grass to grow. A solid drainage system (drainpipe or otherwise) would require a reservoir of some type at one end, but Grasscrete doesn't. It's both an aesthetically pleasing and functional system.

Construction Notes — It's important to determine the proper subgrade for expected loads and drainage. A Geotechnical Engineer may be required, depending on the applications. However, in most applications (except for very heavy loads), native soil with a minimum Resistance Value of 30 and a compaction of 95 percent will provide a suitable subgrade. A testing lab would need to test the subgrade to determine these values, taking into consideration the traffic load that will be using the Grasscrete. In areas with inappropriate soil or very heavy anticipated loads, you'd need to excavate 4 inches or more of soil and replace it with compacted base rock.

Here are some other construction requirements:

1. Slab design needs a minimum $5^{1}/_{2}$-inch thickness. Where used for emergency vehicle access roads, all edges must have solid concrete bands or borders a minimum of 6 inches wide. These are normally painted red.

2. Six-gauge welded wire mesh is required between 2 and 3 inches above the subgrade.

3. Physical expansion joints are necessary where the system abuts other concrete or structures.

4. Forms are placed on the subgrade, then concrete is placed level to the top of forms. Finish the surface with a rough broom texture (Figure 12-23). Notice the Grasscrete forms in the lower section of the photo. A grout pump and brooms are used to evenly distribute the concrete into the holes of the forms. Concrete can also be poured from a cement truck depending on the location of the project.

5. Remove forms after the concrete has hardened sufficiently. In Figure 12-24, a couple of workers in the back lift the forms using specialty tools designed for this purpose. The front displays the finished system.

6. No traffic of any kind should be allowed on the slab until 14 days after the concrete is placed, and only after soil is replaced in holes.

7. Soil movement and seeding should normally be done by a landscaping contractor. In the finished product, only the grass is visible.

Manhours — Again, Bomanite Grasscrete is installed by factory-trained, licensed contractors, so hire a subcontractor to handle this type of installation. For those who are interested, from a manhour standpoint figure 8 manhours for 2,000 square feet with a crew of six. This figure is based on all the prep work being completed first, and doesn't included soil movement or seeding.

Miscellaneous

Have you ever approached the end of a project with the feeling that it just doesn't seem complete? It needs that little something to give it a finished look. I guess that's similar to my experience as I finish this chapter. Things seem to have collected along the way that have no particular category to fill — except "Miscellaneous." I've discovered some products that just don't seem to have a home, but it's amazing what you can accomplish with them. While some can help make a system continue to work efficiently, like rain gutters, others can blend into the surrounding environment, like a screen-covered overhead structure over a deck. Perhaps you can incorporate these products into your next project.

Gutter Guard

You might be able to earn a good living by installing a gutter shield of some type. Genova Products offers a couple of items that help prevent leaves and other debris from entering the gutter system. Gutter Guard slides under the gutter brackets and Debri-Shield has to be cut around the brackets (Figure 12-25). Both are all-vinyl products available in 5-foot lengths that you can cut with tin snips and install with a crew of one. But you'll find a second pair of hands useful to speed production.

Manhours — Once you're set up, use these manhour estimates, using a crew of two and including moving the ladders for a single-story home.

▌ Gutter Guard: 0.020 manhours per 5-foot section.

▌ Debri-Shield: 0.27 manhours per 5-foot section.

Courtesy: Bomanite Corporation

Figure 12-23
Installing Grasscrete forms

Courtesy: Bomanite Corporation

Figure 12-24
The front displays finished Grasscrete system

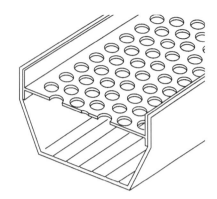

A Gutter Guard fits snugly into metal or vinyl gutters

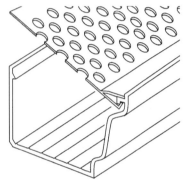

B Debri-Shield fits Genova's Repla K gutter system

Courtesy: Genova Products, Inc.

Figure 12-25
Gutter Guard and Debri-Shield

Courtesy: EnviroWorks Inc.

Figure 12-26
Deck with a Solartex screen for privacy and shade

Solartex

Solartex is a multipurpose outdoor fabric knitted in a lock-stitch construction by EnviroWorks Inc. No matter where you cut it, it doesn't fray or tear, and it stays durable and flexible through hot and cold weather. It resists rot and mildew and is UV stabilized for long-lasting wear. Solartex provides cool relief from the hot sun by reducing the temperature up to 15 degrees F. It's available in a variety of colors (jade, heavy green, and saddle, to name a few), and two quality grades: Premium 70 and Economy 50. Available sizes are 6 × 12 feet (2.844 pounds) or 6 × 25 feet (5.925 pounds).

You can use the Premium 70 fabric to create privacy fencing, deck sunscreens, greenhouses, and playhouses. The black Economy 50 is great for the serious gardener. Accessory items are available, including polyethylene locking clips and ties for attaching Solartex to poles, chain link fences, and other structures. If these accessory items aren't readily available, you can modify Solartex to accommodate common off-the-shelf accessories. The galvanized wood fasteners will spike the fabric securely to wood. This fabric was intended for the do-it-yourself market, but there's no reason why contractors can't use it for the decks they build (Figure 12-26). Solartex makes a perfect setting for a privacy screen, and this deck arbor has been covered to shade out some of the sun.

Manhours — Calculating manhours for this fabric is difficult because there are so many variations in the way it's used. If you're installing it on wooden framing members, a 6 × 25-foot roll will require around 120 wood fasteners. Wooden fasteners begin at the corners and continue in both directions every 10 to 12 inches. In climates where snow and/or leaves might weigh down the cover, place wood fasteners every 4 to 6 inches. For the best overall performance, don't space the framing members more than 6 feet long, or more than 2 feet on center. With a two-man crew, figure 0.075 manhours per square foot to install the fabric with fasteners. This figure is just to install the fabric; it doesn't including framing the project.

Solar SensorLight

The Solar SensorLight (by Alpan, Inc.) relies on the sun instead of electricity to charge the battery. You can use it to solve a lighting situation in a

Courtesy: Alpan, Inc.

Figure 12-27
Solar-powered security light

location where no electric power is available. These are often the very areas around the home, garage, or the tool shed that need security lighting. I wish I'd had a couple of these lights last year when an ice storm hit our area and we were without power for seven days.

The unit comes with a replaceable solar-powered battery designed to last from three to five years. The bulb is an energy-efficient 20-watt halogen that's easily replaced and has a two-year life expectancy. The built-in motion detector activates in response to heat and motion, automatically turning on and then shutting off when the area is clear. Even though the light is made of all-weather UV-stabilized black plastic case, it's still important to protect the main unit from the rain.

It's easy for a homeowner to install, but you might be able to sell and install some lights as part of a larger exterior project. The unit can be installed either horizontally (under the soffit or eaves) as a

convenience light or vertically (on the wall surface) for security (Figure 12-27). When deciding where to mount the main unit, keep two factors in mind. First, the motion sensor has a "field of vision" about 40 feet in front of the light. Second, make sure that the solar module faces south. The solar module is the power source for the motion sensor, converting the sun's energy into electricity that charges the battery. This requires *direct sunlight* falling onto the face of the solar module for as long as possible over the course of the day.

You can adjust the angle of the solar module by moving the adjustment support to the appropriate hook on the mounting bracket. The module comes with a 14-foot cord to help you locate it in direct sun. The unit comes with five fasteners — all you'll need to install the entire unit.

Manhours — It's a simple unit to install. You'll probably spend more time deciding where to install the solar module than actually putting it in. Estimate 0.50 manhours per unit for a one-man crew.

The Final Word

Of course, this isn't the final word on the subject of selecting and installing alternative materials. Actually, I believe it's just the beginning. Why? Because I've only scratched the surface of the alternative materials and methods currently on the market. I expect that new products will enter the market the same day as this book! There's also the possibility that some will depart just as fast as they entered. But I hope I've opened some doors for you, the contractor, and for manufacturers of alternative products to come forward for inclusion in future editions. Let's work together to keep all contractors aware of innovative products that will help them stay on top of the competition.

You've probably noticed that I offered manhour estimates throughout the book. They're for your convenience to help you in the bidding process. But understand that they're just that — estimates. Of course, they won't be accurate for all contractors. There are just too many variables. Your crews may be faster or slower, and only you know your own crews.

It's important to understand your labor charges, whether you're working with alternative materials or traditional building products. If you haven't worked with any of the materials mentioned in this book, carefully read the product description, construction notes, and manhours involved. Visualize yourself installing these products and then check out the manhours to see if you can realistically install this product within those estimates.

Remember that bidding your job correctly can make or break you. Too many mistakes could put you out of business! Why are you in this business in the first place? Because you enjoy what you're doing and because you're making money? The key words here are *making money*.

One way to keep track of your estimating is to itemize the overall project. If you don't have good itemizing forms, you can write to me for my *Contractor's Helping Hands Packet*, or order it from my Web site (www.asktooltalk.com). The cost is $13.50 (which includes postage). The packet contains business forms designed to provide a complete, yet simplified, method of bidding. They'll also help you with organizing your business, accurate price quoting, and legal protections.

There are two purposes to the Itemized Bid Sheets included in the packet:

1. To help your customer better understand the industry. By giving them an itemized breakdown, you instill confidence in your prices. They'll be able to see, in black-and-white, how much materials cost, what labor costs are involved and how much it costs for your subcontractors.

2. To help you be organized. You'll have an inventory of supplies and labor. Your goal will be to do a quality job at an even pace, and keep costs down.

With this new understanding, it'll be easier for you to instill confidence in your customers. They'll understand their project, and see that you are a professional who can accomplish it to their satisfaction.

I encourage you to send for a packet and to write me if for any reason you don't agree with the manhours listed or if you have new information you'd like to share concerning the products after you've installed them. In turn, I can share the information with the manufacturer and add the first-hand experiences back into the book. You can contact me through my Web site or write:

C.R.S., Inc.
P.O. Box 4567
Spokane, WA 99202-0567

With this book, I've tried to open the door to a whole new breed of construction — from materials to installation. Now it's up to you to investigate these products and incorporate them into your projects. To get more information, contact the manufacturers listed in the Appendix. Ask for appropriate information and samples to help you and your customer make intelligent decisions. The only way this will happen is if you make the first move!

Now for the last leg of this journey, turn the page. The final chapter covers the subjects of recycling, reducing waste and "building green."

13

Building to Help
the Environment

▮▮▮

I firmly believe that we can help the environment by sharpening our construction methods and skills, choosing alternative materials, cutting waste and recycling the waste we can't avoid creating, and building "green." The first 12 chapters emphasized alternative materials and the new skills you need to use them. In this chapter, I'll cover the rest of the equation: reducing waste and recycling, and building green.

Reducing Waste and Recycling

Whatever you're building, tighten your material list to reduce waste to the absolute minimum. Order the actual lengths you need instead of ordering by the linear foot. This means you need to sharpen your pencil and spend a little more time on your estimate and the material take-off list. You'll help the environment, and save on your overall material costs into the bargain.

When waste is unavoidable, take time to set up a recycling program on the job site for these materials. But having the site alone won't accomplish the goal — you also need to enlist the cooperation of your employees. The tough job is to educate yourself, your supervisor, and your workers on what materials can and should be recycled. To run efficiently, a recycling program will require constant monitoring and continuing responsibility on the part of your workers. How do you keep them motivated? Try this.

Create a special account for recycling earnings, then divide the proceeds among your employees at the end of the year (a great Christmas bonus) or throw a big party for them.

If you're just too busy to get involved or to teach your employees to separate the waste, then contract with a specialist in recycling who'll haul the materials for you. Whether you'll have to pay for the service may depend on the quality of recyclable waste and how well you negotiate. Whichever direction you take, it's bound to keep your trips to the landfill to a minimum.

Pick up your phone book and check out "Recycling." Learn what materials the companies will accept and what prices they offer. Also look under "Salvage-Merchandise" for companies that may buy the materials from any building you've contracted to take down. If you want to buy salvaged or used materials, look under "Building Materials — Used."

When you're considering today's products, it's important to check with the manufacturer, local distributor, and government agencies to find out how to dispose of any waste. Safe disposal laws are being added to the books all over the country, and enforced more strictly. These disposal regulations and their related costs could have an impact on your material selections — especially when you're dealing with hazardous materials, which nowadays even includes *paint.*

Hazardous Material Handling

New products (as well as existing materials) used in homes and public buildings contain many sources of pollutants that can affect the quality of the air we breathe. If you're trying to create an environmentally safe home, be aware that some of the materials you select could affect the quality of life for the residents. Additionally, some of the existing materials requiring removal may be banned from the landfill — asbestos, for example. Check with local government agencies in your area, such as the County Health District, for information and guidelines on how to safely handle and dispose of asbestos, construction debris, and other polluting or unsafe materials. Dispose of the wrong material in the wrong way or place and you can have the wrath of several governmental agencies descending on you — and it won't be pleasant or cheap. Clearly, you want to know the rules before you bid on any project that may contain polluting materials. Safe handling of these products will protect you, your pocketbook, and your employees, as well as future generations.

Asbestos

By now you're probably well aware of the dangers of asbestos. It's a mineral fiber that's normally found in older homes and commercial buildings, especially in pipe and furnace insulation materials. But it also can be found in ceiling and flooring materials, spackling and plaster compounds made before 1974, asbestos shingles (common in my area), floor tiles including backing, lining felt, or asphaltic "cutback" adhesives, and other building products.

The real threat comes when asbestos fibers become airborne, which happens when materials containing asbestos are disturbed by activities like cutting, drilling, and sanding. So attempting to remove these materials the wrong way will just make the situation worse. It can increase asbestos levels by releasing the fibers into the air — a sure way to endanger people living or working in the building. If you don't know for sure whether there's any asbestos present, then *don't tear anything out*. I recommend that you send a sample to a testing laboratory for analysis. Until you know for sure, it's important to not cut into any existing material with a power tool that could create dust.

Asbestos fibers that are too small to be visible are the most dangerous. Once inhaled, they stay in the lungs and gradually accumulate. Lung cancer, mesothelioma (a cancer of the chest and abdominal linings), and asbestosis (irreversible lung scarring that can be fatal) can all be caused by asbestos. But symptoms of these diseases might not surface until years after exposure. Most individuals with asbestos-related disease were exposed to elevated concentrations on the job. But *others developed the disease from exposure to clothing and equipment brought home from the job sites.*

Various government agencies regulate the removal and safe disposal of in-place asbestos — both residential and commercial. Check with these agencies so you understand the regulations. That's the only way you can know if you'll be in compliance before bidding on any project. Even if your state doesn't require a certified contractor for removing asbestos, I'd consider using one anyway. Let someone with experience and the right safety equipment handle this part of the job!

Lead

Lead is another harmful environmental pollutant. It's widely recognized as the number one environmental threat to the health of our children. So how do we and our children get exposed to lead? Before we knew just how harmful lead was, it was included in paint, lead solder, ceramic glazes, gasoline, water pipes, and many other products. It also found its way into the air, drinking water, and dust. The most significant source of lead exposure in the United States today is old lead-based paint. Most homes built before 1960 contain heavily-leaded paint, and homes built as recently as 1978 may also contain lead-based paint.

As with asbestos, I recommend having old paint tested at a private laboratory to make sure there's no lead present. If you remove the paint from the surface by dry scraping, sanding, or open-flame burning, it sends particles and dust into the air. Airborne lead enters the body when anyone breathes or swallows lead particles or dust. It can also be tracked into a home with contaminated soil.

Exposure to lead can affect practically all bodily systems. At lower levels, it can adversely impact the brain, central nervous system, blood cells and kidneys.

High levels of lead can cause convulsions, coma, and even death. The consequences are especially severe in fetuses, infants, and children. Lead exposure can lead to severe delays in physical and mental development and increased behavioral problems, to name a few. Children are more vulnerable to lead exposure than adults because lead is more easily absorbed into their growing bodies and tissues. Children are also more likely to be exposed because the lead dust on their fingers and/or toys eventually makes it into their mouth.

It's important to minimize dust exposure during a project by keeping the work site as dust-free and clean as possible. During and after your job, advise your customer to mop floors and wipe window ledges and chewable surfaces with a solution of automatic dishwasher detergent (not a multipurpose household cleaner) in warm water. Dishwasher detergents work best because of their high phosphate content. I'd also keep the family, especially children and pregnant women, away from the area until work is completed and cleanup is done. Remember that it'll cling to your clothes, and you'll take it home to your own family. So even where it's not required, consider a subcontractor who specializes in lead paint removal.

Formaldehyde

Formaldehyde can be found in particleboard, fiberboard, plywood, adhesives, and urea-formaldehyde foam insulation (UFFI). In homes, the most significant sources of formaldehyde are likely to be pressed wood products which are used indoors. These products use an adhesive that contains urea-formaldehyde (UF) resins. Products used for exterior applications such as oriented strandboard (OSB) contain the dark or red/black-colored phenol-formaldehyde (PF) resin. While formaldehyde is present in both types of resins, pressed woods that contain PF resin generally leak formaldehyde at considerably lower rates than those containing UF resin.

Since 1985, the Department of Housing and Urban Development (HUD) has permitted the use of plywood and particleboard in prefabricated and mobile homes that conform to specified formaldehyde-emission limits. In the past, some of these homes had high levels of formaldehyde because of the amount of pressed wood products used in their construction. In the '70s, many home wall cavities were filled with UFFI. When tested for formaldehyde soon after being insulated, these homes had high emission levels. Today, very few home are insulated with this product, and the homes built in the '70s are unlikely to have high levels of formaldehyde any more.

Formaldehyde is a colorless gas with an offensive smell. Some individuals who are exposed to elevated levels (above 0.1 parts per million) have difficulty breathing, feel burning sensations in eyes and throat, and other irritating effects. And formaldehyde can trigger attacks in those with asthma when they're exposed to high concentrations.

Check with your customers to see if they've experienced adverse reactions to formaldehyde. If so, avoid using any pressed wood and other formaldehyde-emitting products. Some studies suggest that coating pressed wood products with polyurethane might reduce emissions for a period of time. However, to be effective, the coating must cover all surfaces, especially the edges, and remain intact.

Heat and high humidity levels can accelerate the rate at which formaldehyde is released. With this in mind, suggest that your customer use dehumidifiers and air conditioners to control humidity and to maintain a moderate temperature. Increasing the rate of ventilation in the home will help to reduce formaldehyde levels. There's more information on venting in Chapter 10, Bathrooms and Kitchens.

"Green" Building

As you get deeper into the environmentally-sensitive market, you'll soon discover the phrase *green building*. Nobody knows this approach to building better than the city of Austin, Texas, which runs the Green Builder Program. To them, *green* means *sustainability*. Sustainability is providing for current needs without compromising the ability of future generations to provide for their needs. To do that, construction professionals have to examine the impact of current building projects to decide if their materials and methods are *sustainable* for the long term.

Green Builder Program

What are the criteria for green building? Here's how The City of Austin Green Builder Program defines their criteria:

▌ *Use renewable resources.* This means using solar, wind, and geothermal energy sources as much as possible. Many common building materials come from renewable sources — including wood. But there are other considerations as well, like the harvesting method. Certified lumber from sustainable managed forests is now available at cost-competitive prices.

▌ *Use recycled resources.* Some nonrenewable resources may actually be used many times. This is true, for example, of many metals. Recycling reduces the disruptive and polluting effects of virgin mineral extraction. So do water-reuse technologies and the use of many agricultural and industrial byproducts.

▌ *Reduce "embodied energy."* The total amount of energy required to bring a material into existence is called "embodied energy." This includes all the energy needed to mine, transport, process, distribute, and more. The cumulative energy embodied in creating and transporting the materials and systems of a building may actually exceed the energy required to heat and cool the building for 10 to 30 years.

▌ *Use recyclable materials.* Buildings aren't permanent. Their materials should be a resource, not a waste sent to a landfill after the building's useful life.

▌ *Conserve all resources.* This includes all the resources used in construction and operation of buildings: energy, water and land, as well as materials.

▌ *Avoid toxic materials and systems.* While materials are being produced and transported, toxic chemicals that threaten the sustainability of the planet may be released into the environment. Indoor air quality (IAQ) may be affected by materials and systems. This is especially important since average Americans spend 80 to 90 percent of their time indoors. Many building materials give off unhealthy fumes, like formaldehyde. Additionally, a poorly-designed, installed, or operating mechanical system may result in carbon monoxide poisoning from gas appliances.

▌ *Use climate- and site-responsive design.* This type of design works in harmony with the climate, the sun's path, prevailing air movement, plants, and other natural features on a specific site to achieve comfort. It works with, not against, natural processes.

▌ *Use regional resources.* Solar, wind, and geothermal energy and captured rainwater all come from the building site itself and reduce the need for elaborate energy and water supply systems that increase environmental impacts. Materials from the area where they'll be used reduce the energy needed to transport, and improve regional economy. That's also a goal of sustainability.

The aim of the Green Builder Program is to influence building practices to become sustainable. That includes conserving energy, water and other natural resources, preserving the health of our environment, strengthening the economy and promoting a high quality of life for all citizens.

The Green Builder Program offers ratings of green homes on a scale of one to four stars; the more stars, the more green features and systems found in the home. Building professionals such as builders, architects, engineers, tradespeople, and suppliers receive technical and logistical guidance, as well as marketing assistance, in exchange for agreeing to offer and promote green building practices.

To learn more about sustainable building, get a copy of the *Sustainable Building Sourcebook*, a technical and logistical resource intended for building professionals and designed to fit into a 3-ring binder. Update sheets are available as they're developed. The *Sourcebook* is provided at no charge if you're a member; otherwise, the cost is $25. It's also available on-line with *Infinet*, under Organization (Green Builder Program), and on the Internet at http://www.greenbuilder.com. To learn more, contact them at:

Green Builder Program
City of Austin
Environmental and Conservation Services Dept.
206 E. 9th Street, Suite 17.102
Austin, TX 78701
512-499-7827

You need to ask yourself if green building will work in your area. It probably will, but you may need to take a different approach than they're taking in Austin. Actually, builders around the world can accomplish the same goals — but with different types of materials.

Who Can Help?

This book doesn't have all the answers. It wasn't meant to. What I've tried to do is to bring to your attention alternative products and, in some cases, construction methods to consider. I hope that it moves you toward a new mind-set — an attitude that considers the future impact of decisions made today.

It would be impossible to share all the information that's available on the market today, or to predict what will be there tomorrow. There are individuals, associations, and organizations heavily involved in developing and promoting alternative construction methods and materials, as well as other environmental issues. Some have a long track record and have done extensive research. I suggest that you contact these organizations to gather materials for your library. In some cases, you may even want to join the organization. They can be valuable resources.

Keep in mind that this isn't a complete list of organizations. Probably no author could accomplish that. If you should come across any we missed, I encourage you to contact them to see what they have to offer. Then let me know about them (C.R.S., Inc., P.O. Box 4567, Spokane, WA 99202-0567 or www.asktooltalk.com). I'll check them out and try to include them in future editions.

Also check out the Appendix for a list of building industry associations that can provide reams of information — just for the asking.

American Lung Association, Minneapolis Affiliate

The American Lung Association (ALA) has been fighting lung disease, with assistance from the American Thoracic Society and the medical community, for over 90 years. The ALA offers a variety of health education programs about lung disease and its prevention. They also provide education, community services, advocacy, and research on the subject of good health. To learn more, give them a call at 1-800-LUNG-USA (586-4872).

Up to 30 percent of new and remodeled buildings worldwide may generate excessive complaints related to indoor air quality. A Minneapolis affiliate of the ALA has gone one step further in the fight for good health. From 1992 to 1996 they've been involved in building or remodeling 14 model homes with the goal of providing a healthier indoor environment. These model homes comprise an ongoing project called *Health House*.

The Minneapolis Affiliate offers consulting services through their Healthy Building Consulting Group. Professionals who are serious about building an environmentally-safe home should take a look at their book, *The Health House Workbook: A homeowner's manual for building a healthier home*. It's a workbook that provides an objective approach to targeting and using important information for building a healthier home and eliminating indoor air quality problems. It's geared toward the consumer, but includes an outline of information and questions directed to professional builders. It's a useful resource you could refer to when dealing with a customer who wants to build or remodel to provide a healthier environment.

To order this workbook ($29.95 including shipping and handling) or to find out how the American Lung Association, Minneapolis Affiliate, can help you with your next project, contact:

Health House Project
American Lung Association/MPLS
4220 Old Shakopee Road, Suite 101
Minneapolis, MN 55437-2974
612-885-0338

Center for Resourceful Building Technology (CRBT)

This nonprofit organization was started by Montana homebuilder Steve Loken back in 1990, when he saw the environment around him declining due to the overuse of natural resources. At the same time, he was frustrated by the shortage of quality lumber available. This inspired him to gather information on environmentally-responsible and resource-efficient building materials and technologies.

Through research, education, and demonstration, he's been distributing this information to both the building industry and the general public.

Research

The center is a nationally-recognized source of information on resource-efficient building methods and materials. They continue to collect information on building materials and evaluate them for resource efficiency according to these criteria:

▌ efficient use of limited natural resources

▌ demonstrated recyclability or renewability

▌ lower energy use

They publish a reference directory of manufacturers whose products are resource efficient. *Guide to Resource Efficient Building Elements* ($28.00 including S&H) is now in its sixth edition.

Education

CRBT conducts slide presentations, workshops, and seminars on resource-efficient building. They've also created *Building Our Children's Future*, a 15-part interdisciplinary curriculum on resource-efficient building for grades kindergarten through 12.

Demonstration

In 1992, CRBT completed its first demonstration home building project, called ReCRAFT 90. This 2,400 square foot single-family home featured more than 40 resource-efficient and recycled building materials, with far less dimensional lumber than a conventional wood-framed residence. In 1995, they sponsored a new demonstration project in Missoula, Montana. The Timber-Tech House highlighted advances in efficient wood fiber use, including engineered wood, recovered wood fiber, and salvaged lumber. It shows that builders can build more resource-efficient houses using materials readily available today.

I had the opportunity to interview Steve Loken, who has been designing and building homes for over 20 years that emphasize resource- and energy-efficient design, materials selection, and construction. Here's some of what he had to say:

"The concern with efficient building materials is 'How do you use the least amount of material to accomplish the greatest amount of good in building?' For example, plywood and OSB are more efficient substitutes for 1 × 10 boards, and an I-joist is an efficient replacement for a solid-sawn beam. Overall, efficient building can be defined as being less material-intensive, and engineering for optimized strength-to-weight ratios.

"Alternative products, on the other hand, represent materials and practices not in mainstream use, like replacing a 2 × 6 with straw. In the quest for alternatives, we have to be careful that what we're doing really is best for the building, and isn't just driven by the desire to adopt something new. At the same time, we have to make sure that we don't get locked into using materials just because they're traditional. We need to be open to using new materials that could work better in certain applications within a building, or maybe in certain climates or regions.

"Most areas in the United States tend to overbuild, using an excessive amount of materials, due in part to habits and traditions that are hard to break, as well as code-enforced safety factors. We can decrease our material demands and still make strong and long-lasting buildings. Technology is currently the 'Hamburger Helper' of the construction industry — it's helping us to stretch the resource base. We can build equal or even better structures with less materials and resources. Also, we can decrease labor costs by decreasing the amount of material used, and by using engineered materials that have fewer defects and produce less waste.

"Also, using resource-efficient materials can help keep costs down, by reducing the total amount of material needed in a structure. If you're talking about up-front material cost in dollars, then some alternative products are more expensive. When you factor in labor savings, energy savings, and maintenance savings over the life of a structure, alternative materials often appear as clear winners. And when you include the environmental costs of producing conventional materials — things like loss of biodiversity and water and air quality and preservation of building occupant health and productivity — it throws a whole new light on the conventional cost equation.

"Change in any industry is painful and hard. There will be successes and failures — after all, even conventional materials experience failures. My only hope is that companies will proceed with the research and development of new materials carefully, and not rush to get a product on the market before it's ready. The worst thing that can happen in this market right now is repeated failure of new materials that haven't been completely tested. If that happens, builders start to develop an attitude of 'Oh, I already tried that product and it didn't work,' so it's hard to improve a product once it's already on the market.

"Also, I want to mention that one giant product that's supposed to fit all applications may not be what the industry should be looking for. The most resource-efficient materials could well be the ones that are produced regionally, and are appropriate for the area. It's perfectly acceptable to have a product that performs well in some regions, but shouldn't be used in others.

"I encourage construction professionals to always question what they're doing with materials and applications, and to ask themselves:

▮ Am I being resource efficient?

▮ Am I using the right material the right way?

▮ Am I trying to minimize construction and demolition waste?

▮ Am I being a steward of resources?

"We have an expanding population, an increasing need for affordable housing and a limited resource base. That says it all. We're going to have to enter an age of minimization, in which renovation, salvage, and reuse of materials will all become more and more important as we meet the challenges of housing in the 21st century."

To learn more of the valuable services offered by the Center for Resourceful Building Technology, contact them at:

Center for Resourceful Building Technology
P.O. Box 100
Missoula, Montana 59806
406-549-7678
crbt@montana.com (e-mail)

Environmental Building News

Environmental Building News is a monthly newsletter on environmental issues in both design and construction. The newsletter is packed with information on construction methods, technologies and materials. I found the in-depth feature articles and new product reviews particularly interesting. Each issue also includes a calendar of events where you can get more information. I encourage you to attend some of these events if possible. *EBN's* objectivity has earned the newsletter the unusual distinction of being widely respected both by environmental activists *and* by industry groups. A subscription is $127 per year with a special discount rate of $67 per year for individuals and small businesses with employees of 25 or fewer.

Alex Wilson, the editor and publisher, defines green building like this:

"To me, 'green building' is a way of building that minimizes impact on the environment — including both the outdoor environment and the indoor environment. It is more than construction, however, involving land-use planning, site design, material selection, and building design as well."

You can contact *EBN* at:

Environmental Building News
122 Birge Street
Brattleboro, VT 05301
802-257-7300
ebn@ebuild.com (e-mail)
www.ebuild.com (Web site)

NAHB Research Center

The NAHB (founded in 1942) is an organization of over 185,000 members with more than 800 state and local home builder associations nationwide. Its mission is to represent on Capitol Hill the interests of all members and the housing needs of all Americans. They lobby for housing to remain a priority in legislation and national policy, and work with federal agencies whose regulations affect the housing industry. They also keep both local associations and members informed about changes in building products and techniques, consumer preferences, marketing, finance, regulation, and legislation.

Once a year they stage *The Builders' Show*, with seminars and over 1,000 exhibitors offering the latest in products and techniques. I've attended a few of the shows myself and found them both informative and rewarding. If you've never been to one, I encourage you to go. To inquire about their next show, call them at 800-368-5242.

The NAHB Research Center is a not-for-profit subsidiary of the NAHB, established back in 1964 to carry out research for the home building industry, its related industries and professions, and public sector housing agencies. They focus on studying new technologies, developing techniques for energy and resource conservation, and certifying products used in home building.

One report prepared by the NAHB Research Center under funding from the U.S. Department of Housing and Urban Development (HUD) was the result of a two-year study of alternative materials in residential construction. You can find the results of the first year's study in *Building with Alternatives to Lumber and Plywood* (NAHB). It provides introductory information on alternatives, including basic properties, applicability, and available sources. The second year's study is available in *Alternative Framing Materials in Residential Construction: Three Case Studies*. This report presents three alternative technologies for conventional residential construction: foam-core structural sandwich panels, light-gauge metal framing, and welded-wire sandwich panels. The evaluation included feasibility, quality, and costs associated with each method.

These two publications are a must for your library! You may also find their hotlines useful:

▎ For steel, call 800-79-STEEL (797-8335).

▎ For concrete, call 888-602-HOME (4663).

▎ For general technical questions related to housing, call the HomeBASE hotline at 800-898-2842.

You can reach the NAHB Research Center at:

NAHB Research Center
400 Prince George's Boulevard
Upper Marlboro, MD 20774-8731
301-249-4000
www.nahbrc.com (Web site)

American Fiberboard Association (AFA)

The AFA is a trade organization of manufacturers of fiberboard products used in residential and commercial construction. They provide technical information to the general public as well as professionals, and promote ongoing educational and research activities on the product. Fiberboard can be found in sheathing panels for exterior wall applications, roofing substrate, and other specialty products.

American Fiberboard Association
1210 W. Northwest Hwy.
Palatine, IL 60067
847-934-8394

American Hardboard Association (AHA)

The AHA, located in the same office as the AFA, represents manufacturers of hardboard products used for exterior siding, interior wall paneling, household and commercial furniture, and industrial and commercial products. AHA serves as a central clearinghouse on industry and technical information for trade professionals, government agencies, and the general public. The association publishes a variety of brochures and pamphlets that are free, and videos priced up to $5.00.

American Hardboard Association
1210 W. Northwest Hwy.
Palatine, IL 60067
847-934-8800

APA—The Engineered Wood Association

APA—The Engineered Wood Association, is a nonprofit trade association whose member mills produce approximately 75 percent of the structural wood panel products manufactured in North America. Maybe you've noticed the association's trademark "APA" on some of the plywood sheathing you've used, or on composites and OSB. The trademark appears only on products manufactured by member mills and is the manufacturer's assurance that the products conform to the standard shown on the trademark. The APA EWS trademark appears only on engineered wood products manufactured by members of Engineered Wood Systems (EWS), a related corporation of APA.

They sponsor quality testing and inspection, research and promotion programs. They also offer three publications that I found quite interesting. Use them as handy reference manuals:

Construction Guide — House Building Basics. An elementary guide to wood-frame construction, it illustrates the basic steps to completing the structural shell of a typical single-story house, from the foundation to the roof. In some cases, it describes alternative methods of construction. I firmly believe it doesn't hurt to go back to the beginning for a refresher course.

Design/Construction Guide — Residential & Commercial. This guide contains up-to-date information on panel grades, including APA Performance Rated Panels; specification practices; floor, wall and roof systems; diaphragms and shear walls, fire-rated systems, and methods of finishing.

Product and Application Guide — Glulams. Everything you ever wanted to know about glued laminated timber can be found in this handbook. The nice thing about glulams is that we can use them without worrying about whether they're produced from old-growth timber. Glued laminated timbers are composed of individual pieces of dimension lumber end-jointed together to produce long lengths which are then bonded together with adhesives to create the required beam dimensions. Because of this process, large laminated timbers can be manufactured from smaller trees harvested from second- and third-growth forest and plantations, with a variety of species used.

To get your copy of these or other APA publications, contact them at:

APA—The Engineered Wood Association
7011 S. 19th Street
P.O. Box 11700
Tacoma, WA 98411-0700
253-565-6600

EIFS Industry Members Association (EIMA)

EIMA is a nonprofit trade association representing manufacturers, suppliers, distributors, applicator/contractors and related building professionals involved in the *Exterior Insulation and Finish Systems* (EIFS) industry. The association promotes industry-wide performance standards and develops industry guideline specifications and standards for systems, materials and EIFS application. They sponsor research and testing programs that become the basis for many model building code requirements pertaining to EIFS cladding on exterior wall assemblies.

EIMA publishes specifications and test methods covering such topics as performance, durability, fire testing, application, and use of related materials, including sheathing and sealants. They are committed to enhancing, improving, and promoting the EIFS industry; to advancing the EIFS industry through research and distribution of technical information; and to educating users of EIFS products. To learn more, contact them at:

EIFS Industry Members Association
3000 Corporate Center Drive, Suite 270
Morrow, GA 30260
800-294-3462

Portland Cement Association (PCA)

The PCA has been perfecting concrete for over 75 years. They offer information on virtually all uses of cement and concrete, from casting a sidewalk or patio or laying concrete block, to building complex highways and skyscrapers. Their resources range from do-it-yourself help to sophisticated design software for engineering firms.

The PCA's Library Services department can provide research and information service on cement and concrete technology, materials technology, engineering, and related topics for member companies and others. Its collection of books, government reports, journals, standards, and patents on cement and concrete is one of the most comprehensive anywhere in the world. In addition, the Library offers:

❚ Literature searches on electronic databases

❚ Electronic interlibrary loan capabilities

❚ Document delivery (photocopies of journal articles, out-of-print PCA publications, etc.)

❚ Internet site information

❚ Bimonthly newsletter

Concrete Solutions (their annual catalog) is filled with helpful information on publications, software, audiovisuals, and ongoing seminars. Request a copy from:

Portland Cement Association
5420 Old Orchard Road
Skokie, IL 60077-1083
847-966-6200 (for information)
800-868-6733 (to order publications)

Structural Board Association (SBA)

SBA supports technical and research programs to enhance OSB quality, improve product manufacturers, and provide technical support for market expansion programs. To learn more about the association and the use of OSB in floor, roof, and wall sheathing, contact them and ask for their 28-page booklet, *OSB In Wood Frame Construction*, at:

Structural Board Association
45 Sheppard Ave. E., Suite 412
Willowdale, Ontario, M2N 5W9
Canada
416-730-9090
www.sba-osb.com (Web site)

Structural Insulated Panel Association (SIPA)

This organization promotes the structural insulated panel industry, including panel manufacturers as well as their suppliers and customers. They organize forums, workshops, mailings, and other activities for members, all designed to share information and provide business opportunities. SIPA conducts an ongoing publicity and information dissemination program to gain visibility for the industry and to promote the interests of its members. Their newsletter, *Spotlight On SIPA*, is packed full of information. To learn more about this organization, contact them at:

Structural Insulated Panel Association
1331 H Street, NW, Suite 1000
Washington, D.C. 20005
202-347-7800

Where Do We Go From Here?

Where *do* we go from here? Do we continue to build as though land, material, and energy resources are endless? Do we continue to make trip after trip to the landfill with construction waste? Do we continue to pay higher and higher prices as material supplies dwindle? Or do we move into the next century as confident construction professionals, making profitable use of new technology and techniques?

Does this book have all the answers? No, of course not. But I do hope it moves you toward a new mindset, an attitude that considers the future impact of decisions made today. That's what this book is all about. I hope the new products and construction tips I've included here will help get your creative juices flowing and to put you in the forefront of 21st century construction.

Abbreviations

AAC	autoclaved aerated concrete	**DBH**	dead plow hammer
ABS	acrylonitrile butadiene styrene	**EBN**	Environmental Building News
ACAA	American Coal Ash Association	**EIFS**	Exterior Insulation and Finish Systems
ACQ	ammoniacal copper quaternary	**EIMA**	EIFS Industry Members Association
ACZA	ammoniacal copper zinc arsenate	**EPA**	Environmental Protection Agency
ADA	Americans with Disabilities Act	**EPS**	expand polystyrene
AFA	American Fence Association	**EWP**	Engineered Wood Products
AFA	American Fiberboard Association	**FAS**	First And Second
AHA	American Hardboard Association	**FEQ**	First Export Quality
AISI	American Iron & Steel Institute	**FiRP**	fiber reinforced product
ALA	American Lung Association	**FRTW**	fire retardant treated wood
APA	American Plywood Association	**gpf**	gallons per flush
ASTM	American Society for Testing Materials	**gpm**	gallons per minute
AWP	all weather primer	**GS**	granulate surface
AWPA	American Wood-Preservers' Association	**HCFCs**	hydrochlorofluorocarbons
AWPI	American Wood Preservers Institute	**HDF**	high density fiberboard
BOCA	Building Officials & Code Administrators International	**HDM**	Holz Dammers Moers
Btu	British thermal units	**HDPE**	high-density polyethylene
CABO	Council of American Building Officials	**HRV**	recovery ventilator
CCA	chromated copper arsenate	**HUD**	U.S. Department of Housing and Urban Development
CCBs	coal combustion by-products	**IAQ**	indoor air quality
CEM	carbon enhanced modified	**IBAMA**	Brazilian Institute for the Environment and Renewable Natural Resources
CFCs	chlorofluorocarbons	**ICBO**	International Conference of Building Officials
cfm	cubic feet per minute		
CMUs	concrete masonry units	**ICF**	Insulated Concrete Forms
CRBT	Center for Resourceful Building Technology	**IMSI**	Insulated Reinforced Masonry System
csf	100 square feet	**ITTO**	International Tropical Timber Organization
CSPC	Consumer Products Safety Commission	**KDAT**	kiln dried after treatment

LDPE	low-density polyethylene		**PVC**	polyvinyl chloride
L-P	Louisiana-Pacific		**RSP**	Residential Steel Partnership
LSL	laminated strand lumber		**SAFB**	sound attenuation fire batt insulation
LVL	laminated veneer lumber		**SBA**	Structural Board Association
MDF	medium density fiberboard		**SBCCI**	Southern Building Code Congress International
MDI	methylenediphenyl diisocyanate			
MEC	Model Energy Code		**SBS**	styrene-butadiene-styrene
MPV	multi-port ventilators		**sf**	square foot
MSDS	material safety data sheet		**SIP**	Structural Insulated Panel
NAHB	National Association of Home Builders		**SIPA**	Structural Insulated Panel Association
NES	National Evaluation Services, Inc.		**sq**	per square
NFPA	National Fire Protection Association		**T&G**	tongue-and-groove
NSPI	National Spa & Pool Institute		**UBC**	Uniform Building Code
NWG	New West Gypsum Recycling, Inc.		**UF**	urea-formaldehyde
OSB	oriented strandboard		**UFFI**	urea-formaldehyde foam insulation
PB	polymer based		**UL**	Underwriters Laboratories
PCA	Portland Cement Association		**ULF**	ultra-low flushing
pcf	per cubic foot		**USG**	United States Gypsum Company
PET	polyethylene terephthalate		**USVS**	United States Vinyl Shakes
PF	phenol-formaldehyde		**UV**	ultraviolet
PM	polymer modified		**VHI**	very high impact
psf	pounds per square foot		**WFK**	warm floor kit
PSL	parallel strand lumber		**XPS**	extruded polystyrene

Who to Contact

Contributing Organizations and Companies

Build Smarter with Alternative Materials would not have been possible without the information, photos, and drawings provided by the organizations and companies listed here. I am grateful for their cooperation and assistance.

Please note that while the companies are listed under specific chapters, they also market products discussed in other chapters.

Chapter 1   Alternative Building Materials

National Evaluation Service, Inc.
900 Montclair Road, Suite A
Birmingham, AL 35213-1206
205-599-9888
jheaton@sbcci.org (e-mail)
www.nateval.org

Building Officials and Code Administrators International (BOCA)
4051 West Flossmoor Road
Country Club Hills, IL 60478-5795
708-799-2300
www.bocai.org

BOCA Evaluation Services, Inc.
708-799-2305
boca@aecnet.com (e-mail)
www.boca-es.com/~boca-es

International Conference of Building Officials (ICBO)
5360 Workman Mill Road
Whittier, CA 90601
562-699-0541

ICBO Evaluation Service, Inc.
562-699-0543
es@icbo.org (e-mail)
www.icbo.org

Southern Building Code Congress International (SBCCI)
900 Montclair Road
Birmingham, AL 35213
205-591-1853

SBCCI Public Safety Testing and Evaluation Services, Inc.
205-599-9800
rfazel@sbcci.org (e-mail)

Chapter 2   Foundations

American Wood Preservers Institute
2750 Prosperity Avenue, Suite 550
Fairfax, VA 22031-4312
703-204-0500
www.awpi.org

American Wood-Preservers' Association
P.O. Box 5690
Granbury, TX 76049
817-326-6300
817-326-6306 (Fax)
www.anpa.com

Simpson Strong-Tie Company, Inc.
260 North Palm
Brea, CA 92820
800-999-5099
www.strongtie.com

MiraDRI Moisture Protection Products
3500 Parkway Lane, Suite 500
Norcross, GA 30092
800-234-0484
www.miradri.com

Hickson Corporation
1955 Lake Park Drive, Suite 250
Smyrna, GA 30080
770-801-6600
ptwinfo@hicksoncorp.com

American Forest & Paper Association
1111 19th Street NW, Suite 800
Washington, DC 20036
202-463-2700
503-721-5826 (Fax on demand, index 101)
www.afandpa.org

Southern Pine Council
c/o Southern Forest Products Assn.
P.O. Box 641700
Kenner, LA 70064-1700
504-443-4464
www.southernpine.com

American Coal Ash Association, Inc.
2760 Eisenhower Avenue, Suite 304
Alexandria, VA 22314-4554
703-317-2400
ACAA-USA@msn.com (e-mail)
www.ACAA-USA.org

Rousseau Co.
1712 Thirteenth Street
Clarkston, WA 99403
800-635-3416
www.rousseau.com

Wind-Lock Corporation
1055 Leisz's Bridge Road
Leesport, PA 19533
800-521-9255
www.wind-lock.com

Flexible Products
1007 Industrial Park Dr.
Marietta, GA 30061
800-800-3626
www.flexibleproducts.com
www.itsgreatstuff.com

American ConForm Industries, Inc.
1820 South Santa Fe St.
Santa Ana, CA 92705
800-266-3676
www.smartblock.com

Reward Wall Systems, Inc.
4115 S. 87th Street
Omaha, NE 68127
402-592-7077
www.rewardwallsystem.com

AFM Corporation
P.O. Box 246
Excelsior, MN 55331
800-255-0176
www.r-control.com

Lite-Form International
P.O. Box 774
Sioux City, IA 51102
800-551-3313
www.liteform.com

Chapter 3   Wall and Roof Systems

Hebel Building Systems
6600 Highlands Parkway
Smyrna, GA 30082
800-994-3235
www.hebel.com

Insulated Masonry Systems International, Ltd.
165 South Mountain Way Drive
Orem, UT 84058
801-235-1565
corpd@imsi-ltd.com (domestic e-mail)
corpi@imsi-ltd.com (foreign e-mail)

Structural Insulated Panel Association
1331 H Street NW, Suite 1000
Washington, DC 20005
202-347-7800
www.sips.com

AFM Corporation
P.O. Box 246
Excelsior, MN 55331
800-255-0176
www.r-control.com

ICS 3-D Panel Works, Inc.
2610 Sidney Lanier Drive
Brunswick, GA 31525
912-264-3772
www.3-dpanelworks.com

Thermastructure XT, Corp.
609 Rock Road
Radford, VA 24141
888-633-5001
www.infoatthermasteel.com

Royall Wall Systems, Inc.
801 Pike Road
West Palm Beach, FL 33411
561-689-5398

Chapter 4   Framing Materials

Anthony Log Homes
P.O. Box 787
Mountain Home, NC 28758
800-837-8786

GRABBER Construction Products
205 Mason Circle
Concord, CA 94520
800-869-1375
www.grabberman.com

NAHB Research Center
1133 15th Street, NW
Washington, DC 20005-4700
202-452-7100

HL Stud Corporation
5516 Caplestone Lane
Dublin, OH 43017
800-457-8837

Plastic Components, Inc.
9051 N.W. 97th Terrace
Miami, FL 33178
800-327-7077

Tri-Steel Structures, Inc.
5400 S. Stemmons Frwy.
Denton, TX 76205
800-874-7833
www.tri-steel.com

Advanced Wood Resources
34363 Lake Creek Drive
Brownsville, OR 97327
800-533-3374

Hickson Corporation
1955 Lake Park Drive, Suite 250
Smyrna, GA 30080
770-801-6600
ptwinfo@hicksoncorp.com

Homasote Company
P.O. Box 7240
West Trenton, NJ 08628-0240
800-257-9491
www.homasote.com

Louisiana-Pacific Corporation
111 SW Fifth Avenue
Portland, OR 97204-3601
800-648-6893 (tech support)
www.lpcorp.com

Trus Joist MacMillan
200 E. Mallard Drive
Boise, ID 83706
800-628-3997
800-338-0515
www.TJM.com

American Laminators
P.O. Box 858
Drain, OR 97435
541-836-2026

Simpson Strong-Tie Company, Inc.
260 N. Palm
Brea, CA 92820
800-999-5099
www.strongtie.com

Boise Cascade
Timber and Wood Products Division
P.O. Box 2400
White City, OR 97503-0400
800-232-0788
www.bc.com

Truswal Systems Corporation
1101 N. Great Southwest Parkway
Arlington, TX 76011
800-521-9790

Arlington Industries, Inc.
1 Stausser, Industrial Park
Scranton, PA 18517
800-233-4717

Chapter 5   Roofing Materials

Cedar Shake & Shingle Bureau
P.O. Box 1178
Sumas, WA 98295-1178
604-462-8961
www.cedarbureau.org

Monier, Inc.
1 Park Plaza, Suite 900
Irvine, CA 92614
800-224-2024
www.monierlifetile.com

Re-New Wood, Incorporated
104 NW 8th Street
Wagoner, OK 74467
800-420-7576

Gerard Roofing Technologies
955 Columbia Street
Brea, CA 92821-2923
800-841-3213
www.gerardusa.com

Louisiana-Pacific Corporation
111 SW Fifth Avenue
Portland, OR 97204-3601
800-648-6893 (tech support)
www.lpcorp.com

Met-Tile Inc.
P.O. Box 4268
Ontario, CA 91761
909-947-0311
www.met-tile.com/roof

MaxiTile, Inc.
849 E. Sandhill Avenue
Carson, CA 90746
800-338-8453
www.maxitile.com

Nuline Industries
4900 Ondura Drive
Fredericksburg, VA 22407-8773
800-777-7663
www.nuline.com

Tropic Top
2028-3 Eastbourne Way
Orlando, FL 32812
407-273-0069
www.thatchart.com

Benjamin Obdyke Incorporated
65 Steamboat Drive
Warminster, PA 18974-4889
800-346-7655
www.obdyke.com

Owens Corning Fiberglas Corporation
Roofing & Asphalt Division
1 Owens Corning Parkway
Toledo, OH 43659
800-GET PINK (438-7465)
www.owenscorning.com

ado Products
21800 129th Ave. N.
P.O. Box 236
Rogers, MN 55374
800-666-8191
www.adoproducts.com

TrimLine Roof Ventilation Systems Inc.
705 Pennsylvania Avenue
Minneapolis, MN 55426
800-438-2920
www.trimline.com

Johns Manville
P.O. Box 5108
Denver. CO 80217-5108
800-654-3103
www.jm.com

Chapter 6   Siding Materials

Shakertown 1992, Inc.
1200 Kerron Street
P.O. Box 400
Winlock, WA 98596-0400
800-426-8970
www.shakertown.com

Louisiana-Pacific Corporation
111 SW Fifth Avenue
Portland, OR 97204-3601
800-648-6893 (tech support)
www.lpcorp.com

Masonite Corporation
Building & Industrial Products
1 South Wacker Drive
Chicago, IL 60601
800-323-4591
www.masonite.com

ABT Building Products
10115 Kincey Avenue, Suite 150
Huntersville, NC 28078
800-566-2282
www.abtco.com

MaxiTile, Inc.
849 E. Sandhill Avenue
Carson, CA 90746
800-338-8453
www.maxitile.com

S-B Power Tool Company
4300 West Peterson Avenue
Chicago, IL 60646-5999
877-267-2499
www.boschtools.com

James Hardie Building Products, Inc.
26300 La Alameda, Suite 250
Mission Viejo, CA 92691
800-942-7343
www.jameshardie.com

United States Vinyl Shakes, Ltd.
P.O. Box 690
Rockville Centre, NY 11571-0690
516-536-5237

Nailite International
1251 Northwest 165th Street
Miami, FL 33169-5871
305-620-6200
www.nailite.com

Wolverine Vinyl Siding
750 E. Swedesford Road
P.O. Box 860
Valley Forge, PA 19482
888-838-8100

Style-Mark Inc.
960 West Barre Road
Archbold, OH 43502
800-446-3040
www.style-mark.com

Cultured Stone Corporation
P.O. Box 270
Napa, CA 94559-0270
800-255-1727
www.culturedstone.com

EIFS Industry Members Association (EIMA)
3000 Corporate Center Drive, Suite 270
Morrow, GA 30260
800-294-3462

Acrocrete, Inc.
1259 NW 21st Street
Pompano Beach, FL 33069-1417
305-592-5000

Dryvit Systems, Inc.
One Energy Way
West Warwick, RI 02893
800-556-7752
www.dryvit.com

Omega Products Corporation
P.O. Box 1889
Orange, CA 92856
714-556-3830
www.omegaproducts.com

Parex, Inc.
1870 Stone Mountain Lithonia Rd.
P.O. Box 189
Redan, GA 30074
800-537-2739
www.parex.com

Pleko System International, Inc.
P.O Box 98360
Tacoma, WA 98498
253-472-9637

Retro Tek
584 Progress Way
Sun Prairie, WI 53590
800-225-9001

Senergy Division
Harris Specialty Chemicals
10245 Centurian Parkway North
Jacksonville, FL 32256-0564
904-996-6000

Simplex Products (Finestone)
1801 West U.S. Route 223
P.O. Box 10
Adrian, MI 49221
800-545-6555
www.simplex-products.com

Sto Corporation
6175 Riverside Drive SW
Atlanta, GA 30331
800-221-2397
www.stocorp.com

TEC Incorporated
315 S. Hicks Road
Palatine, IL 60067
847-358-9500

Texas EIFS, LLC.
220 Burleson Street
San Antonio, TX 78202
800-358-4785

Universal Polymers, Inc.
319 N. Main Avenue
Springfield, MO 65806
800-752-5403
www.universalpolymers.com

Plastic Components, Inc.
9051 NW 97th Terrace
Miami, FL 33178
800-327-7077

Chapter 7   Insulation Materials and Radiant Heat

Johns Manville Corp.
Product Information Center
P.O. Box 5108
Denver, CO 80127-5108
800-654-3103
www.jm.com

Waterhill Products Inc.
P.O. Box 633
Gorham, ME 04038
207-929-8796

Owens Corning World Headquarters
1 Owens Corning Way
Toledo, OH 43659
800-GET-PINK (438-7465)
www.owenscorning.com

Fibrex Insulations, Inc.
561 Scott Road
Sarnia, Ontario
Canada N7T 7L4
800-265-7514
www.fibrex.on.ca

Tenneco Building Products
2907 Log Cabin Drive
Smyrna, GA 30080-7013
800-241-4402

The Dow Chemical Company
Construction Materials
2020 Dow Center
Midland, MI 48674
800-441-4369 (tech line)
800-232-2436 (sales)

Ark-Seal International, Inc.
2190 S. Kalamath Street
Denver, CO 80223
800-525-8992
www.arkseal.com

Louisiana-Pacific Corp.
111 SW Fifth Avenue
Portland, OR 97204-3601
800-648-6893
www.lpcorp.com

Electro Plastics, Inc.
4406 St. Vincent Ave.
St. Louis, MO 63119
314-781-2121
www.warmfloor.com

Maxxon Corporation
920 Hamel Road
Hamel, MN 55340
800-588-4470
www.maxxon.com

Chapter 8   Doors, Windows, and Trim

JELD-WEN, Inc.
P.O. Box 1329
Klamath Falls, OR 97601
800-535-3462
www.doors-windows.com

Therma-Tru Corporation
P.O. Box 8780
1687 Woodlands Drive
Maumee, OH 43537
800-537-8827
www.thermatru.com

Traco
A Three Rivers Aluminum Company
71 Progress Avenue
Cranberry Township, PA 16066
800-837-7003

Comfort Line Inc.
5500 Enterprise Blvd.
Toledo, OH 43612
800-522-4999

Cedar Shake & Shingle Bureau
P.O. Box 1178
Sumas, WA 98295-1178
604-462-8961
www.cedarbureau.com

Wenco, A Division of JELD-WEN, Inc.
3250 Lakeport Blvd.
Klamath Falls, OR 97601
800-535-3462
www.doors-windows.com

Stillwater Products, Inc.
400 Commerce Street
Brevard, NC 28712
800-326-5355
www.decoroof.com

CeeFlow, Inc.
5334 South Lake Shore Drive
Harbor Springs, MI 49740
616-526-5579

The Sun Tunnel
786 McGlincey Lane
Campbell, CA 95008
800-369-3664
www.suntunnel.com

Focal Point Architectural Products, Inc.
P.O. Box 93327
Atlanta, GA 30377-0327
888-534-5196

TRIMTRAMP Ltd.
151 Carlingview Drive, Unit #11
Toronto, Ontario
Canada M9W 5S4
800-387-8746
www.trimtramp.com

Noxon Sales & Marketing
101 East Nora
Spokane, WA 99207
888-741-7419

Style-Mark Inc.
960 West Barre Road
Archbold, OH 43502
800-446-3040
www.style-mark.com

ABTco, Inc.
3250 West Big Beaver Road, Suite 200
Troy, MI 48084
800-521-4250
www.abtco.com

ResinArt East, Inc.
17 Continuum Drive
Fletcher, NC 28732
800-497-4376
www.resinart.com

Fen-Tech, Inc.
1510 North 5th Street
Superior, WI 54880
715-392-9500
www.fentechinc.com

Chapter 9   Interior Products

New West Gypsum Recycling Inc.
5620 - 198th Street
Langley, BC
Canada V3A 7C7
800-929-1817 (US)
604-534-9925 (Corporate)
nwgypsum@direct.ca (e-mail)

The Nailer
The Millennium Group, Inc.
141 E. Madison Street
Waterloo, WI 53594-1206
800-280-2304

United States Gypsum Company
125 South Franklin Street
Chicago, IL 60606-4678
800-874-4968
www.usg.com

Takagi Tools, Inc.
337-A Figueroa Street
Wilmington, CA 90744
800-891-7855

Unimast Incorporated
4825 North Scott Street, Suite 300
Schiller Park, IL 60176
800-654-7883
www.unimast.com

Ultraflex
GRABBER Construction Products
205 Mason Circle
Concord, CA 94520
800-869-1375
www.grabberman.com/nocoat.htm

Ultraflex
NO COAT Products
P.O. Box 5937
Bend, OR 97708
888-662-6281

Flexi-Wall Systems
208 Carolina Drive
P.O. Box 89
Liberty, SC 29657-0089
800-843-5394

Outwater Plastic Industries, Inc.
Architectural Products
4 Passaic Street
Wood-Ridge, NJ 07075
800-631-8375
www.outwater.com

Pole-Wrap
P.O. Box 1523
Troy, MI 48099
800-241-7653

Natural Impressions
HDM USA
4200 Northside Parkway
Building 4, Suite 100
Atlanta, GA 30327
404-842-0077

Cedarline
Giles & Kendall
P.O. Box 188
Huntsville, AL 35804
800-225-6738
www.aromaticcedar.com

Homasote Company
P.O. Box 7240
West Trenton, NJ 08628-0240
800-257-9491
www.homasote.com

WECork, Inc.
P.O. Box 276
Exeter, NH 03833
800-666-CORK (666-2675)
www.wecork.com

W.R. Bonsal Company
PO Box 241148
Charlotte, NC 28224-1148
800-738-1621
www.bonsal.com

Natural Cork, Ltd. Co.
1825 Killingsworth Road
Augusta, GA 30904
800-404-2675

Smith & Fong Company
601 Grandview Drive
S. San Francisco, CA 94080
650-872-1184
www.plyboo.com

QuikJack
Cepco Tool Company
P.O. Box 153
Spencer, NY 14883
800-466-9626
cepcotool@netzero.net

National Oak Floor Manufacturers Association
P.O. Box 3009
Memphis, TN 38173
901-526-5016
www.nofma.org

Wilsonart International Inc.
2400 Wilson Place
P.O. Box 6110
Temple, TX 76503-6110
800-435-9109
www.wilsonart.com

Image Carpets, Inc.
P.O. Box 5555
Armuchee, GA 30105
800-722-2504

Hinge-It Corporation
3999 Millersville Road
Indianapolis, IN 46205
800-599-6328
www.hingeit.com

Best Dressed Homes Company
P.O. Box 2447
Shelby, NC 28151
800-411-5738
bdh@unidial.com

Vance Industries, Inc.
250 Wille Road
Des Plaines, IL 60018-1866
847-375-8900

RACO/Hubbell Electrical Products
3902 W. Sample Street
P.O. Box 4002
South Bend, IN 46634
800-722-6462 (Fax)

Chapter 10   Bathroom and Kitchens

Re-Bath Corporation
1055 S. Country Club Drive, Bldg. 2
Mesa, AZ 85210-4613
800-426-4573
www.re-bath.com

Unique Refinishers, Inc.
5171 Nelson Brogdon Blvd. (Hwy. 20)
Sugar Hill, GA 30518
800-332-0048
www.uniquerefinishers.com

The Swan Corporation
1 City Centre, Suite 2300
St. Louis, MO 63101
800-325-7008
www.theswancorp.com

Sachwin Products, Inc.
P.O. Box 1366
Torrance, CA 90505
310-378-3800

Wilsonart International Inc.
2400 Wilson Place
P.O. Box 6110
Temple, TX 76503-6110
800-435-9109
www.wilsonart.com

Kuehn Bevel
111 Canfield Ave., A-17
Randolph, NJ 07869
800-862-3835

TerraGreen Ceramics, Inc.
1650 Progress Drive
Richmond, IN 47374
765-935-4760
www.terragreenceramics.com

Makita U.S.A., Inc.
14930 Northam Street
La Mirada, CA 90638-5753
800-462-5482
www.makitatools.com

Custom Building Products
13001 Seal Beach Blvd.
Seal Beach, CA 90740
800-272-8786

James Hardie Building Products, Inc.
26300 La Alameda, Suite 250
Mission Viejo, CA 92691
800-942-7343
www.jameshardie.com

ABTco, Inc.
3250 West Big Beaver Road, Suite 200
Troy, MI 48084
800-521-4250
www.abtco.com

Sloan Flushmate
51155 Grand River Avenue
Wixom, MI 48393
800-533-3460
www.flushmate.com

Mister Miser Urinal
4901 North Twelfth Street
Quincy, IL 62301
888-228-6900

International Cushioned Products Inc.
9505 Haldane Road
Kelowna, BC
Canada V4V 2K5
800-882-7638

The Product Source
302 Bell Park Drive
Woodstock, GA 30188
770-592-3145

Sterling Plumbing Group, Inc.
2900 Golf Road
Rolling Meadows, IL 60008
800-783-7546

Gloucester Co., Inc.
P.O. Box 428
Franklin, MA 02038
800-343-4963
www.phenoseal.com

Vance Industries, Inc.
250 Wille Road
Des Plaines, IL 60018-1866
847-375-8900

American Aldes Ventilation Corporation
4537 Northgate Court
Sarasota, FL 34234-2124
800-255-7749
www.oikos/aldes.com

Acme Brick
IBP
P.O. Box 425
Fort Worth, TX 76101
800-932-2263
www.IBPglassblock.com

HY-LITE Block Windows
101 California Avenue
Beaumont, CA 92223-2812
800-827-3691
www.hy-lite.com

Chapter 11   Decking

Makita U.S.A., Inc.
14930 Northam Street
La Mirada, CA 90638-5753
800-462-5482
www.makitatools.com

TREX Company, LLC
20 South Cameron Street
Winchester, VA 22601-9917
800-289-8739
www.trex.com

TimberTech
Crane Plastics Company
P.O. Box 1047
Columbus, OH 43216-1047
800-307-7780
www.timbertech.com

Re-Source Building Products, Ltd.
1685 Holmes Road
Elgin, IL 60123
800-231-9721
www.plastival.com

BoWrench
Cepco Tool Company
P.O. Box 153
Spencer, NY 14883
800-466-9626
cepcotool@netzero.net

ZCL Composites Inc.
E-Z Deck
2305 8th Street
Nisku, Alberta
Canada T9E 7Z3
800-990-3099
www.Ezdeck.com

Phoenix Recycled Plastics, Inc.
225 Washington Street
Conshohocken, PA 19428
610-940-1590
www.plasticlumberyard.com

American Ecoboard, Inc.
200 Finn Court
Farmingdale, NY 11735
516-753-5151
www.americanecoboard.com

Carefree Building Products
U.S. Plastic Lumber, Ltd.
2600 W. Roosevelt Rd.
Chicago, IL 60608
800-653-2784
www.ecpl.com

Brock Deck
Royal Crown Limited
P.O. Box 360
Milford, IN 46542-0360
800-365-3625
www.royalcrownltd.com

Supreme Decking Inc.
P.O. Box 1459
Lorton, VA 22079
800-532-1323

Iron Woods
Timber Holdings, Ltd.
2400 West Cornell Avenue
Milwaukee, WI 53209
414-445-8989
www.ironwoods.com

Greenheart-Durawoods, Inc.
665 Route 9 North
Bayville, NJ 08721
800-783-7220

ACQ Preserve
Chemical Specialties, Inc.
200 East Woodlawn Road, Suite 250
Charlotte, NC 28217
800-421-8661
www.treatedwood.com

DeckMaster
GRABBER Construction Products
205 Mason Circle
Concord, CA 94520
800-869-1375
www.deckmaster.com

EB-TY
Blue Heron Enterprises L.L.C.
P.O. Box 414
Califon, NJ 07830
888-438-3289
www.ebty.com

Dec-Klip
Ben Manufacturing, Inc.
21229 Cypress Way
Lynnwood, WA 98036
425-776-5340
ben69@premier1.net (email)

O'Berry Enterprises, Inc.
664 Exmoor Court
Crystal Lake, IL 60014
800-459-8428
globery@mc.net

P. A. Stratton & Co., Inc.
P.O. Box 436
Milan, OH 44846-0436
800-768-8880

Crawford Products, Inc.
301 Winter Street
West Hanover, MA 02339-0702
800-523-9382

Canadian Dekbrands
P.O. Box 1027
Cobourg, Ontario
Canada K9A 4W5
800-263-7903

Gaco Western, Inc.
P.O. Box 88698
Seattle, WA 98138-2698
800-456-4226
www.gaco.com

Versadek Industries
P.O. Box 73073
San Jose, CA 95173
800-497-3325

Feeney Wire Rope & Rigging, Inc.
2603 Union Street
Oakland, CA 94607-2420
800-888-2418
www.cablerail.com

Cross Industries, Inc.
6685 Jimmy Carter Blvd.
Norcross, GA 30071
800-521-9878
www.arcat.com/sd/clients/crossind.cfm
www.sweets.com/index/mfg.htm?id=447

Thompson Industries, Inc.
4260 Arkansas Ave. South
Russellville, AR 72801
800-621-2960

Chapter 12   Outdoor Products

Style-Mark Inc.
960 West Barre Road
Archbold, OH 43502
800-446-3040
www.style-mark.com

Re-Source Building Products, Ltd.
1685 Holmes Road
Elgin, IL 60123
800-231-9721
www.plastival.com

Triple Crown Fence
Royal Crown Limited
P.O. Box 360
Milford, IN 46542-0360
800-365-3625
www.royalcrownltd.com

Genova Products, Inc.
7034 E. Court Street
Davison, MI 48423-0309
800-521-7488
www.genovaproducts.com

Column Concepts
451 W. Channel Road
Benicia, CA 94510
707-747-0400
www.forcolumn.com

D&D Technologies, Inc.
1590 Sunland Lane
Costa Mesa, CA 92626-1515
800-716-0888
www.Ddtech.com

TimberTech
Crane Plastics Company
P.O. Box 1047
Columbus, OH 43216-1047
800-307-7780
www.timbertech.com

Keystone Retaining Wall Systems, Inc.
4444 West 78th Street
Minneapolis, MN 55435
800-747-8971
www.keystonewalls.com

Carlisle Tire & Wheel Company
P.O. Box 99
Carlisle, PA 17013
717-249-1000
www.carlisletire.com

Bomanite Corporation
P.O. Box 599
Madera, CA 93639
559-673-2411
www.bomanite.com

Grasspave2
Invisible Structures, Inc.
20100 E. 35th Drive
Aurora, CO 80011-8160
800-233-1510

EnviroWorks Inc.
3000 West Orange Avenue
Apopka, FL 32703
800-621-4253
www.enviroworksfiscars.com

Alpan, Inc.
425-I Constitution Ave.
Camarillo, CA 93012
800-233-1106
www.Alpan.com

C.R.S., Inc.
P. O. Box 4567
Spokane, WA 99202-0567
www.asktooltalk.com

Chapter 13 Building for the Environment

Health House Project
American Lung Association/MPLS
490 Concordia Ave.
St. Paul, MN 55103-2441
651-227-8014
www.alamn.org

Center for Resourceful Building Technology
P.O. Box 100
Missoula, MT 59806
406-549-7678
crbt@montana.com (e-mail)
www.montana.com/crbt

Environmental Building News
122 Birge St.
Brattleboro, VT 05301
802-257-7300
ebn@ebuild.com (e-mail)
www.ebuild.com

Green Builder Program
City of Austin
Environmental & Conservation Services Dept.
206 E. 9th Street, Suite 17.102
Austin, TX 78701
512-499-7827
www.greenbuilder.com

American Fiberboard Association
1210 W. Northwest Hwy.
Palatine, IL 60067
847-934-8394
aha@ahardbd.org

American Hardboard Association
1210 W. Northwest Hwy.
Palatine, IL 60067
847-934-8800
aha@ahardbd.org

APA—The Engineered Wood Association
7011 S. 19th Street
P.O. Box 11700
Tacoma, WA 98411-0700
253-565-6600
www.apawood.org

EIFS Industry Members Association (EIMA)
3000 Corporate Center Drive, Suite 270
Morrow, GA 30260
800-294-3462
www.eifsfacts.com

NAHB Research Center
400 Prince George's Boulevard
Upper Marlboro, MD 20774-8731
301-249-4000
www.nahbrc.org

Portland Cement Association
5420 Old Orchard Road
Skokie, IL 60077-1083
847-966-6200
www.portcement.org
www.concretehomes.com

Structural Board Association
45 Sheppard Ave. E., Suite 412
Willowdale, Ontario
Canada M2N 5W9
416-730-9090
www.sba-osb.com

Structural Insulated Panel Association
1331 H Street, NW, Suite 1000
Washington, D.C. 20005
202-347-7800
www.sips.org

Organizations

■■

The AFA is a national trade organization of manufacturers of cellulosic fiberboard products used in residential and commercial construction. They provide product literature for individuals engaged in distribution, specification, and construction fields. They work with government and private organizations to develop and improve industry standards and specifications and to update building codes. They also provide technical information to the general public as well as professionals. Of course, their objective is to promote the use of fiberboard, but they also promote ongoing educational and research activities on the product.

Fiberboard can be found in sheathing panels for exterior wall applications, roofing substrate, and other specialty products.

American Fiberboard Association
1210 W. Northwest Hwy.
Palatine, IL 60067
847-934-8394
aha@ahardbd.org

American Hardboard Association (AHA)

The AHA is located in the same office as the AFA. Also a national trade organization, the AHA represents manufacturers of hardboard products used for exterior siding, interior wall paneling, household and commercial furniture, and industrial and commercial products. Hardboard (generic term) can be found in such products as lap siding and siding accessories such as corners; paneling — smooth or textured surfaces; doorskins; garage door panels; and the list

goes on. Hardboard is made from wood chips converted to fibers and permanently bonded under heat and pressure into a panel. The wood fibers are combined with natural and synthetic binders and other additives that improve certain properties.

AHA serves as a central clearinghouse on industry and technical information for trade professionals, government agencies, and the general public. The association publishes a variety of brochures and pamphlets that are free, and videos that cost up to $5.00 each. They are concerned about the environment, and promote educational and promotion programs. With the help of independent researchers, they conduct timely research activities relating to hardboard use and performance; they improve industry standards and specifications, and they work to update building codes.

American Hardboard Association
1210 W. Northwest Hwy.
Palatine, IL 60067
847-934-8800
aha@ahardbd.org

APA—The Engineered Wood Association

Formerly known as the American Plywood Association, APA—The Engineered Wood Association is a nonprofit trade association whose member mills produce approximately 75 percent of the structural wood panel products manufactured in North America. You may have noticed the association's trademark "APA" on some of the plywood sheathing you've used. The same trademark can be found on composites (veneer faces bonded to wood strand cores) or on OSB. The trademark appears only on products manufactured by member mills and is the manufacturer's assurance that the products conform to the standard shown on the trademark. The APA EWS trademark appears only on engineered wood products manufactured by members of Engineered Wood Systems (EWS), a related corporation of APA.

Besides quality testing and inspection, research and promotion programs play important roles in developing and improving plywood and other panel construction systems, and in helping contractors, architects, and specifiers better understand and apply products affiliated with the association. They offer three publications that you can use as handy reference manuals:

Construction Guide — House Building Basics. This handbook is designed as an *elementary* guide to wood-frame construction. It illustrates the basic steps to completing the structural shell of a typical single-story house, from the foundation to the roof. In some cases, alternative methods of construction are explained or described. I firmly believe it doesn't hurt to go back to the beginning for a refresher course. This isn't meant to insult your professionalism, but merely to put you back on track. After all, some of us get sloppy with experience. Anyone who plans to enter the construction market could use this guide for a simplified, yet detailed, understanding of construction and framing methods.

Design/Construction Guide — Residential & Commercial. This publication contains up-to-date information on panel grades, including APA Performance Rated Panels; specification practices; floor, wall and roof systems; diaphragms and shear walls, fire-rated systems, and methods of finishing.

Product and Application Guide — Glulams. Everything you ever wanted to know about glued laminated timber can be found in this handbook — and then some. A nice thing about this product is no old-growth timber was cut down to produce it. Glued laminated timbers are composed of individual pieces of dimension lumber end-jointed together to produce long lengths, which are then bonded together with adhesives to create the required beam dimensions. Because of this process, large laminated timbers can

be manufactured from smaller trees of a variety of species harvested from second- and third-growth forest and plantations.

To get your copy of this or other APA publications, contact them at:

APA — The Engineered Wood Association
7011 S. 19th Street
P.O. Box 11700
Tacoma, WA 98411-0700
253-565-6600

EIFS Industry Members Association (EIMA)

EIFS Industry Members Association (EIMA), founded in 1981, is a nonprofit trade association representing manufacturers, suppliers, distributors, applicator/contractors and related building professionals involved in the *Exterior Insulation and Finish Systems* (EIFS) industry. The association promotes industry-wide performance standards and develops industry guideline specifications and standards for systems, materials and EIFS application. They sponsor research and testing programs that become the basis for many model building code requirements pertaining to EIFS cladding on exterior wall assemblies.

EIMA publishes specifications and test methods covering such topics as performance, durability, fire testing, application, and use of related materials including sheathing and sealants. They are committed to enhancing, improving, and promoting the EIFS industry; to advancing the EIFS industry through research and distribution of technical information; and to educating users of EIFS products. To learn more, contact them at:

EIFS Industry Members Association
3000 Corporate Center Drive, Suite 270
Morrow, GA 30260
800-294-3462
www.eifsfacts.com

Portland Cement Association (PCA)

The PCA has been perfecting concrete for over 75 years. They offer information on virtually all uses of cement and concrete — casting a sidewalk or patio, laying concrete block, or building complex highways and skyscrapers. Their resources range from do-it-yourself help to sophisticated design software for engineering firms. A fascinating organization — headquartered in Skokie, Illinois — they conduct research, market development, and educational work on behalf of their members. And, of course, they promote the use of cement and concrete. Other associations that help make the PCA strong include:

▌ The American Portland Cement Alliance in Washington, D.C., a sister organization that represents the cement industry in legislative and regulatory affairs.

▌ The Canadian Portland Cement Association, headquartered in Ottawa, with offices in Toronto, Halifax, Montreal, and Vancouver.

▌ Construction Technology Laboratories, Inc., a wholly-owned subsidiary of PCA. They offer engineering, testing, and research services to the construction, transportation, mineral, and related industries.

The PCA's Library Services department can provide research and information service on cement and concrete technology, materials technology, engineering, and related topics for member companies and others. Its collection of books, government reports, journals, standards, and patents is one of the most comprehensive on cement and concrete anywhere in the world. The library systematically exchanges technical literature with institutions around the world. In addition, the library offers:

▌ Literature searches on electronic databases

▌ Electronic interlibrary loan capabilities

▌ Document delivery (photocopies of journal articles, out-of-print PCA publications, etc.)

▌ Internet site information

▌ Bimonthly newsletter

The Library's schedule of member and nonmember fees for services is available by calling 847-966-6200, ext. 530. *Concrete Solutions* (a yearly catalog) is filled with helpful information on publications, software, audiovisuals, and ongoing seminars. I reviewed three products that I felt would be beneficial to readers of this book:

▌ The first is a 5-part video training series *(Building with Insulating Concrete Forms Video Training Series — Code # VC500)* where contractors can learn to build beautiful and marketable concrete homes using insulating concrete forms (ICFs). It takes you through the entire ICF building process discussed in Chapter 2 of this book. This 80-minute video set shows the entire process, starting with planning and design, through setting the forms and placing and finishing the concrete, to installing utilities.

▌ The other two products seem to go hand in hand. *Building Concrete Homes with Insulating Concrete Forms* is a 30-minute video that gives the nuts and bolts on ICF (insulating concrete form) systems. This hands-on video was shot out at the job site and just seems to make you feel right at home.

▌ To go more in-depth on the subject of ICF and to complement the video, there's a construction manual: *Insulating Concrete Forms — Successful Methods & Techniques.* This is a step-by-step manual that walks you through the entire process in 150 pages packed with photographs, diagrams, and other useful information. Also included is a directory of products and resources to help you get started in this business.

These three resource materials are a must for your library! To order copies, call them at 800-868-6733. To learn more about the PCA, contact them at:

Portland Cement Association
5420 Old Orchard Road
Skokie, IL 60077-1083
847-966-6200
www.portcement.org
www.cementhomes.com

Structural Board Association (SBA)

The SBA represents a group of international OSB producers. The association was founded in 1976 representing five companies and six mills. Their mission was to coordinate research, establish a distinct identity for waferboard as a commodity building panel, promote acceptance in building codes, and develop technical information in support of market end uses. The association has quite a history and has gone through some name changes. It finally incorporated in the U.S. in 1994. But it was in 1978 and 1979, when SBA coordinated the development of the first North American standards for waferboard, that OSB distinguished itself from particleboard. Then in 1980, the U.S. Model Codes and the National Building Code of Canada specified separate and distinct waferboard end uses.

SBA undertakes and supports short- and long-term technical and research programs which will enhance OSB quality, improve product manufacturers, and provide technical support for market expansion programs. They continue to promote the recognition of OSB as a structural-use wood panel product in worldwide building codes and standards.

To learn more about the association and the use of OSB in floor, roof, and wall sheathing, write to them for their 28-page booklet, *OSB In Wood Frame Construction*, at:

Structural Board Association
45 Sheppard Ave. E., Suite 412
Willowdale, Ontario
Canada M2N 5W9
416-730-9090
http://www.sba-osb.com

Structural Insulated Panel Association (SIPA)

This organization was established to promote the structural insulated panel industry, including panel manufacturers as well as their suppliers and customers. It maintains ongoing communications between its members and other interested parties, including government agencies, trade associations, industry professionals, and code agencies, to ensure that SIPA interests are recognized and its members are kept abreast of relevant issues. They organize forums, workshops, mailings, and other activities for members, all designed to share information and provide business opportunities.

SIPA conducts an ongoing publicity and information program to gain visibility for the industry and to promote the interests of its members. Press releases, articles, directory listings, and a national workshop series all generate inquiries and direct leads for SIPA and its members. They assist the structural insulated panel industry in promoting professional business ethics among its members. Their newsletter, *Spotlight On SIPA*, is packed full of information. To learn more about this organization, contact them at:

Structural Insulated Panel Association
1511 K. Street, N.W., Suite 600
Washington, D.C. 20005
202-347-7800
www.sips.org

Index

Practical References for Builders

Construction Forms & Contracts

125 forms you can copy and use — or load into your computer (from the FREE disk enclosed). Then you can customize the forms to fit your company, fill them out, and print. Loads into *Word* for *Windows*™, *Lotus 1-2-3*, *WordPerfect*, *Works*, or *Excel* programs. You'll find forms covering accounting, estimating, fieldwork, contracts, and general office. Each form comes with complete instructions on when to use it and how to fill it out. These forms were designed, tested and used by contractors, and will help keep your business organized, profitable and out of legal, accounting and collection troubles. Includes a CD-ROM for *Windows*™ and Mac. **400 pages, 8¹/₂ x 11, $39.75**

CD Estimator

If your computer has *Windows*™ and a CD-ROM drive, *CD Estimator* puts at your fingertips 85,000 construction costs for new construction, remodeling, renovation & insurance repair, electrical, plumbing, HVAC and painting. You'll also have the *National Estimator* program — a stand-alone estimating program for *Windows*™ that *Remodeling* magazine called a "computer wiz." Quarterly cost updates are available at no charge on the Internet. To help you create professional-looking estimates, the disk includes over 40 construction estimating and bidding forms in a format that's perfect for nearly any word processing or spreadsheet program for *Windows*™. And to top it off, a 70-minute interactive video teaches you how to use this CD-ROM to estimate construction costs. **CD Estimator is $68.50**

Contractor's Survival Manual

How to survive hard times and succeed during the up cycles. Shows what to do when the bills can't be paid, finding money and buying time, transferring debt, and all the alternatives to bankruptcy. Explains how to build profits, avoid problems in zoning and permits, taxes, time-keeping, and payroll. Unconventional advice on how to invest in inflation, get high appraisals, trade and postpone income, and stay hip-deep in profitable work. **160 pages, 8¹/₂ x 11, $22.25**

National Construction Estimator

Current building costs for residential, commercial, and industrial construction. Estimated prices for every common building material. Provides man-hours, recommended crew, and gives the labor cost for installation. Includes a CD-ROM with an electronic version of the book with *National Estimator*, a stand-alone *Windows*™ estimating program, plus an interactive multimedia video that shows how to use the disk to compile construction cost estimates. **592 pages, 8¹/₂ x 11, $47.50. Revised annually**

Residential Steel Framing Guide

Steel is stronger and lighter than wood — straight walls are guaranteed — steel framing will not wrap, shrink, split, swell, bow, or rot. Here you'll find full page schematics and details that show how steel is connected in just about all residential framing work. You won't find lengthy explanations here on how to run your business, or even how to do the work. What you will find are over 150 easy-to-read full-page details on how to construct steel-framed floors, roofs, interior and exterior walls, bridging, blocking, and reinforcing for all residential construction. Also includes recommended fasteners and their applications, and fastening schedules for attaching every type of steel framing member to steel as well as wood. **170 pages, 8¹/₂ x 11, $38.80**

Builder's Guide to Accounting Revised

Step-by-step, easy-to-follow guidelines for setting up and maintaining records for your building business. This practical, newly-revised guide to all accounting methods shows how to meet state and federal accounting requirements, explains the new depreciation rules, and describes how the Tax Reform Act can affect the way you keep records. Full of charts, diagrams, simple directions and examples, to help you keep track of where your money is going. Recommended reading for many state contractor's exams. **320 pages, 8¹/₂ x 11, $26.50**

Contractor's Guide to the Building Code Revised

This new edition was written in collaboration with the International Conference of Building Officials, writers of the code. It explains in plain English exactly what the latest edition of the *Uniform Building Code* requires. Based on the 1997 code, it explains the changes and what they mean for the builder. Also covers the *Uniform Mechanical Code* and the *Uniform Plumbing Code*. Shows how to design and construct residential and light commercial buildings that'll pass inspection the first time. Suggests how to work with an inspector to minimize construction costs, what common building shortcuts are likely to be cited, and where exceptions may be granted. **320 pages, 8¹/₂ x 11, $39.00**

How to Succeed With Your Own Construction Business

Everything you need to start your own construction business: setting up the paperwork, finding the work, advertising, using contracts, dealing with lenders, estimating, scheduling, finding and keeping good employees, keeping the books, and coping with success. If you're considering starting your own construction business, all the knowledge, tips, and blank forms you need are here. **336 pages, 8¹/₂ x 11, $24.25**

Building Contractor's Exam Preparation Guide

Passing today's contractor's exams can be a major task. This book shows you how to study, how questions are likely to be worded, and the kinds of choices usually given for answers. Includes sample questions from actual state, county, and city examinations, plus a sample exam to practice on. This book isn't a substitute for the study material that your testing board recommends, but it will help prepare you for the types of questions — and their correct answers — that are likely to appear on the actual exam. Knowing how to answer these questions, as well as what to expect from the exam, can greatly increase your chances of passing. **320 pages, 8¹/₂ x 11, $35.00**

Contractor's Index to the 1997 *Uniform Building Code*

Finally, there's a common-sense index that helps you quickly and easily find the section you're looking for in the *UBC*. It lists topics under the names builders actually use in construction. Best of all, it gives the full section number and the actual page in the *UBC* where you'll find it. If you need to know the requirements for windows in exit access corridor walls, just look under *Windows*™. You'll find the requirements you need are in Section 1004.3.4.3.2.2 in the *UBC* — on page 115. This practical index was written by a former builder and building inspector who knows the *UBC* from both perspectives. If you hate to spend valuable time hunting through pages of fine print for the information you need, this is the book for you. **192 pages, 8¹/₂ x 11, paperback edition, $26.00**
192 pages, 8¹/₂ x 11, looseleaf edition, $29.00.

Blueprint Reading for the Building Trades

How to read and understand construction documents, blueprints, and schedules. Includes layouts of structural, mechanical, HVAC and electrical drawings. Shows how to interpret sectional views, follow diagrams and schematics, and covers common problems with construction specifications. **192 pages, 5¹/₂ x 8¹/₂, $14.75**

Basic Engineering for Builders

If you've ever been stumped by an engineering problem on the job, yet wanted to avoid the expense of hiring a qualified engineer, you should have this book. Here you'll find engineering principles explained in non-technical language and practical methods for applying them on the job. With the help of this book you'll be able to understand engineering functions in the plans and how to meet the requirements, how to get permits issued without the help of an engineer, and anticipate requirements for concrete, steel, wood and masonry. See why you sometimes have to hire an engineer and what you can undertake yourself: surveying, concrete, lumber loads and stresses, steel, masonry, plumbing, and HVAC systems. This book is designed to help the builder save money by understanding engineering principles that you can incorporate into the jobs you bid. **400 pages, 8¹/₂ x 11, $34.00**

Basic Lumber Engineering for Builders

Beam and lumber requirements for many jobs aren't always clear, especially with changing building codes and lumber products. Most of the time you rely on your own "rules of thumb" when figuring spans or lumber engineering. This book can help you fill the gap between what you can find in the building code span tables and what you need to pay a certified engineer to do. With its large, clear illustrations and examples, this book shows you how to figure stresses for pre-engineered wood or wood structural members, how to calculate loads, and how to design your own girders, joists and beams. Included FREE with the book — an easy-to-use version of NorthBridge Software's *Wood Beam Sizing* program. **272 pages, 8¹/₂ x 11, $38.00**

Contractor's Guide to QuickBooks Pro 99

This user-friendly manual walks you through QuickBooks Pro's detailed setup procedure and explains step-by-step how to create a first-rate accounting system. You'll learn in days, rather than weeks, how to use QuickBooks Pro to get your contracting business organized, with simple, fast accounting procedures. On the CD included with the book you'll find a full version of QuickBooks Pro, good for 25 uses, with a QuickBooks Pro file preconfigured for a construction company (you drag it over onto your computer and plug in your own company's data). You'll also get a complete estimating program, including a database, and a job costing program that lets you export your estimates to QuickBooks Pro. It even includes many useful construction forms to use in your business.
296 pages, 8¹/₂ x 11, $42.00
Also available: **Contractor's Guide to QuickBooks Pro *version 6*. $39.75**

Electrician's Exam Preparation Guide

Need help in passing the apprentice, journeyman, or master electrician's exam? This is a book of questions and answers based on actual electrician's exams over the last few years. Almost a thousand multiple-choice questions — exactly the type you'll find on the exam — cover every area of electrical installation: electrical drawings, services and systems, transformers, capacitors, distribution equipment, branch circuits, feeders, calculations, measuring and testing, and more. It gives you the correct answer, an explanation, and where to find it in the latest *NEC*. Also tells how to apply for the test, how best to study, and what to expect on examination day.
352 pages, 8¹/₂ x 11, $28.00

Craftsman's Illustrated Dictionary of Construction Terms

Almost everything you could possibly want to know about any word or technique in construction. Hundreds of up-to-date construction terms, materials, drawings and pictures with detailed, illustrated articles describing equipment and methods. Terms and techniques are explained or illustrated in vivid detail. Use this valuable reference to check spelling, find clear, concise definitions of construction terms used on plans and construction documents, or learn about

little-known tools, equipment, tests and methods used in the building industry. It's all here. **416 pages, 8¹/₂ x 11, $36.00**

Estimating & Bidding for Builders & Remodelers w/ CD-ROM

If your computer has a CD-ROM drive, the *CD Estimator* disk enclosed in this book could change forever the way you estimate construction. You get over 2,500 pages from six current cost databases published by Craftsman, plus an estimating program you can master in minutes, plus a 70-minute interactive video on how to use this program, plus an award-winning book. This package is your best bargain for estimating and bidding construction costs. **272 pages, 8¹/₂ x 11, $69.50**

Contracting in All 50 States

Every state has its own licensing requirements that you must meet to do business there. These are usually written exams, financial requirements, and letters of reference. This book shows how to get a building, mechanical or specialty contractor's license, qualify for DOT work, and register as an out-of-state corporation, for every state in the U.S. It lists addresses, phone numbers, application fees, requirements, where an exam is required, what's covered on the exam and how much weight each area of construction is given on the exam. You'll find just about everything you need to know in order to apply for your out-of-state license. **416 pages, 8¹/₂ x 11, $36.00**

Illustrated Guide to the 1999 *National Electrical Code*

This fully-illustrated guide offers a quick and easy visual reference for installing electrical systems. Whether you're installing a new system or repairing an old one, you'll appreciate the simple explanations written by a code expert, and the detailed, intricately-drawn and labeled diagrams. A real time-saver when it comes to deciphering the current *NEC*.
384 pages, 8¹/₂ x 11, $38.75

Fences & Retaining Walls

Everything you need to know to run a profitable business in fence and retaining wall contracting. Takes you through layout and design, construction techniques for wood, masonry, and chain link fences, gates and entries, including finishing and electrical details. How to build retaining and rock walls. How to get your business off to the right start, keep the books, and estimate accurately. The book even includes a chapter on contractor's math. **400 pages, 8¹/₂ x 11, $23.25**

CD Estimator Heavy

CD Estimator Heavy has a complete 780-page heavy construction cost estimating volume for each of the 50 states. Select the cost database for the state where the work will be done. Includes thousands of cost estimates you won't find anywhere else, and in-depth coverage of demolition, hazardous materials remediation, tunneling, site utilities, precast concrete, structural framing, heavy timber construction, membrane waterproofing, industrial windows and doors, specialty finishes, built-in commercial and industrial equipment, and HVAC and electrical systems for commercial and industrial buildings. **CD Estimator Heavy is $69.00**

The Contractor's Legal Kit

Stop "eating" the costs of bad designs, hidden conditions, and job surprises. Set ground rules that assign those costs to the rightful party ahead of time. And it's all in plain English, not "legalese." For less than the cost of an hour with a lawyer you'll learn the exclusions to put in your agreements, why your insurance company may pay for your legal defense, how to avoid liability for injuries to your sub and his employees or damages they cause, how to collect on lawsuits you win, and much more. It also includes a FREE computer disk with contracts and forms you can customize for your own use. **352 pages, 8¹/₂ x 11, $59.95**

Contractor's Year-Round Tax Guide Revised

How to set up and run your construction business to minimize taxes: corporate tax strategy and how to use it to your advantage, and what you should be aware of in contracts with others. Covers tax shelters for builders, write-offs and investments that will reduce your taxes, accounting methods that are best for contractors, and what the I.R.S. allows and what it often questions. **192 pages, 8¹/₂ x 11, $26.50**

Construction Estimating Reference Data

Provides the 300 most useful manhour tables for practically every item of construction. Labor requirements are listed for sitework, concrete work, masonry, steel, carpentry, thermal and moisture protection, doors and windows, finishes, mechanical and electrical. Each section details the work being estimated and gives appropriate crew size and equipment needed. Includes a CD-ROM with an electronic version of the book with *National Estimator*, a stand-alone *Windows*™ estimating program, plus an interactive multimedia video that shows how to use the disk to compile construction cost estimates. **432 pages, 11 x 8¹/₂, $39.50**

Land Development

The industry's bible. Nine chapters cover everything you need to know about land development from initial market studies to site selection and analysis. New and innovative design ideas for streets, houses, and neighborhoods are included. Whether you're developing a whole neighborhood or just one site, you shouldn't be without this essential reference.
360 pages, 5¹/₂ x 8¹/₂, $40.70

Home Inspection Handbook

Every area you need to check in a home inspection — especially in older homes. Twenty complete inspection checklists: building site, foundation and basement, structural, bathrooms, chimneys and flues, ceilings, interior & exterior finishes, electrical, plumbing, HVAC, insects, vermin and decay, and more. Also includes information on starting and running your own home inspection business. **324 pages, 5¹/₂ x 8¹/₂, $24.95**

Estimating Excavation

How to calculate the amount of dirt you'll have to move and the cost of owning and operating the machines you'll do it with. Detailed, step-by-step instructions on how to assign bid prices to each part of the job, including labor and equipment costs. Also, the best ways to set up an organized and logical estimating system, take off from contour maps, estimate quantities in irregular areas, and figure your overhead. **448 pages, 8¹/₂ x 11, $39.50**

Quicken for Contractors

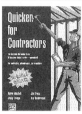

Most builders, contractors, and remodelers came up through the ranks, learning their craft "in the trenches." They know how buildings go together, but office management skills like bookkeeping, accounting, payroll and job costing are often a new, and dangerous, challenge. *Quicken for Contractors* explains how to use *Quicken*, an affordable, easy-to-understand computer program published by Intuit. It shows step-by-step how *Quicken* can work in the builder's office, putting accurate bookkeeping within easy reach. This manual does not include the program, but has instructions and examples, with onscreen "pictures" of familiar forms like checks, deposit slips and time cards that you can create with *Quicken*. Even if you only have a few minutes a week, this book will help you use *Quicken* to set up a basic checkbook system that will quickly tell you your cash position and job costs. You'll learn how to set up your financial files and create valuable reports, including profit & loss statements, payroll reports and job cost reports. The companion diskette included with the book consists of a sample construction company that you can use as a tutorial, or as a template for you to plug in your own data. **240 pages, 8¹/₂ x 11, $32.50**

Profits in Building Spec Homes

If you've ever wanted to make big profits in building spec homes yet were held back by the risks involved, you should have this book. Here you'll learn how to do a market study and feasibility analysis to make sure your finished home will sell quickly, and for a good profit. You'll find tips that can save you thousands in negotiating for land, learn how to impress bankers and get the financing package you want, how to nail down cost estimating, schedule realistically, work effectively yet harmoniously with subcontractors so they'll come back for your next home, and finally, what to look for in the agent you choose to sell your finished home. Includes forms, checklists, worksheets, and step-by-step instructions. **208 pages, 8¹/₂ x 11, $27.25**

Estimating Home Building Costs

Estimate every phase of residential construction from site costs to the profit margin you include in your bid. Shows how to keep track of man-hours and make accurate labor cost estimates for footings, foundations, framing and sheathing finishes, electrical, plumbing, and more. Provides and explains sample cost estimate worksheets with complete instructions for each job phase. **320 pages, 5¹/₂ x 8¹/₂, $17.00**

Rough Framing Carpentry

If you'd like to make good money working outdoors as a framer, this is the book for you. Here you'll find shortcuts to laying out studs; speed cutting blocks, trimmers and plates by eye; quickly building and blocking rake walls; installing ceiling backing, ceiling joists, and truss joists; cutting and assembling hip trusses and California fills; arches and drop ceilings — all with production line procedures that save you time and help you make more money. Over 100 on-the-job photos of how to do it right and what can go wrong. **304 pages, 8¹/₂ x 11, $26.50**

Craftsman Book Company
6058 Corte del Cedro
P.O. Box 6500
Carlsbad, CA 92018

☎ **24 hour order line**
1-800-829-8123
Fax (760) 438-0398

In A Hurry?
We accept phone orders charged to your
○ Visa, ○ MasterCard, ○ Discover or ○ American Express

Card#_____

Exp. date_____Initials_____

Total enclosed_____(In California add 7.25% tax)
We pay shipping when your check covers your order in full.

Tax Deductible: Treasury regulations make these references tax deductible when used in your work. Save the canceled check or charge card statement as your receipt.

Order online http://www.craftsman-book.com
Free on the Internet! Download any of Craftsman's estimating costbooks for a 30-day free trial! http://costbook.com

Name_____

Company_____

Address_____

City/State/Zip_____

○ This is a residence

10-Day Money Back Guarantee

- ○ 34.00 Basic Engineering for Builders
- ○ 38.00 Basic Lumber Engineering for Builders
- ○ 14.75 Blueprint Reading for the Building Trades
- ○ 26.50 Builder's Guide to Accounting Revised
- ○ 35.00 Building Contractor's Exam Preparation Guide
- ○ 68.50 CD Estimator
- ○ 69.00 CD Estimator Heavy
- ○ 39.50 Construction Estimating Reference Data with FREE *National Estimator* on a CD-ROM.
- ○ 39.75 Construction Forms & Contracts with a CD-ROM for *Windows*™ and Macintosh.
- ○ 36.00 Contracting in All 50 States
- ○ 42.00 Contractor's Guide to QuickBooks Pro 99
- ○ 39.75 Contractor's Guide to QuickBooks Pro *version 6*

- ○ 39.00 Contractor's Guide to the Building Code Rev.
- ○ 26.00 Contractor's Index to the *UBC — Paperback*
- ○ 29.00 Contractor's Index to the *UBC — Looseleaf*
- ○ 59.95 Contractor's Legal Kit
- ○ 22.25 Contractor's Survival Manual
- ○ 26.50 Contractor's Year-Round Tax Guide Revised
- ○ 36.00 Craftsman's Illustrated Dictionary of Construction Terms
- ○ 28.00 Electrician's Exam Preparation Guide
- ○ 69.50 Estimating & Bidding for Builders & Remodelers w/ CD-ROM
- ○ 39.50 Estimating Excavation
- ○ 17.00 Estimating Home Building Costs
- ○ 23.25 Fences and Retaining Walls

- ○ 24.95 Home Inspection Handbook
- ○ 24.25 How to Succeed w/Your Own Construction Business
- ○ 38.75 Illustrated Guide to the 1999 *NEC*
- ○ 40.70 Land Development
- ○ 47.50 National Construction Estimator with FREE *National Estimator* on a CD-ROM.
- ○ 27.25 Profits in Building Spec Homes
- ○ 32.50 *Quicken* for Contractors
- ○ 38.80 Residential Steel Framing Guide
- ○ 26.50 Rough Framing Carpentry
- ○ 34.75 Build Smarter with Alternative Materials
- ○ FREE Full Color Catalog

Prices subject to change without notice

Craftsman Book Company
6058 Corte del Cedro
P.O. Box 6500
Carlsbad, CA 92018

☎ **24 hour order line**
1-800-829-8123
Fax (760) 438-0398

In A Hurry?
We accept phone orders charged to your
○ Visa, ○ MasterCard, ○ Discover or ○ American Express

Card#_____

Exp. date_____Initials_____

Total enclosed_____(In California add 7.25% tax)
We pay shipping when your check covers your order in full.

Tax Deductible: Treasury regulations make these references tax deductible when used in your work. Save the canceled check or charge card statement as your receipt.

Order online http://www.craftsman-book.com
Free on the Internet! Download any of Craftsman's estimating costbooks for a 30-day free trial! http://costbook.com

Name_____

Company_____

Address_____

City/State/Zip_____

○ This is a residence

10-Day Money Back Guarantee

- ○ 34.00 Basic Engineering for Builders
- ○ 38.00 Basic Lumber Engineering for Builders
- ○ 14.75 Blueprint Reading for the Building Trades
- ○ 26.50 Builder's Guide to Accounting Revised
- ○ 35.00 Building Contractor's Exam Preparation Guide
- ○ 68.50 CD Estimator
- ○ 69.00 CD Estimator Heavy
- ○ 39.50 Construction Estimating Reference Data with FREE *National Estimator* on a CD-ROM.
- ○ 39.75 Construction Forms & Contracts with a CD-ROM for *Windows*™ and Macintosh.
- ○ 36.00 Contracting in All 50 States
- ○ 42.00 Contractor's Guide to QuickBooks Pro 99
- ○ 39.75 Contractor's Guide to QuickBooks Pro *version 6*

- ○ 39.00 Contractor's Guide to the Building Code Rev.
- ○ 26.00 Contractor's Index to the *UBC — Paperback*
- ○ 29.00 Contractor's Index to the *UBC — Looseleaf*
- ○ 59.95 Contractor's Legal Kit
- ○ 22.25 Contractor's Survival Manual
- ○ 26.50 Contractor's Year-Round Tax Guide Revised
- ○ 36.00 Craftsman's Illustrated Dictionary of Construction Terms
- ○ 28.00 Electrician's Exam Preparation Guide
- ○ 69.50 Estimating & Bidding for Builders & Remodelers w/ CD-ROM
- ○ 39.50 Estimating Excavation
- ○ 17.00 Estimating Home Building Costs
- ○ 23.25 Fences and Retaining Walls

- ○ 24.95 Home Inspection Handbook
- ○ 24.25 How to Succeed w/Your Own Construction Business
- ○ 38.75 Illustrated Guide to the 1999 *NEC*
- ○ 40.70 Land Development
- ○ 47.50 National Construction Estimator with FREE *National Estimator* on a CD-ROM.
- ○ 27.25 Profits in Building Spec Homes
- ○ 32.50 *Quicken* for Contractors
- ○ 38.80 Residential Steel Framing Guide
- ○ 26.50 Rough Framing Carpentry
- ○ 34.75 Build Smarter with Alternative Materials
- ○ FREE Full Color Catalog

Prices subject to change without notice

Mail This Card Today For a Free Full Color Catalog

Over 100 books, annual cost guides and estimating software packages at your fingertips with information that can save you time and money. Here you'll find information on carpentry, contracting, estimating, remodeling, electrical work, and plumbing.

All items come with an unconditional 10-day money-back guarantee. If they don't save you money, mail them back for a full refund.

Name_____

Company_____

Address_____

City/State/Zip_____

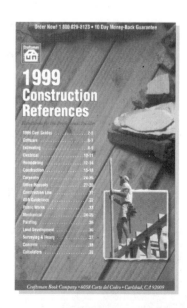

Craftsman Book Company / 6058 Corte del Cedro / P.O. Box 6500 / Carlsbad, CA 92018